IMS Multimedia Telephony over Cellular Systems

IMS Multimedia Telephony over Cellular Systems

VoIP Evolution in a Converged Telecommunication World

Edited by

Shyam Chakraborty and Janne Peisa
Ericsson Research, Finland

Tomas Frankkila and Per Synnergren
Ericsson Research, Sweden

1807
WILEY
2007

BICENTENNIAL

John Wiley & Sons, Ltd

Telephone   (+44) 1243 779777

Email (for orders and customer service enquiries): cs-books@wiley.co.uk
Visit our Home Page on www.wiley.com

Other Wiley Editorial Offices

John Wiley & Sons Inc., 111 River Street, Hoboken, NJ 07030, USA

Jossey-Bass, 989 Market Street, San Francisco, CA 94103-1741, USA

Wiley-VCH Verlag GmbH, Boschstr. 12, D-69469 Weinheim, Germany

John Wiley & Sons Australia Ltd, 42 McDougall Street, Milton, Queensland 4064, Australia

John Wiley & Sons (Asia) Pte Ltd, 2 Clementi Loop #02-01, Jin Xing Distripark, Singapore
129809

John Wiley & Sons Canada Ltd, 6045 Freemont Blvd, Mississauga, ONT, L5R 4J3, Canada

Wiley also publishes its books in a variety of electronic formats. Some content that appears
in print may not be available in electronic books.

British Library Cataloguing in Publication Data

A catalogue record for this book is available from the British Library

ISBN 978-0-470-05855-8 (HB)

Typeset by Sunrise Setting Ltd, Torquay, Devon, UK.
Printed and bound in Great Britain by Antony Rowe Ltd, Chippenham, England.
This book is printed on acid-free paper responsibly manufactured from sustainable forestry
in which at least two trees are planted for each one used for paper production.

Contents

5   Media Flow   115
Daniel Enström, Tomas Frankkila, Per Fröjdh, Janne Peisa, Krister Svanbro

6   Security   209
Rolf Blom, Yi Cheng, Vesa Lehtovirta, Karl Norrman, Göran Schultz

Preface

This preface is somewhat different from prefaces found in similar books because it does not focus so much on the content of the book. We have instead chosen to write a few words about our own experiences from working with telephony services over Internet Protocol (IP). Here are our stories.

Shyam Chakraborty

In my childhood, black ebonite telephones were a rare commodity and a status symbol. When I made my first telephone call, after a lot of tries and shouting hellos, I could hear a metallic voice through sharp hissing and 'click/clack' sounds. My father told me that it was an art to converse over the telephone, and that it may even be possible to recognize a few voices with sufficient practice. Telephony as an art and as a technology fascinated me. Over the years, I could manage to call effortlessly and talk and chat for hours. And not only identify voices clearly . . . it has even been possible to understand emotions over the telephone.

During the late 1980s the extensive proliferation of computers fueled the growth of data communications at a fast pace. Though the present prevalence of the Internet was not then fully understood, forecasts were aplenty that market of data communications would exceed that of voice communications by leaps and bounds. I wondered, even if these predictions are valid, would voice communications take a back seat? It did not. The basic need for telephony got tremendous support from cellular systems due to the offered mobility, portability, good voice quality and wide coverage. Mobile telephony has reached the pinnacle of consumer items, with both grace and utility.

The concept of a converged network has been on the drawing board for quite some time. With meticulous provisioning, the packet switched Internet gains an increasingly convincing role for such a converged network. Rather than talking of voice and data networks separately, a broader concept of services with different quality of service requirements has emerged. A few years back, I became curious whether the wireless interface, despite its 'limited bandwidth', would be adequate for providing real-time services in a packet switched mode, given the different aspects – mobility, security and latency issues – to be satisfied. These thoughts were primarily studied in a more academic setup, somewhat different from that of the rest of my co-editors and authors, who have been studying the design of the radio interface and VoIP services in an industrial research environment. The preliminary results showed me that, as the offered bit rates over the radio interface increased, packet switched real-time services would in general be feasible. This, of course, calls for a clever design of the associated protocol stacks. When I joined Ericsson Research and discussed my thoughts with my colleagues here, I had full corroboration from them.

Mobility and portability have provided fertile ground for a number of conversational and interactive services that are provided more flexibly over a packet switched network. These services allow a richer experience for users in communicating with more information and even personal closeness. Surely, not only the networking paradigms are converging, but a convergence of service paradigms also looms large. I hope this will redefine interpersonal communications in the future.

Tomas Frankkila

During my years within the company I have mainly worked with speech coding for Circuit Switched (CS) cellular systems. This work includes fixed-point and DSP implementation, research, verification of speech quality and standardization. I started working with Voice over IP (VoIP) issues during 2001 and have worked with VoIP ever since. During these years, I have had three 'Aha! experiences' and I will try to describe these here.

When I started with VoIP, most people working in this area were focusing on VoIP over the fixed Internet. VoIP over wireless had of course started but it did not really seem to be realistic to deploy it for a few reasons, mainly these:

1. For wireless systems, one cannot waste half of the resources or more on transmitting the IP, UDP and RTP headers. It is possible to reduce the overhead (per frame) by packing several speech frames into each RTP packet. However, due to the tough latency requirements for full-duplex, real-time voice services, this aggregation needs to be limited to two or maybe three frames per packet, which still gives too much overhead. It was quite clear that, for successful VoIP deployment, header compression would be needed.

2. The Packet Switched (PS) radio bearers were far from optimal for VoIP. For both GPRS and UMTS PS bearers, the latencies were too long. Acknowledged Mode (AM) could not be used because of the quite long retransmission time between the mobile terminal and the RNC, which would give very problematic jitter behavior. And Unacknowledged Mode (UM) bearers were either not available or were too limited to take advantage of the flexibility in IP services.

3. VoIP could not use the radio bearers as efficiently as CS because unequal error protection would not be as optimal as for CS. UDPlite was of course available but it was not as optimal as the super-optimized channel coding and interleaving schemes used on CS bearers.

It was obvious that significant improvements were required. There was also ongoing work to solve these issues, but the work was far from completion in most areas.

One of the most important features that would eventually make VoIP over wireless realistic was the ongoing work with header compression and especially with RObust Header Compression (ROHC). The introduction of ROHC made it clear that the overhead due to protocol headers was manageable. Since ROHC also provided good resilience against packet losses, much better than other header compression schemes, it was quite clear that packet loss due to the air interface would not be a big problem. The problems with inefficient and non-flexible radio bearers still remained.

After working with VoIP for a little while, it became clear to me that VoIP over wireless will actually be better than CS voice. My thinking at that time was that the sound quality of the VoIP service will be better because the great flexibility in IP makes it very easy to introduce wideband speech codecs in the systems. With the development and standardization of AMR-WB it also became clear to me that wideband codecs do not need to have a much higher bit rate than the narrowband codecs used in the existing systems. Previous wideband speech codecs had bit rates in the 32–64 kbps range, which is too high to be useful in wireless systems. With AMR-WB however it became obvious that good wideband speech quality could be achieved at about 12–16 kbps. The complexity of the AMR-WB codec was also manageable, making it realistic to implement the codec in mobile phones.

The first 'Aha!'

My first Aha! experience came when I realized that the quality could be improved by combining:

1. the flexibility of IP, which makes it very easy to introduce AMR-WB;

2. AMR-WB, which gives much better quality than narrowband codecs at a bit rate that is not much higher than for the codecs used for CS, i.e. AMR 12.2 kbps;

3. ROHC, which compresses the headers to reasonable sizes.

Even though radio bearers optimized for VoIP were still not available, and even though unequal error protection was not as optimal as for CS, it was clear to me that the users would appreciate the great quality improvements with wideband speech. In fact I believed that the users would like this so much that they would be willing to pay more for the service and this would compensate for the inefficiencies of the existing radio bearers.

During this time, we were also studying time scaling of speech. This worked quite well, at least for moderate amount of scaling. It became clear to me that a reasonable amount of jitter would not be a big problem.

The second 'Aha!'

The second Aha! experience came in 2003–04 when I learned about the ongoing discussions for high-speed channels. At that time, the general thinking in the high-speed field was focusing on data services and it seemed like they thought that there will be two general types of channels:

- One type of channel is optimized for Transmission Control Protocol (TCP) traffic. This channel type would have short Transmission Time Intervals (TTI), short round-trip time (RTT) and fast retransmissions, which would give low packet loss rates.

- The other type of channel would be specially designed for VoIP. The idea was that this is needed because voice has, as it was said to me, constant requirements for bit rate, packet rate, Frame Erasure Rate (FER) and delay. Since it was also realized that voice is one very important service, one will need radio bearers that are optimized for these requirements.

The short round-trip time and the low packet loss rates were needed to make it possible for the TCP rate control to reach data rates up to the several megabits per second. This actually gave tougher latency and packet loss rate requirements for data than for real-time voice services.

When hearing about this, however, I stated that it is not true that voice has constant FER and delay requirements. The reasons why one uses constant requirements in the CS system is more a design choice than an actual speech property. We had been studying different redundancy schemes for a while and it was quite clear that the quality degradation due to packet losses were much worse for some speech frames than for others. Packet losses gave much larger distortions for onset frames and frames with discontinuities than for steady-state frames. This is because the error concealment, which typically uses repeat-and-mute, works much better for steady-state periods than for transitions regions.

Learning about the short end-to-end delays made me realize that the latency problem was going to be solved for data services, and the transport functions that accomplished the low delay could of course be used also for VoIP. One therefore no longer needed the great quality improvement with wideband speech to compensate for long delays. In addition, it seemed realistic that the low packet loss rates could also be achieved for voice.

Improved service quality would, however, still be needed because VoIP still required more resources than CS because of the non-zero header and since unequal error protection was not as optimal for VoIP as for CS, which gave lower capacity than for CS. Another factor that could probably also compensate for the reduced capacity was the fact that all-IP networks are typically less expensive to operate since one only has one network, the PS network, to manage instead of two networks, PS and CS.

These things made me realize, for the second time, that VoIP will be better than CS voice, even with narrowband voice.

The third 'Aha!'

My third Aha! experience came when learning more about the Hybrid Automatic Repeat reQuest (HARQ) performance and when I was involved in discussions and evaluations on the delay scheduler. When using HARQ, the delay scheduler, and a few other improvements, the capacity for VoIP in High Speed Packet Access (HSPA) was significantly increased and VoIP over HSPA now showed at least as good capacity figures as CS.

So now all components were in place for claiming that VoIPoHS will be better than CS. The quality of the sound will be as good as for CS, since the same codec is used. The performance will actually be a little better for most cases since most users will have lower FER than what they would have for CS voice in UTRAN. Using the same codec as in CS also makes it possible to do Tandem-Free Operation (TFO) with CS, which gives great backwards compatibility and maximizes the quality for interworking scenarios. And none of these optimizations reduced the flexibility, which means that it will still be very easy to improve the quality by introducing AMR-WB.

The end-to-end delay is also not going to be a big issue. The requirements for high bit rates for data services means that short delays are required because of the TCP rate control. The delays actually need to be shorter for data services than for voice, if one wants TCP to reach data rates up to several megabits per second. So data services will actually be the driver for shorter delays and it is natural to use the same transport mechanisms also for VoIP. Thereby the delays will be shorter than for CS for most users under most operating conditions. It is only for the very high loads that the users will experience delays that might be a little worse than for CS.

Conclusion

It is my opinion that VoIP over HSPA will be better than CS for the following reasons:

- The sound quality will be at least as good as for CS voice since the same codec is used and most users will have close to zero FER. The sound quality can also be significantly improved by introducing AMR-WB.

- The end-to-end delay will be about the same as or even shorter than for CS voice.

- The capacity will be at least as good as for CS.

These properties are, in my mind, the most important ones that will enable a successful launch of VoIP in HSPA.

It is my hope that this book will show how to do VoIP over HSPA and also that one should expect as good performance as for CS, or even slightly better, regarding both quality and capacity. Maybe the reader will even experience the same 'Aha! experiences' as I have experienced while working with VoIP?

Janne Peisa

Unlike Tomas, I have spent most of my career in telecommunications optimizing the air interface for IP-based applications. While doing so the focus was (almost) always on the applications using TCP. We quickly realized that one of the fundamental problems with the first cellular packet data access systems (especially GPRS) was the round-trip time (which was close to one second), and we became almost obsessed with reducing the air interface round-trip time. This culminated in the work for High Speed Packet Access (which introduced two millisecond transmission time interval) and Long Term Evolution of UTRAN (which will introduce an even shorter TTI).

It never occurred to me that there would be any interest in providing a high capacity voice service over the HSPA channels we had created. The design goal of the HSPA had always been interactive applications, and we explicitly ruled out any conversational services over HSPA. But suddenly this changed. Preliminary analysis showed that it was theoretically possible to reach or exceed the CS capacity for voice service, and I spent a lot of effort trying to understand how this is possible (for curious readers, the reasons are explained in Section 7.3). The outcome was surprising: when designing the HSPA we had accidentally designed an air interface that was capable of supporting voice applications with higher efficiency than the existing CS bearers could.

As soon as I understood that it would be better to provide the voice service with HSPA access, it also become apparent that suddenly we have *both* the flexibility of the IP-based applications, allowing one to quickly introduce new codecs, add new modes of communications, such as video calls or instant messaging, *and* the efficient performance the CS service.

After redesigning the air interface, it was time to redesign the basic telephony application – to replace the voice telephony with Multimedia Telephony.

I hope the reader can appreciate both the flexibility and the efficiency of the IMS Multimedia Telephony.

Per Synnergren

During my rather brief career in telecommunications I have had the pleasure to work with various nodes belonging to almost all the layers in the ISO/OSI reference model. But the common denominator has always been the end goal of realizing working packet switched communication services.

I started out working with speech coding during the early part of this decade. At that time, much work was being performed in my company, in universities and in the industry in general to optimize the operation of the speech codecs and de-jitter buffering algorithms to secure the voice quality for a voice service running over IP and Internet. Soon companies with Internet telephony as their main business sprung up and released products based on some of the ideas developed during this time frame. For some of the companies the timing was excellent and today we see the success of the IP and Internet telephony business. For me as for many others, it was obvious that IP-based telephony was going to be big business and the discussions about fixed–mobile convergence started to gain momentum.

IMS was the new thing everyone talked about! IMS had been specified in 3GPP release 5 and the first releases of the important base specifications were developed during the time period of 2001–02. However, IMS lacked services. IMS was built to be a general service platform that in theory didn't need any standardization of services. The thinking was that services could be developed by third party companies and just implemented on top of IMS using the ISC interface. But it was soon realized that in practice interoperability could only be achieved by standardization of the services. The first service was PoC (Push-to-talk over Cellular). In 2002, many companies in the telecommunication industry struggled. The operators lost money due to expensive 3G license fees and an increased price pressure on mobile phone calls. It was noticed that one operator seemed to handle the 'bad times' better than the rest, at least in the US. It was NEXTEL, and the specific thing with NEXTEL was their offerings to small and medium businesses. They had rugged phones for the blue collar segment, and they had services that no one else could offer. One such service was Push-to-talk, the cellular walkie-talkie with nationwide coverage. The operators and vendors were desperate to find a new blockbuster application that could help turn the tide around. Maybe PoC on IMS was the savior? Soon an industry consortium was formed that contained Ericsson, Nokia, Motorola and Siemens as the leading players. I ended up as one of many people that worked in this industry consortium producing the set of pre-OMA PoC specifications. This was a really fun time and we all had great hope that PoC was going to be the 'smash-hit' that was to promote IMS. During this time and during the time period I followed the PoC work in OMA I had the opportunity to work with and learn a lot about both the IMS control plane and IMS user plane.

PoC was soon surrounded by hype, but commercially it struggled. The reason soon became obvious. In 2003 and 2004 the commercial mobile networks that were deployed were not good enough to handle the real-time packet switched voice the PoC service produced. At this time the deployment of WCDMA had just started and market penetration was low. Thus PoC had to work over GSM/GPRS to be a success. PoC was designed in such a way that it could be used in a GSM/GPRS network even in situations when only one timeslot was assigned to the mobile terminal. At least it should work in theory, or maybe in a well-planned GSM network that was compliant to the latest 3GPP release. However, the GSM packet switched radio bearer suffered from significantly larger overhead than the CS radio bearer (the LLC and SNDCP overhead). Therefore, the coverage radius of the packet switched voice was significantly less than for CS voice. In reality the commercial GSM networks didn't support all standardized features that were beneficial for PoC and most often the cell planning was optimized for the CS voice service, leading to quality issues for PoC. From that experience I got interested in the radio related issues and I started to work with radio access functionality for IP multimedia.

WCDMA High Speed Packet Access (HSPA) is the most promising way forward. It is certainly not impossible to make packet switched voice services work well over GSM/GPRS and EDGE. For instance, the 3GPP work item EDGE continued evolution may secure the performance needed for the packet switched voice service over EDGE. Another alternative is WCDMA using dedicated channels. But neither of the alternatives above has the same potential as WCDMA HSPA to offer a versatile radio bearer that can deliver the service quality, system capacity and flexibility that allow the operator to do IP multimedia service offerings.

In this book we present the Multimedia Telephony communication service being standardized by 3GPP and promote the idea that Multimedia Telephony has the technological potential to beat the legacy CS telephony service when it comes to capacity and quality at least when utilizing the WCDMA HSPA air interface. I sincerely hope that this will be true also in real implementations. Then maybe in 10 years time we may be able to conclude that the introduction of WCDMA HSPA made IMS and its services become a commercial success.

Acknowledgments

This book is a joint effort, and the editors would like to thank all of our co-authors, Rolf Blom, Gonzalo Camarillo, Yi Cheng, Daniel Enström, Per Fröjdh, Vesa Lehtovirta, Karl Norrman, Göran Schultz, and Krister Svanbro, for their hard work.

The idea for writing this book was conceived while most of the editors were working at Ericsson Research. The editors would like to thank the personnel and management of Ericsson Research for providing exciting research topics to work on as well as the possibility to spend a small part of our working time actually preparing the book. Shyam Chakraborty wishes to thank Raimo Vuopionpera and Johan Torsner of Nomadic Lab, Ericsson Finland, for picking up the potential of this book at the first glance and providing the necessary support. In addition to Ericsson Research, Janne Peisa would like to thank the Mobile Media Gateway unit of Design Unit Core Network Evolution, which has fostered an atmosphere of innovativeness and research even as part of their normal design process. The support of Raul Söderström, Ari Jouppila and Johan Fagerström has been vital for the success of this book.

We would like to express our gratitude to Anders Nohlgren, Martin Körling, Sara Mazur, Hans Hermansson, Mats Nordberg, Håkan Olofsson, Lars Bergenlid, Krister Svanbro, Lotta Voigt, Stefan Håkansson, Fredrik Jansson and Torbjörn Einarsson for reading the manuscript and providing valuable comments.

We would also like to thank all our colleagues, with whom we have had many insightful discussions. We would especially like to thank Rickard Sjöberg, Mårten Ericson, Stefan Wänstedt and Stefan Wager, who have kindly allowed us to use their data as part of our performance evaluation chapter.

Last we would like to thank our families, for whom the process of writing the book has surely been stressful, for their support. Janne Peisa would like to thank Duyên and Duy. Per Synnergren would like to thank his children Johan and Rebecka for just being part of his life. Kids, this book was written during an extremely stressful period for us all, but for whatever it is worth I'll always love you! Shyam Chakraborty embraces Milan and Vikram for providing constant trouble and Joanna for providing boundless joy. Tomas Frankkila would like to thank Lars, Tyra and Kristina for their patience during this very busy period.

Glossary

3GPP	3rd Generation Partnership Project. An international forum responsible for standardizing the GSM and UMTS systems
3GPP2	3rd Generation Partnership Project 2
A-BGF	Access Border Gateway Function
AAC	Advanced Audio Coding
ACR	Anonymous Communication Rejection
ACS	Active Codec Set
ADPCM	Adaptive Differential PCM waveform codec
AEC	Acoustic Echo Cancellation
AF	Application Function
AGC	Automatic Gain Control
AKA	Authentication and Key Agreement
AL-SDU	Adaptation Layer SDU
AM	Acknowledged Mode. One of the modes of the UMTS RLC protocol
AMR	Adaptive Multi-Rate. Speech codec used in GSM and UMTS networks
AMR-WB	AMR wideband. 16 kHz speech codec specified for GSM and UMTS networks
APN	Access Point Name
ARQ	Automatic Repeat Request
AS	Application Server
ATM	Asynchronous Transfer Mode
AV	Authentication Vector
AVC	Advanced Video Coding
AVP	Audio-Video Profile. An RTP profile
BGCF	Breakout Gateway Control Function
BICC	Bearer Independent Call Control
BLER	BLock Error Rate
BSC	Base Station Controller
BSS	Base Station Subsystem
BTS	Base Transceiver Station

C/I	Carrier to Interference ratio. Indication of the link quality
CB	Communication Barring
CCDF	Complementary Cumulative Distribution Function
CCPCH	Common Control Physical CHannel
CCS	Composite Character Sequence
CD	Communication Deflection
CDF	Charging Data Function
CDIV	Communication DIVersion
CDMA	Code Division Multiple Access
CDMA2000	A family of third-generation (3G) mobile telecommunications standards that use CDMA specified by 3GPP2
CDR	Charging Data Record
CELP	Codebook Excited Linear Prediction. General terminology for a group of speech codecs
CFB	Communication Forwarding on Busy user
CFNL	Communication Forwarding on Not Logged-in
CFNR	Communication Forwarding on No Reply
CFNRc	Communication Forwarding on Mobile Subscriber Not Reachable
CFU	Communication Forwarding Unconditional
CGF	Charging Gateway Function
CIF	Common Intermediate Format
CMC	Codec Mode Command
CMI	Codec Mode Indication
CMR	Codec Mode Request
CN	Core Network
CONF	CONFerence
CQI	Channel Quality Indicator. Measurement of the downlink channel quality used for HS-DSCH in UMTS
CRC	Cyclic Redundancy Check
CRT	Cathode Ray Tube
CRTP	Compressed RTP
CS	Circuit Switched
CSCF	Call Session Control Function
CSICS	Circuit Switched IMS Combinational Service
CSQ	Circuit Switched Quality
DCCH	Dedicated Control CHannel. Logical channel used in UMTS
DCH	Dedicated CHannel. Transport channel used in UMTS
DCT	Discrete Cosine Transform

DL	DownLink
DNS	Domain Name System
DPCCH	Dedicated Physical Control CHannel
DPCH	Dedicated Physical CHannel
DPDCH	Dedicated Physical Data CHannel
DRX	Discontinuous Reception
DTCH	Dedicated Traffic CHannel. Logical channel used in UMTS
DTX	Discontinuous Transmission
DVD	Digital Versatile Disc or Digital Video Disc
E-AGCH	Absolute Grant CHannel. Control channel used to schedule transmissions on E-DCH
E-DCH	Enhanced Dedicated CHannel. Improved version of the dedicated transport channels in UMTS systems
E-DPCCH	Enhanced Dedicated Physical Control CHannel. Physical channel used to carry control information for E-DCH
E-DPDCH	Enhanced Dedicated Physical Data CHannel. Physical channel used to carry E-DCH
E-HICH	HARQ Indicator CHannel. Control channel used to for E-DCH
E-RGCH	Relative Grant CHannel. Control channel used to schedule transmissions on E-DCH
EC	Echo Cancellation
ECT	Explicit Communication Transfer
ECU	Error Concealment Unit
EDGE	Enhanced Data rates for GSM Evolution. An updated air interface for GPRS
EEP	Equal Error Protection
EFR	Enhanced Full Rate. Terminology commonly used for improved versions of full rate speech codecs
EIA	Electronic Industries Alliance
ENUM	Electronic NUMbering
EQ	Economy Quality
ESP	Encapsulating Security Payload
ETSI	European Telecommunication Standards Institute
EUL	Enhanced UpLink
FACH	Forward Access CHannel
FBC	Flow Based Charging
FEC	Forward Error Correction
FER	Frame Erasure Rate
FMC	Fixed Mobile Convergence
FR	Full Rate. Often used in combination with a speech codec operating on a full rate channel

FTP	File Transfer Protocol
Gb	Interface between GSM/EDGE Radio Access Network and Core Network
GERAN	GSM/EDGE Radio Access Network
GGSN	Gateway GPRS Support Node
GIF	Graphics Interchange Format
GOB	Group of Blocks
GPRS	General Packet Radio Service
GPS	Global Positioning System
GRX	GPRS Roaming eXchange
GSM	Global System for Mobile communications
GSM-EFR	Enhanced Full Rate speech codec for GSM
GSMA	GSM Association
GTP	GPRS Tunneling Protocol
HARQ	Hybrid Automatic Repeat reQuest
HDTV	High-Definition TV
HLR	Home Location Register
HOLD	Offline Communication Hold
HQ	High Quality
HS-DPCCH	High Speed Dedicated Physical Control CHannel. Special physical control channel used for HSDPA
HS-DSCH	High Speed Downlink Shared CHannel. Transport channel used for HSDPA
HS-PDSCH	High Speed Physical Downlink Shared CHannel. Physical channel used for HSDPA
HS-SCCH	High Speed Shared Control CHannel
HSDPA	High Speed Downlink Packet Access. An improved air interface for UMTS downlink transmission
HSPA	High Speed Packet Access. Enhancement of the UMTS packet acces. HSPA consists of HSDPA and E-DCH
HSS	Home Subscriber Server
HTTP	HyperText Transfer Protocol
I-BGF	Interconnect Border Gateway Function
I-CSCF	Interrogating Call Session Control Function
IBCF	Interconnect Border Control Function
ICB	Incoming Communication Barring
IDEN	Integrated Digital Enhanced Network
IDR	Instantaneous Decoder Refresh
IEC	International Electrotechnical Commission
IETF	Internet Engineering Task Force
IKE	Internet Key Exchange
IM	Instant Messaging

IM-MGW	IP Multimedia Media GateWay function
IMPI	Private User Identity
IMPU	Public User Identity
IMS	IP Multimedia Subsystem
IMS-GWF	IMS GateWay Function
IMSI	International Mobile Subscriber Identity
IP	Internet Protocol
IPv4	Internet Protocol version 4
IPv6	Internet Protocol version 6
IRS	Intermediate Reference System
ISDN	Integrated Service Digital Network
ISIM	IP multimedia Service Identity Module
ISO	International Organization for Standardization
ISP	Internet Service Provider
ISUP	ISDN User Part
ITU	International Telecommunication Union
Iu	Interface between UMTS Radio Access Network and Core Network
Iub	Interface between UMTS RNC and Node B
JFIF	JPEG File Interchange Format
JPEG	Joint Photographic Experts Group
JVT	Joint Video Team
L1	Layer 1
L2	Layer 2
LAN	Local Area Network
LCD	Liquid Crystal Display
LLC	Logical Link Control. Protocol layer used in GPRS
LPC	Linear Predictive Coding
LTE	UTRAN Long Term Evolution
MAC	Medium Access Control
MAC-d	Dedicated MAC entity. Link layer entity used for Medium Access Control in UMTS
MAC-e	Enhanced MAC entity. Link layer entity used for Medium Access Control in UMTS when using E-DCH
MAC-hs	High Speed MAC entity. Link layer entity used for Medium Access Control in UMTS when using HS-DSCH
MGCF	Media Gateway Control Function
MGW	Media GateWay
MIDI	Musical Instrument Digital Interface
ModIRS	Modified IRS
MOS	Mean Opinion Score

MPEG	Moving Picture Experts Group
MRF	Media Resource Function
MRFC	Media Resource Function Controller
MRFP	Media Resource Function Processor
MSC	Mobile Switching Center
MSC Server	Mobile Switching Center Server
MSRP	Message Session Relay Protocol
MT	Mobile Terminal
MTP	Message Transfer Part
MTU	Maximum Transmission Unit
MVC	Multiview Video Coding
MWI	Message Waiting Indication
NAI	Network Access Identifier
NAL	Network Abstraction Layer
NASS	Network Attachment Sub-System
NDS	Network Domain Security
NDS/AF	Network Domain Security Authentication Framework
NDS/IP	Network Domain Security for IP
NE	Network Element
NGN	Next Generation Network
Node B	UMTS radio base station
NRSPCA	Network Requested Secondary PDP Context Activation
NSP	Network Service Provider
OCB	Outgoing Communication Barring
OCS	Online Charging System
OFCS	OFfline Charging System
OFDM	Orthogonal Frequency Division Multiplexing
OIP	Originating Indication Presentation
OIR	Originating Indication Restriction
OMA	Open Mobile Alliance
OSI	Open System Interconnect
P-CCPCH	Primary Common Control Physical CHannel
P-CSCF	Proxy Call Session Control Function
PC	Personal Computer
PCC	Policy Control and Charging
PCEF	Policy and Charging Enforcement Function
PCM	Pulse Code Modulation waveform codec
PCRF	Policy and Charging Rules Function
PDC	Personal Digital Cellular
PDC-EFR	Enhanced Full Rate codec in PDC
PDCP	Packet Data Convergence Protocol

PDF	Policy Decision Function
PDP	Packet Data Protocol
PDU	Protocol Data Unit
PEP	Policy Enforcement Point
PESQ	Perceptual Evaluation of Speech Quality
PFC	Packet Flow Context
PLC	Packet Loss Concealment
PLMN	Public Land Mobile Network
PLR	Packet Loss Rate
PMM	Packet Mobility Management
PNG	Portable Networks Graphics
PoC	Push-to-talk over Cellular
PRACH	Physical Random Access CHannel
PS	Packet Switched
PSI	Public Service Identity
PSTN	Public Switched Telephone Network
QCIF	Quarter CIF
QoS	Quality of Service
QVGA	Quarter VGA
RAB	Radio Access Bearer
RACH	Random Access CHannel
RACS	Resource and Admission Control Sub-system
RAN	Radio Access Network
RB	Radio Bearer
RFC	Request For Comments
RGB	Red–Green–Blue
RLC	Radio Link Control. Link layer protocol used in UMTS
RNC	Radio Network Controller
ROHC	RObust Header Compression
RRC	Radio Resource Control protocol
RTCP	RTP Control Protocol
RTCP XR	RTCP eXtended Reports
RTP	Real-time Transport Protocol
S-CCPCH	Secondary Common Control Physical CHannel
S-CSCF	Serving Call Session Control Function
SA	Security Association
SB-ADPCM	Sub-Band ADPCM waveform codec
SBLP	Service Based Local Policy
SC-FDMA	Single Carrier Frequency Division Multiple Access
SCTP	Stream Control Transmission Protocol
SDP	Session Description Protocol
SDU	Service Data Unit

SEG	SEcurity Gateways
SGSN	Serving GPRS Support Node
SGW	Signalling GateWay function
SigComp	Signalling Compression
SIM	Subscriber Identity Module
SIMPLE	SIP for Instant Messaging and Presence Leveraging Extensions
SIP	Session Initiation Protocol
SLF	Subscription Location Function
SM	Session Management
SMIL	Synchronised Multimedia Integration Language
SNDCP	Sub Network Dependent Convergence Protocol
SPIT	SPam over Internet Telephony
SPR	Subscription Profile Repository
SRTP	Secure Real-time Transport Protocol
SVG	Scalable Vector Graphics
TAS	Telephony Application Server
TBCP	Talk Burst Control Protocol
TCP	Transmission Control Protocol
TDMA	Time Division Multiple Access
TDMA-EFR	Enhanced Full Rate codec in TDMA
TE	Terminal Equipment
TETRA	TErrestrial Trunked RAdio
TFO	Tandem Free Operation
TFT	Traffic Flow Template
THIG	Topology Hiding Interwork Gateway
TIA	Telecommunications Industry Association
TIP	Terminating Indication Presentation
TIR	Terminating Indication Restriction
TM	Transparent Mode. One of the modes of the UMTS RLC protocol
TPF	Traffic Plane Function
TrFO	Transcoder Free Operation
TrGW	Transition GateWay
TTI	Transmission Time Interval
TTL	Time To Live
UDP	User Datagram Protocol
UDVM	Universal Decompressor Virtual Machine
UE	User Equipment
UEP	Unequal Error Protection
UICC	Universal Integrated Circuit Card
UL	UpLink
UM	Unacknowledged Mode. One of the modes of the UMTS RLC protocol

UMTS	Universal Mobile Telecommunications System
URI	Uniform Resource Identifier
URL	Uniform Resource Locator
UTRAN	UMTS Terrestial Radio Access Network
VCD	Video CD
VCL	Video Coding Layer
VCR	VideoCassette Recorder
VGA	Video Graphics Array
VHS	Video Home System
VoIP	Voice over IP
WCDMA	Wideband Code Division Multiple Access. The air interface technology used for UMTS
WLAN	Wireless LAN
WWW	World Wide Web
XCAP	XML Configuration Access Protocol
XDM	XML Document Management
XML	eXtensible Markup Language
YCbCr	Luminance and (two) color difference signals used for video processing

Chapter 1

Introduction

Shyam Chakraborty, Tomas Frankkila

The innovation of voice communication with electrical means by Alexander Graham Bell initiated the era of Public Switched Telephone Networks (PSTN), and revolutionized the way people in modern civilization would communicate. The important elements of PSTN are full-duplex conversational service, narrowband speech and Circuit Switching (CS). The success of PSTN provided a strong impetus for seeking improvements and modifications to the basic conversational service to provide a richer experience, better convenience and wider availability to the user, in a variety of ways. The evolution of digital signal processing allowed conversion of analog signals to digital format to take advantage of digital communication and switching techniques. The development of the Integrated Service Digital Network (ISDN) provisioned different services, speech telephony, video telephony and data, over the same interface of a single network. Intelligent networking provided enhancement to basic services and cellular systems added the freedom from desktop telephony. Needless to say, these innovations have also been strong stimulants for industrial growth.

The scope of non-conversational data communications that was catering primarily for telegraphy and facsimile expanded very much with the proliferation of computers. The information generated by computers is highly bursty in nature and communicated either one way (simplex) or interactively (half-duplex). This necessitated a totally new switching paradigm, known as packet switching, which essentially stemmed from a US Department of Defense (DoD) research network ARPANET. Computers proliferated from research laboratories to work places as daily working tools, and ultimately to homes as consumer goods providing information with web (World Wide Web, WWW) browsing, communication with emails and entertainment with a variety of games, music, videos, etc., over a single user interface. This growth is actually aided by the evolution of data networks from the ARPANET to the Internet, a huge collection of heterogeneous networks glued together by the TCP/IP protocol suite, and reaching practically all the offices and homes of a modern society.

Despite the provision of mobility, the first generation analog systems, for example, NMT, TACS and AMPS, met with limited market success, mainly because of limited coverage, bulky terminals and high cost. The potentials of mobile communications were better exploited with the digital second generation Global System for Mobile communication (GSM) that provides circuit switched connections to carry low bit rate coded speech or data. The GSM systems have enjoyed enormous market response due to the highly portable terminals, thanks

IMS Multimedia Telephony over Cellular Systems   S. Chakraborty, T. Frankkila, J. Peisa and P. Synnergren
© 2007 John Wiley & Sons, Ltd

to component miniaturization and improved battery technology, a voice quality that matches closely the quality of PSTN, reduced cost due to scale-factor cost benefits and rapidly expanding coverage. These attributes have transformed cellular mobile systems from business tools to consumer items with a much bigger market potential. An important dimension added to conversational telephony was the introduction of Short Message Service (SMS). Akin to email, the SMS is a much simpler, informal and user friendly service. Though SMS is not a truly conversational service, its high interactive feature is extensively used in near-real-time chatting, and often complements a voice conversation with such additional text information as exchange of address, telephone number, jokes etc. Not only has this enhanced interpersonal communications to a new height, but also it has been a major revenue churner for the operators. With advanced display devices, processors and networking technology, cellular systems are increasingly equipped with data services, Multimedia Message Service (MMS), video telephony and other non-communication facilities, such as camera, MP3 player, etc. An up-to-date mobile terminal is thus a highly sophisticated user equipment with multi-service capabilities.

1.1 Convergence of Networking Paradigms

The concept of an integrated network that provides the user with a variety of services over a single interface has a number of advantages for both the users and the operators. This is well reflected in the developments of both circuit switched narrowband ISDN (N-ISDN) and further packet switched broadband ISDN (B-ISDN) networks. However, a homogeneous, 'end-to-end Asynchronous Transfer Mode (ATM)' concept of a B-ISDN had some obvious difficulties in taking off, and the Internet as a heterogeneous collection of subnetworks has become increasingly dominant. The Internet provides an abundance of bandwidth owing to its core optical fiber network. This factor makes a good case for an integrated network. However, the inherent design principles of the Internet provide only a 'best effort' service that is typically not optimized for conversational services. To provide a variety of services over the Internet, a few improvements in traffic management and handling within the network are beneficial. The first of them is to define Quality of Services (QoS) requirements for the different services. The second is to develop means within the Internet to provide 'better than best-effort' services that would suit the different QoS requirements of different conversational, streaming and interactive classes of services. Examples of such improvements are provisioning of Diffserv (differentiated services) and Multi-Protocol Label Switching (MPLS) that handles the different QoS classes with a statistical guarantee rather than an absolute guarantee. Equipped with these techniques, the Internet can provide the basis of an efficient packet switched voice, commonly known as Voice over IP (VoIP).

The explosion of the Internet into the consumer market has also influenced the cellular systems. While GSM provided data communications over fixed rate circuit switched links at 9.6 kbps, packet services with higher data rate capabilities are increasingly offered. This is reflected in such enhancements of the second generation GSM systems as General Packet Radio Services (GPRS) and Enhanced Data rates for GSM Evolution (EDGE), and the third generation system Universal Mobile Telecommunication Systems (UMTS). These systems offer increasingly higher speed packet access to the Internet, in parallel to the circuit switched links to be provided for narrowband voice telephony. One can therefore truthfully conclude that fixed and mobile networks have converged for data communication.

The fixed–mobile convergence of real-time services has however not been realized so far. Real-time services may have bit rate requirements that are significantly lower than high-speed data services, but the real-time services have the special requirement that the transport of the packets must be constant. For data services, interruptions of for example 100–200 ms or even up to 500 ms are insignificant, even if they occur quite frequently. Interruptions of 100–200 ms, in addition to the delay jitter, can only be allowed for real-time services if the interruptions are very rare. And longer interruptions can never be allowed. The reason is that the receiving client cannot buffer enough frames to survive such interruptions without causing interruptions in the produced sound, because long buffering times increase the response time from the users, which reduces the conversational quality for the users that are involved in the conversation. The alternative, to maintain a short buffering time and accept silence periods in the produced sound, is even worse since this reduces the listening quality even faster. Another fundamental requirement for successful convergence between real-time and data services is the fact that the capacity for the real-time services must be as good for the packet switched system as for the corresponding circuit switched system.

IMS is the system that enables fixed–mobile convergence, as shown in Figure 1.1, and bridges the gap between these two environments. The IMS system will control the session and route the media stream regardless of the access types that are used and regardless of what operators are involved. A key benefit of the IMS system is that it is also backwards compatible since signaling and media gateways will be allocated when at least one of the users is using a legacy circuit switched system.

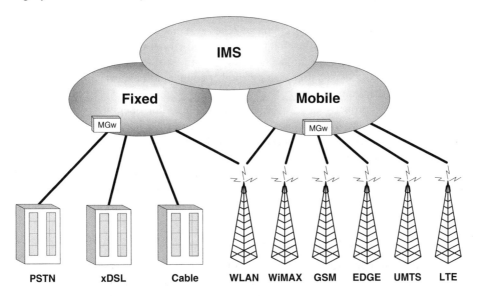

Figure 1.1: Fixed–mobile convergence with IMS as the platform that enables communication regardless of access type and regardless of whether the users have subscriptions with the same or different operators.

1.2 IMS and the IMS Multimedia Telephony Service

The IMS Multimedia Telephony service is seen as the next step in telecommunications and the services which would finally harmonize telephony with data communications. The service will offer enriched communication with real-time speech, video and text communication. In addition, it also offers file sharing and media sharing capabilities that allow users to send, for example, images and video clips to other users. The development of the Multimedia Telephony service is driven by the fact that end-users desire to communicate in new ways, while still requiring a telecom grade service for the traditional real-time voice and video telephony service components, i.e. the same quality, reliability and security.

For enterprise users, Multimedia Telephony will also offer integration of email, support for remote workers, conferencing and collaboration features. Another important property of Multimedia Telephony is personal mobility. Professional users want to use the telephony and data communication services in the same way while traveling as when being in the office. A third important feature for enterprise users is the possibility to control what calls and sessions should be allowed at any given point in time. For example, when attending a business meeting or a conference, the users may want to allow only the most important calls or sessions to reach the receiver. Other, less important, calls may be routed to an answering machine where the sound is recorded and attached in an e-mail to the subscriber's e-mail address.

IMS and the IMS Multimedia Telephony service are interesting also for the operators for several reasons. The IMS system incorporates the control mechanisms that are required in order to ensure that they can meet the users' expectations on quality of the service, reliability and security. The generic architecture and flexibility of the IMS platform also offers simple development and deployment of new services that can be added to the existing services, which gives the operators new revenue opportunities. In addition, the operator also requires an efficiency that is on a par with existing circuit switched systems and also interoperability with legacy systems. IMS also supports developing standardized services, such as IMS Multimedia Telephony. The standardization is required to make the core set of service components behave the same for all operators since a user visiting another country still expects the same service behavior as in his or her home network.

Another attractive property of the IMS is that it allows for operating a single all-IP network, thereby removing the need for maintaining a circuit switched system in parallel with the IP network. This reduces the need for capital expenditure (CAPEX) as well as the operational expenditure (OPEX).

The Multimedia Telephony service is also transport agnostic. This means that service providers only need to implement one version of the service, which significantly reduces the implementation, testing and verification efforts and allows for faster time to market.

1.3   Requirements and Challenges

In all cellular telephony systems up to 3G, voice communication is the service that has defined the toughest requirements on the system. These requirements are related to: capacity, quality, error rates, end-to-end delay and consistent delivered bit rate. Before the introduction of GPRS, voice communication even defined the bit rate requirements.

This changed with the introduction of GPRS, EDGE and data services in 3G. For future systems, it is also the data services that will put the highest requirements on the system regarding error rates (packet loss rates) and end-to-end delay. This is because low packet loss rates and short round-trip times are required if rate control in TCP is to be able to adapt to data rates of several megabits per second. Thereby, it is no longer voice communication that defines the requirements for throughput and delay, but rather data communication.

The requirements that IMS Multimedia Telephony has to fulfill are still challenging. To be able to replace legacy circuit switched services, the capacity and the coverage need to be at least as good. The capacity and coverage requirements must be fulfilled while still delivering the same quality, or better, even when the system is loaded to the capacity limit.

In addition, the system design must also be more flexible than the existing circuit switched systems. This is needed in order to allow for developing and deploying different service variants and to have a future-proof solution that allows for simple introduction of new service components.

During the development of the IMS Multimedia Telephony service, one important property has been that the quality experienced should be consistent. The service should deliver similar performance, in terms of both capacity and quality, regardless of network, operating condition, equipment, etc.

Another requirement is that the IMS Multimedia Telephony service must show consistent behavior for different networks. Equipment from different vendors must also give comparable performance. This is needed because the end-users want the freedom to purchase phones from different vendors and they expect that these will show consistent behavior. Inconsistent performance would also give problems for the operators, if the inconsistencies had an impact on the cell planning, since the operator then might not know if his cell planning gives sufficient coverage or not. This would, for example, be the case if one phone gives adequate performance while another phones gives too much frame erasures, for the same delay jitter.

The IMS Multimedia Telephony service must therefore be standardized in sufficient detail so that the vendors can know that their products will fulfill the performance requirements. This requirement is a real challenge since different transport networks may have quite different characteristics and capabilities.

1.4 Outline of this Book

A prerequisite for IMS Multimedia Telephony to become successful is, as described above, that the basic voice service must be able to replace the traditional circuit switched voice communication in existing cellular systems. To fulfill this requirement, the VoIP in Multimedia Telephony must deliver the same quality as CS voice, while still matching the capacity of the legacy systems. In this book we describe how this requirement can be fulfilled for HSPA systems. We try to cover all areas needed to understand how the IMS Multimedia Telephony service works and how the performance requirements can be fulfilled.

In Chapter 2, the IMS Multimedia Telephony service is discussed in more detail. A few show examples how the service may be used. Requirements are also briefly discussed.

Chapter 3 describes the architecture of the IMS system and the service realization for the Multimedia Telephony. Interworking with legacy system is also an important aspect, since one can expect that the legacy systems will not be replaced overnight. Interconnecting to legacy systems is therefore also briefly described.

Session control is an important part of IMS services and is discussed in Chapter 4. The session setup obviously sets up the session between the users, but it also defines the service components that can be used in the session. One can expect that different users have different subscriptions, which will allow different service components. Session control is also used for routing (finding the other user), setting up radio bearers, QoS, charging, policing, etc.

Chapter 5 describes the media flow between the users. This includes: how the media is encoded, the protocols that are needed for the media, the transmission impairments that may occur in packet switched networks, and how these transmission impairments are managed in the receiving client. This chapter furthermore outlines solutions for interworking with legacy systems. Most of these things are not unique for HSPA or other cellular systems but rather generic for all IP systems. The last section in this chapter therefore presents the special considerations that have been made during the development of the IMS Multimedia Telephony service.

Chapter 6 provides an overview of the security components and mechanisms defined for IMS. The access domain and IMS domain security solutions are discussed, and the outlook of the currently discussed extensions is presented.

In Chapter 7, the performance of the IMS Multimedia Telephony system over HSPA is addressed. The chapter shows that the capacity for voice service is as good, or even better than for the circuit switched service with similar quality requirements. The possibilities to enhance quality are discussed. In addition to the capacity, the coverage for Multimedia Telephony service is addressed, and the network characteristics are presented. Finally the call setup delays are briefly examined.

Chapter 8 discusses services that are related to IMS Multimedia Telephony. These services are: the Circuit Switched IMS Combinational Service (CSICS), Push-to-talk over Cellular (PoC), weShare, Instant Messaging (IM), presence and group list management.

The summary of the book is found in Chapter 9.

The focus of this book is IMS Multimedia Telephony over HSPA, and especially the real-time speech, video and text media included in this service. There is nothing that prohibits using the IMS Multimedia Telephony service on other access methods like EDGE, TISPAN (land-line IP) networks, WLAN or WiMAX, but these access methods are outside the scope of this book.

A few related services, for example PoC, instant messaging and presence, are briefly discussed. Supplementary services like Message Waiting Indication (MWI), Originating Identification Presentation (OIP), Communication Hold (HOLD), etc., are also described.

Chapter 2

The Multimedia Telephony Communication Service

Daniel Enström, Krister Svanbro, Per Synnergren

Since the introduction of IMS in 3GPP release 5, VoIP over cellular access has been discussed in the mobile industry. However, the early 3GPP specifications did not specify any VoIP service in detail and thus there was an ambiguity in the industry regarding the technical realization of an interoperable VoIP service. To address this issue, Ericsson initiated an activity to define an IMS-based Multimedia Telephony communication service in 3GPP release 7. The aim was to create specifications defining a minimum set of capabilities required to secure a multi-vendor and multi-operator inter-operable Multimedia Telephony communication service, which could be supplemented by richer media types and classical telephony-type features such supplementary services (call forwarding, etc.).

This chapter discusses the benefits of an all-IP cellular system that uses IMS as the service platform and thus discusses the importance of being able to realize an IP-based telephony service. Further, the concept of standardized IMS communication services is explained, and the relationship between Multimedia Telephony and other standardized IMS services like Open Mobile Alliance (OMA) Push-to-talk over Cellular (PoC) is discussed. The main part of the chapter is devoted to outline how Multimedia Telephony could be envisioned by a user through the presentation of a service scenario. From this service scenario some of the requirements that a Multimedia Telephony communication service must meet are derived. It should however be noted that 3GPP only specifies the behavior of the Multimedia Telephony client software toward the IMS-based network. The exact user experience resulting from the Multimedia Telephony client software implementation toward the user is not standardized. It is thus important to understand that there is not a one-to-one mapping between the standardized Multimedia Telephony communication service and the resulting user experience that is outlined in this chapter.

2.1 Benefits with IMS

The vision of an all-IP cellular system has been present within the industry for several years. The vision has included different phases, from introducing IP in fixed parts of radio

---

IMS Multimedia Telephony over Cellular Systems   S. Chakraborty, T. Frankkila, J. Peisa and P. Synnergren
© 2007 John Wiley & Sons, Ltd

access and core network for simplified transport to introducing full-fledged support for all types of services also over the air interface. The latter would then include real-time services like Voice over IP (VoIP). On the other hand, the circuit switched voice service has been a tremendous success and continues to make new progress. It provides a reliable, efficient voice service with acceptable quality and is deployed in cellular networks all over the world today. Any new cellular voice service realization will thus be compared with the circuit switched voice service. The circuit switched voice service is today complemented with e.g. SMS for messaging and mobile data but the creation of new service offerings by the introduction of new media components to complement the circuit switched voice service is not a straightforward process.

There is certainly something very compelling about the notion of having one network and one architecture for all traffic and all types of services. Most likely this network would be a packet switched network based on IP. This is the essence of the all-IP network vision. It might not be feasible to quickly migrate to an all-IP network from a deployment point of view and it is important to look at a phased migration, but this vision is still valid, yesterday, today, as well as tomorrow. The idea would be to create a network that supports the introduction of a multitude of interoperable services based on one coherent set of technologies. This all-IP network could also be a cornerstone in a convergence of fixed and mobile systems. For the voice or telephony service over cellular systems in this context, it is only natural that the performance of this service is compared with today's circuit switched alternatives. A future packet-based telephony service over cellular systems must thus show the technology potential to at least the same performance as voice services of today but also provide something extra. This extra can be summarized with multimedia capabilities. Multimedia capabilities in this context imply that it should be possible to create rich communication services by combining e.g. video, audio, messaging and data capabilities to enable a communication service suited for every need. Figure 2.1 illustrates some possible components of such an enriched communication service.

This new telephony or communication service must thus be able to be part of complex multimedia application scenarios while at the same time provide circuit switched equivalent performance over cellular access. This is a challenge that calls for a structured way of realizing these new services. The IMS (IP Multimedia Subsystem) as specified by 3GPP provides the required functionality to both cellular and fixed network technologies.

IMS is the means to provide interoperable multimedia services over cellular systems. IMS can be seen as the technology platform that makes it possible to reuse parts of the protocol stack that are common to all services. For example, addressing or routing is typically common to all IP multimedia services. Hence, IMS can be seen as a horizontally layered technology platform that enables different (vertical) service realizations. This is further visualized in Figure 2.2. For each application there are of course some parts of the application logic that are unique for the service and some parts such as handling of contact lists or use of codecs may be common among the different applications. This fact is utilized in IMS to provide a common base for several services.

The requirements on IMS have been further specified in 3GPP TS22.228 [43]. These requirements conclude that support for IP multimedia sessions shall be provided in a flexible manner to allow operators to differentiate services in the market place and also to customize the services to fulfill specific needs. This is provided by the usage of service capabilities in both networks and terminals. The requirements on IMS also include the principle of access independence even though it should be noted that, even if this principle applies in general,

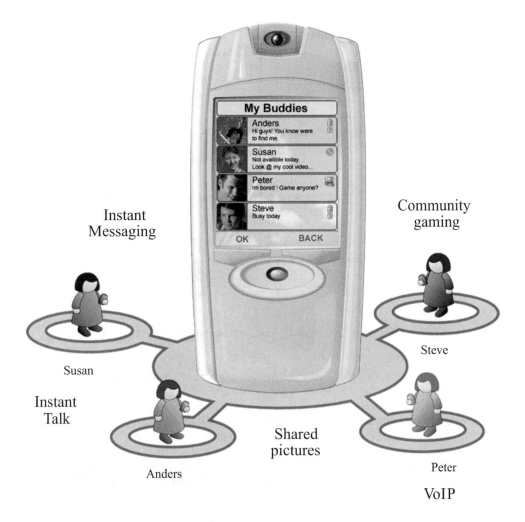

Figure 2.1: Example of components in an enriched communication service scenario.

it can only be assured by 3GPP for the access technologies that 3GPP specifies or has specific interworking with.

The high level requirements on IMS systems for IP multimedia applications include the following:

- Possibility to negotiate QoS at session setup as well as during the session.

- End-to-end voice quality at least as good as that achieved by the circuit switched (e.g. AMR codec based) telephony services.

- Support of roaming and negotiation between operators for QoS and service capabilities.

- Possibility of policy control of IP multimedia applications.

- Support for a variety of different media types but also a defined set of default media types to ensure interoperability.

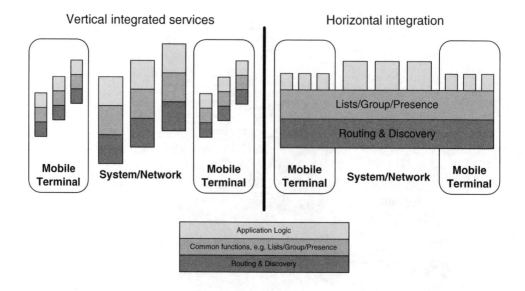

Figure 2.2: IMS as a technology platform supporting different services.

- Possibility to have several IP multimedia applications within each IP multimedia session.

- Maintain the amount of privacy, security, or authentication protection compared to corresponding GPRS and circuit switched services.

- The principle of access independence shall be supported. It should be possible to offer services to users regardless of how they achieve connectivity (e.g. WCDMA, GPRS, fixed lines, LAN).

- Possibility to support applications that have been developed outside the 3GPP community.

Fulfilment of these requirements provides benefits to the end-user, the operator and the vendors.

The end-user will benefit from rich multimedia interoperable communication services, maintained security and reliability of services also in all-IP networks in addition to a service transparency from today's circuit switched voice services to tomorrow's VoIP/IMS-based services. The end-user will also benefit from the possibility to have interoperability between operators and terminals.

The operator will benefit from an improved common control of services as well as an improved efficiency of IP multimedia services. An improved control of offered services from an operator include:

- radio bearer control with service controlled and optimized QoS;

- common authentication and authorization mechanisms;

- service control and fraud management;

- charging mechanisms;

- legacy telephony interworking.

The operator will also benefit from improved efficiency with IP technology and the IMS:

- IMS can help improve radio efficiency by providing services information to the radio bearer establishment process. This helps the system to choose suitable radio bearers for services like VoIP that include the usage of performance boosters such as header compression (e.g. ROHC).

- Radio access and core network transport efficiency is improved by using IP technology.

The vendors will benefit from building one technology platform for supporting a variety of services compared to developing new technology for each new service. Interoperability and possibility for larger markets for standard-based technology will also be beneficial for development of system infrastructure and terminals.

Hence, IMS as specified by 3GPP is a strong alternative for meeting the challenges of the all-IP vision. The voice service will of course also in future systems be of vital importance. Multimedia Telephony is the IMS way of providing IP-based telephony over cellular systems.

2.2 IMS Communication Services

IMS was already from the beginning seen as a service platform that should be able to support a multitude of different service types including telephony kind of services. The 3GPP provided the first set of IMS specifications in 3GPP release 5 in 2002. But the first set of IMS specifications lacked standardized services. In fact, the early view was that IP multimedia applications should not be standardized and existing legacy telephony and supplementary services should not be restandardized as IP multimedia applications. Instead, the start of the IMS standardization effort became tightly coupled to a set of specified signaling mechanisms, protocols and media types that for instance could be used to realize a rudimentary single multimedia conversational service using the call control, codecs and transport protocols specified in 3GPP TS 24.229 [25], 3GPP TS26.235 [37] and 3GPP TS26.236 [38]. The main focus of IMS according to the original view was to:

- allow service differentiation between operators on a single market;

- allow a large number of different services to be developed and deployed;

- design its network nodes in a service agnostic way.

Other standardization communities (e.g. 3GPP2, OMA and TISPAN) incorporated IMS solutions in their specification making it possible to have true interoperable IP services within managed networks stemming from different standardized technologies. In this process, the view on standardization of IMS services changed. The shift of view on IMS and IMS services triggered new standardization activities after the first release of IMS in 2002. After 3GPP release 5, the foundation had been laid which included the usage of IETF protocols, adopting them to the IMS usage and business model, and getting the overall infrastructure in place. After this period, new standardization efforts were started which focused upon specific

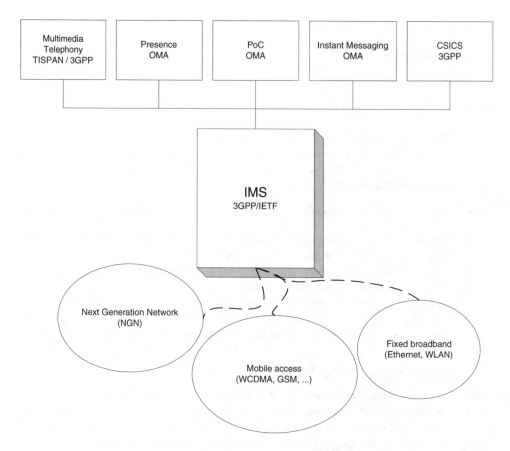

Figure 2.3: Standardization overview of IMS-based services.

service standardization on top of the common IMS specification. Several standardization bodies have been involved, some of which are shown in Figure 2.3.

Today, the efforts made in OMA, TISPAN and 3GPP has evolved IMS into a broad service platform that has a number of specified services that can be offered to customers. The evolved view on standardized IMS services has made the definition of the purpose of IMS somewhat more complicated. For instance, a multimedia communication service would benefit from service interoperability and service predictability. Therefore, in later years much focus has been put on:

- service interoperability;

- uniform service behavior;

- harmonized, single service design on user terminals;

- optimizations for a single service.

This expanded view has made it necessary to enable IMS to support a basic set of functionalities upon which applications can be based which then offer interoperability on a technical level. But still, IMS has to support both operator cooperation as well as operator differentiation. Therefore, IMS services should not be specified to such a detailed level that operator differentiation is not possible. It is within this scope that the IMS communication services should be viewed.

As mentioned above a number of different IMS-based services have emerged from the respective standardization processes. At first the different IMS services were viewed in the standardization processes as almost completely autonomous services on top of the IMS service layer. The consequence was that each of these services was designed to be deployed alone in the service layer network. Hence, none of the services has within their original scope to coexist with other services or to provide functionality to share resources with other services. This could lead to duplication of functions and a need to update each standard with new functions. As a consequence, there may be an inefficient use of resources and market fragmentation.

The service standardization was thus somewhat conflicting with the view that IMS should be seen as a multi-environment enabling multimedia, multi-session and multi-service. This 'multi view' on IMS certainly satisfies the need to get a flexible toolbox to build services and applications, but what about the vision of an IMS-based real-time multimedia communication service? The answer is that some sort of service component bundling, harmonizing with other standardized services, should be used to realize such a service offering.

An IMS communication service (Figure 2.4) is a way of identifying the respective components in one service and bundling them together into a set of service functionalities. However, an IMS communication service does not define a specific application; it only defines a set of functionalities associated with certain service requirements. In the end, an end-user application might utilize several different IMS communication services at the same time and perhaps only also a subset of each IMS communication service.

Hence, from an end-user perspective, it is ultimately the design of the application which governs in what way the user uses a specific functionality of an IMS communication service. In this way, all interoperability issues are handled via standardized IMS communication services while still keeping the flexibility of a dynamic service development and deployment environment. Any interoperability agreement signed between operators is done based on the IMS communication services.

An IMS communication service is defined by its behavior, its network model and the media combinations and formats it supports. It is the combined characteristics of these contributions that make out the definition; support for a specific media type does not constitute a separate IMS communication service. Although an IMS communication service can be proprietary, the current industry view is that there are a few different IMS communication services existing in standards today:

- Multimedia Telephony;

- Push-to-talk over Cellular (PoC);

- Presence (and list management);

- Instant Messaging (IM);

- Circuit Switched IMS Combinational Service (CSICS).

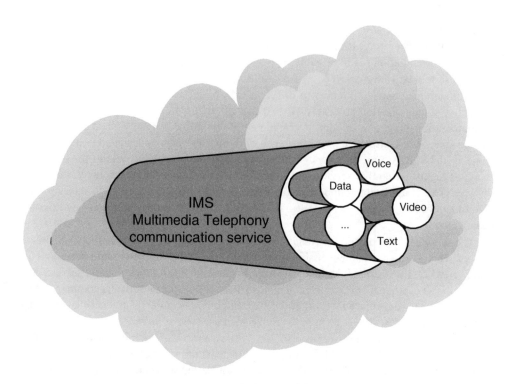

Figure 2.4: Multimedia Telephony, an IMS communication service.

Although CSICS can be seen as another IMS communication service, it can also be a special case of the Multimedia Telephony communication service (basically a CSICS session can be viewed as a Multimedia Telephony session without any voice component that occurs simultaneously as a CS voice call). Presence can be offered as a standalone communication service with no relation to other communication services only indicating user specific states such as 'happy', 'sad' and 'do-not-disturb'. However, presence and list management are also designed to be common service enablers that have relations to the other communication services. For instance, presence can be interfaced to Multimedia Telephony and support service specific presence states (e.g. 'in a call') while group list management may be used to create PoC and IM groups.

2.2.1   An IMS Application Example

In order to better illustrate the concept of an IMS application making use of several different IMS communication services, let us consider Figure 2.5, which illustrates the relation between an IMS application and IMS communication services. To further illustrate how the concept of IMS communications services is used by an IMS application a gaming example is given. In this example a multi-user, science fiction game called 'Neo' is using several different IMS communication services to enable full mobile support of the game. The user Peter wants to play it while commuting on the train. The following takes place.

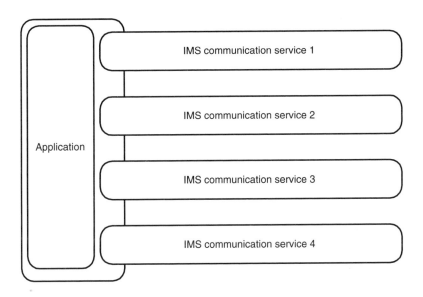

Figure 2.5: IMS communication services as application components.

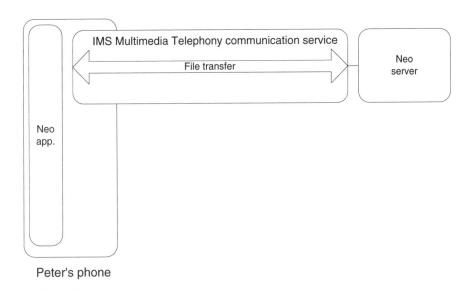

Figure 2.6: Using the IMS Multimedia Telephony communication service for file transfer.

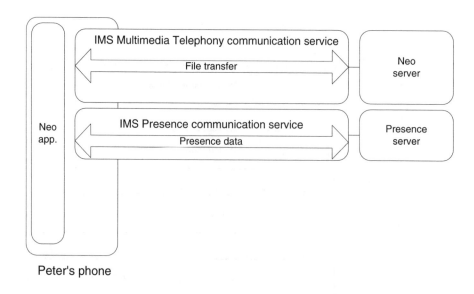

Peter's phone

Figure 2.7: Adding the IMS Presence communication service for locating his friends to help Peter raid the enemy space ship.

1. Peter starts the gaming application on his mobile terminal. The first thing that happens is that an IMS Multimedia Telephony session starts to set up a data channel for file transfer to download the latest updates to the game. This is illustrated in Figure 2.6.

2. After downloading any late patches, Peter enters the Neo space and starts playing. After a while, he has the possibility to raid an enemy space ship in the game. However, this is too difficult to manage all by himself. He activates his presence feature in the gaming application and checks if any of his Neo buddies are online and can assist him. The IMS Presence communication service is added in Figure 2.7.

3. After checking the presence state of his Neo buddies, Peter invites four of his Neo buddies, that are online, into the raid and initiates a PoC session where they can coordinate their actions in the game; see Figure 2.8.

4. After the successful raid, the raid party dissolves and Peter returns to the home space ship trade market to sell items which he looted during the raid on the space ship. He meets a potential buyer and needs to negotiate the price. A chat session is started by starting the IMS Messaging communication service; see Figure 2.9.

5. After a successful deal, Peter meets a long-time friend, Maya, in the open market space in the game. Since they haven't seen each other for quite some time, Peter initiates a voice session with Maya. The voice session uses the full-duplex voice component of the Multimedia Telephony communication service (see Figure 2.10).

6. After terminating the call with Maya, Peter's train arrives at its destination. Peter exits the game and closes the gaming application. Right after doing that, he receives an incoming call from his brother. The phone application is started on Peter's mobile

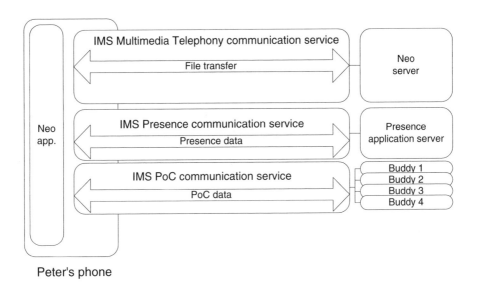

Figure 2.8: Adding the IMS PoC communication service to coordinate the space ship raid.

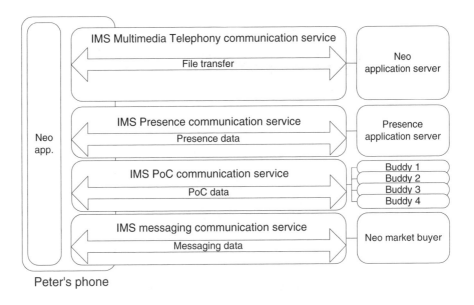

Figure 2.9: Adding the IMS Messaging communication service to negotiate the price of the goods for sale.

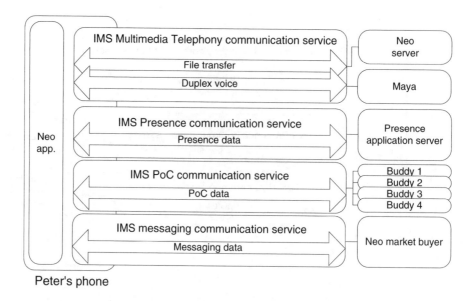

Figure 2.10: Adding another component of the IMS Multimedia Telephony communication service to start a full-duplex voice communication session.

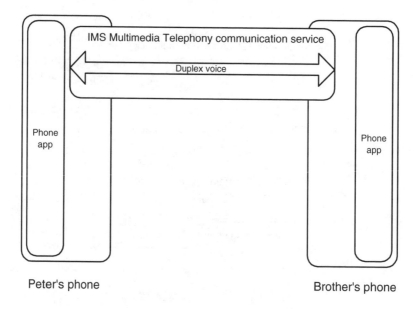

Figure 2.11: The Neo application is terminated at this stage. Peter receives an incoming call and that triggers the execution of another application, the phone application, which also uses the Multimedia Telephony communication service.

terminal. Peter answers and agrees to meet him in 20 minutes at the local pub. The same IMS Multimedia Telephony communication service is used to realize the call as Peter used when playing the Neo game, talking with Maya. It is used within a different application (the phone application) though; see Figure 2.11.

In this example, three IMS communication services are used to realize a gaming scenario. The scenario itself exists in many online games on the Internet today. Here, IMS communication services provides a toolbox of functions to realize this scenario leaving the actual gaming application as an operator differentiator but making inter-operator scenarios including interoperability agreements possible even though the inter-operator agreements do not target the actual gaming application itself, only the IMS communication services used in the application.

2.3 Multimedia Telephony Service Scenario

A service scenario defines how an end-user may want to use a service to fulfill a particular communication need. The scenario tries to explain the context of use and the concept of the service from an end-user perspective. By developing scenarios for a service under development a good foundation for requirement derivation is laid. In this section a Multimedia Telephony communication service scenario is presented in order to highlight end-user and operator requirements and benefits of a Multimedia Telephony service offering based on the standardized Multimedia Telephony communication service. It describes how a new subscriber, Rebecka, is using her Multimedia Telephony enabled mobile phone to make a number of different types of calls. Whenever the scenario describes certain interesting benefits or requirements of the Multimedia Telephony communication service the scenario text stops and there is text describing that specific benefit/requirement.

Rebecka is a great communicator who uses the telephone a lot. She spends hours every day making telephone calls and messaging to friends and family. Rebecka has just bought a new mobile phone that enables her to use the new IP-based telephony service. The IP-based telephony is of no significance to her, but the important thing for her is that the service is affordable and makes her life a bit easier since she now only needs one telephony subscription for all her devices. The new subscription makes it possible to communicate via her mobile telephone, the PC and her old fixed telephone.

The fixed–mobile convergence vision, i.e. the introduction of IMS in both the fixed and mobile domains, makes it possible to have one subscription for a number of communication devices that are using different access methods but all are connected to IMS. This makes life simpler for the end-users and also reduces the cost for the operator since the number of subscriptions the operator needs to handle will be lower and the amount of bills sent out is reduced.

It is Friday, Rebecka is at home. She is on her way out with a couple of friends to celebrate a friend's birthday at a pub downtown. She suddenly remembers that she promised to call her father before the weekend. So before she leaves home, she takes her newly bought mobile phone, plugs in the hands-free and starts to call her dad. Now she discovers that there are a number of options that her old phone didn't have. She can actually choose from a number of methods to communicate.

The most apparent thing that differentiates Multimedia Telephony from the CS telephony service is the possibility of flexible selection of media types. The Multimedia Telephony client is free to support many different types of media but 3GPP specifies a core set of media capabilities communication components that the Multimedia Telephony communication service should support. The following media capabilities should be included in a Multimedia Telephony service offering to an end-user (see 3GPP TS22.173 [27]).

- Speech: Full duplex voice between two Multimedia Telephony clients. The voice component is regarded as the most important of the media capabilities. Still the voice component does not need to be present in a Multimedia Telephony session (i.e. a call); it can be added or dropped dynamically during the session.

- Video: Real-time video from the camera used by Multimedia Telephony client A to the display used by Multimedia Telephony client B. The devices used must support full-duplex video but the video component can also be used to set up one-way flows. When the voice component is present lip synchronization is required. It must be possible to add/drop the video component during the session.

- Text: Text typed on the keyboard used by Multimedia Telephony client A to the display used by Multimedia Telephony client B. It must be possible to do full-duplex text communication, but it is also required that the text communication can be set up as a one-way flow. It must be possible to add/drop the text component during the session.

- File transfer: A file sent from one Multimedia Telephony client to another. The receiving user should be offered the possibility to either save or open the received file.

- Video clip sharing, still image sharing, and audio clip sharing: A file of predefined format is sent from one Multimedia Telephony client to another. The received file is displayed or replayed directly at reception by the receiving Multimedia Telephony client without interactions by the user.

To ensure interoperability at least one common standardized media format per media type is standardized in 3GPP. One example is that the Multimedia Telephony client is required to support the transmission and reception of AMR narrowband and AMR wideband encoded voice. In addition to the core set of media capabilities a Multimedia Telephony client may provide the possibility to communicate using other media components like whiteboard sharing or various types of gaming.

When Rebecka purchased the subscription she was asked by the salesperson whether she wanted a more expensive alternative that gave her the possibility to download more data every month and a higher voice quality. Rebecka didn't buy the more expensive subscription, but she got a trial period for free. So for the next month she will experience the high-quality voice service at least if the person she is calling is also allowed to use the higher speech quality setting. Fortunately, her dad also uses IMS-based telephony via a business subscription, which also allows higher voice quality.

IP technology provides great flexibility. This flexibility allows not only for using a wide range of media types but also for using different variants of each media type. The flexibility can, for example, be used to create different voice quality classes.

- High Quality (HQ): A quality class for users who spend lot of time on the phone and want better sound quality. This variant should use a wideband speech codec that allows the users to listen to sampled audio information at a rate of approximately 7 kHz instead of approximately 3.5 kHz that narrowband codecs provide.

- Circuit Switched equivalent Quality (CSQ): This quality class would replace the current circuit switched voice service. This variant should have very similar performance as CS voice to give a seamless transition from Circuit Switched (CS) to Packet Switched (PS) networks. For this service to be truly equivalent, one would need to have the same codec as in CS UTRAN and CS GERAN, i.e. the AMR 12.2 kbps codec mode, which is equivalent to the GSM-EFR codec. The end-to-end delay should be in the order of 180–230 ms and the frame erasure rate should be less than 2% for the vast majority of the users (95%).

- Economy Quality (EQ): A quality class that reduces the voice quality by using a lower rate codec. The use of a lower rate codec reduces the amount of (e.g. radio) resources needed per user, which allows for more active users. Since more users can use the same system simultaneously, it is also natural that this service variant could be provided less expensively for the end-users. Another possibility for the operator to increase the capacity is to allow for longer delay, perhaps in the 250–350 ms one-way delay (a CS call has a one-way delay of 180–230 ms). A third possibility is to also allow for operating with higher frame erasure rates, for example 3–5% instead of 1–2%.

The requirement on the Multimedia Telephony communication service should be that it enables the deployment of several variants of a voice service. It should, of course, be possible for an operator to provide a variant that gives the same user experience as for CS, but it should also be possible to have a variant that gives a better experience, i.e. a better quality. This in turn may put requirements on the mobile terminal to support wideband speech codecs. It should also be possible to have an inexpensive version where one possibly sacrifices quality in order to be able to increase the capacity so that the cost can be shared among more users. It may also be possible for the operator to force the use of the inexpensive version at 'busy hour' to temporarily increase network capacity.

Rebecka chooses her father from the address book and presses the dial button. A brief moment later she gets an indication that the call is connected and that her father's phone is ringing. Seconds later, her father answers the call. The conversation starts and Rebecka thinks that this new phone is really good, as it feels almost as if her father is next to her in her apartment.

End-users are accustomed to the service quality of the CS voice service provided by most cellular networks. It is important that basic voice services based on the Multimedia Telephony communication service have the potential to be as good as the CS service. Therefore, a realization of the Multimedia Telephony communication service over a modern air interface such as WCDMA HSPA must meet a number of strict requirements derived from the legacy CS voice service. Examples of such performance requirements are 'mouth-to-ear delay' (one-way delay) for the speech in the range of 200 ms. The call setup phase must also feel fast. The faster the better, but if a call setup is completed in about 4–5 seconds or less it is probable that an end-user is satisfied. A third performance requirement is the frame erasure rate. The speech codecs used in CS networks are in general designed to work best if the frame erasure rate is less than 2% end-to-end. Hence, for a 'telecommunication grade' service offering the requirement on frame erasure should be in the region 1% per air link. However, another requirement is that the speech quality of the Multimedia Telephony service should be kept fairly high even for severely degraded channel conditions. When using IP, smart algorithms that reduce the speech quality impact of lost packets can more easily be deployed than for CS. So it may be possible to run Multimedia Telephony on links with higher frame erasure rates without any drastic reduction of the perceived speech quality. Thus, appropriate mechanisms are needed so that the service can adapt to different operating conditions and always deliver the optimum compromise between quality and error resilience.

> *During the conversation, Rebecka gets an indication of another incoming call. Rebecka looks on her screen and sees that the calling party is her brother Johan. Earlier during the day Johan had said that he also wanted to go to their mutual friend's birthday but he didn't know when and where the party was to take place. Rebecka tells her dad that she will put him on hold and quickly switches to Johan and tells him about the plans for the evening. After finishing talking to Johan, Rebecka switches back to the call with her dad. They talk for another 5 minutes before hanging up.*

The Multimedia Telephony communication service must include supplementary services similar to the supplementary services for CS voice. The scenario mentions Originating Indication Presentation that is activated when Johan calls Rebecka and his number/nickname is presented on the display. The scenario also mentions Communication Hold that is used when she switches to the other call. These are examples of the supported supplementary services, but a service based on the Multimedia Telephony communication service must support at least the following set of supplementary services.

- Originating Indication Presentation (OIP): This supplementary service provides the terminating user with the possibility to receive identity information about the user who originates the call.

- Originating Indication Restriction (OIR): This supplementary service is used to restrict the presentation of the identity of the originating user.

- Terminating Indication Presentation (TIP): This supplementary service provides the originating user with the possibility to get identity information about the terminating user.

- Terminating Indication Restriction (TIR): This supplementary service enables the terminating party to prevent the presentation of the identity information.

- Communication DIVersion (CDIV): A set of supplementary services that enable a user to divert the communication to another destination either via network or client settings. The CDIV set of supplementary services includes Communication Forwarding on Mobile Subscriber Not Reachable (CFNRc) that targets a mobile specific scenario. The diversion of the call is made when the user is not reachable due to e.g. being out of coverage.

- Communication Hold (HOLD): This supplementary service enables a user to suspend the media stream(s) in a session and resume the media stream(s) at a later time.

- Communication Barring (CB): A set of supplementary services that is used to reject communications.

- Message Waiting Indication (MWI): This supplementary service is used to notify a user that a message has been received by e.g. a voice mail box.

- CONFerence (CONF): This supplementary service enables a user to participate and control a communication session involving a group of users.

- Explicit Communication Transfer (ECT): This supplementary service provides the users in a call to transfer the communication to a third party.

It should be noted that the supported supplementary services are sometimes referred to as simulation services. The reason for the name 'simulation services' is that the IMS-based supplementary services do not work in exactly the same way as the supplementary services offered in ISDN/PSTN. Instead the IMS solution tries to simulate the supplementary services of ISDN/PSTN.

> *The week after her friend's party, Rebecka goes on a leisure trip to Gotland. She will be away from her family for five days. She makes a telephone call to her brother Johan in downtown Visby. Rebecka and Johan talk for a while before Johan starts to ask what it is like in Visby. Rebecka switches to video mode and starts to film the sunny landscape. Johan watches his screen and is pleased with what he sees. Then he starts to play around with his phone and manages to enable two-way video and thus the telephone call becomes a video telephony call. Rebecka and Johan turn their cameras so that they can film their faces instead of the surroundings. Rebecka and Johan communicate for a while before she needs to hang up.*

Maybe the main difference the end-users will notice after having introduced IP services is the increased service flexibility. In this part of the service scenario the possibility to enhance a communication session by adding media components during the communication session is highlighted. The adding of video may fulfill two types of communication needs. In the first case Rebecka adds video to share her experience. This type of communication should in most cases use one-way video and give the viewer the possibility to 'see what I see'. The second communication need is video telephony where the persons who communicate want to see each other by using full-duplex video. It is required that Multimedia Telephony can handle both cases. The flexibility of IP also gives the possibility to toggle between voice only, 'see what I see' and video telephony within a communication session.

As mentioned earlier, the Multimedia Telephony communication service must support a number of different media. These media have different characteristics. Video and voice are the real-time media types that are delay sensitive but robust enough to be able to lose a certain percentage of the packets without affecting the media quality too much. The real-time media often uses Real-time Transport Protocol (RTP) and User Datagram Protocol (UDP) as transport protocols. Text and files, sent using Transmission Control Protocol (TCP), are non-real-time media types. This means that they are not very delay sensitive, but packets cannot be lost in order to avoid e.g. file corruption. In an optimized system, the different media types should utilize different radio bearers with different radio bearer settings. Therefore, the introduction of the Multimedia Telephony requires the Radio Access Network (RAN) to support a suitable set of PS radio bearers and Radio Access Bearer (RAB) combinations for all the media combinations possible. Another requirement is that the RAN should be able to switch e.g. from a RAB combination for a plain voice call to a RAB combination for a video telephony call in a sufficiently short time. The time for doing the service change must not take more time than the call setup and should be done seamlessly, without interruption of the ongoing media flows.

Negotiable QoS must be supported. The individual media components should be treated differently in radio access and in the transport network to secure a good end-user perception of the service. One example is the case of a call with video and voice where the user moves further away from the base station. At some point, the radio conditions will be so poor that the radio link cannot maintain the bit rate needed for the voice and video streams. In that case, from a user perspective, it would be beneficial if the video froze and the playback of the voice stream maintained high quality. This can be achieved by using different QoS for the individual media components and maybe by distribution the media streams on several radio bearers.

> *Later that evening, Rebecka calls her mother. Rebecka knows that her mother has an old phone, so most probably she doesn't have any multimedia capabilities. So Rebecka thinks that there is no point in trying to do a video call of some sort so she just chooses her mom from the phone book and calls her using the plain voice service. Rebecka and her mother talk for about one and a half hours before they hang up.*

Interworking with the CS domain and other external networks is a key feature to get faster deployment of the Multimedia Telephony service. The Multimedia Telephony communication service will support interworking since it is a requirement for IMS to have support to other networks; see 3GPP TS22.228 [43]. Interworking with CS networks using Tandem Free Operation (TFO) should be possible so that the optimum quality can be provided to the end-user. The interworking function should not only work for the plain voice service, but should also be able to handle the CS video telephony interworking cases. Multimedia services may also be realized using 3GPP CSICS (see Section 8.1), which means that the voice uses the CS infrastructure and the multimedia extensions use the PS infrastructure. The introduction of 3GPP CSICS creates yet another interworking case that the Multimedia Telephony communication service should be able to handle.

> *Disaster! Rebecka left her charger at home. Her battery has gone flat and she needs to call her father. Fortunately, she travels with a friend who has a similar phone but from another brand. She cannot borrow her charger but can borrow her phone. Rebecka removes her SIM card and puts it into her friend's phone so she doesn't add to her friend's phone bill. She starts the phone and it registers her to the network. Funnily enough the phone asks her if she would like to have her phone book downloaded. Well, she knows the number for her dad, but it couldn't hurt to download the phone book. After a brief moment she sees the same buddy list that she had on her phone. The list shows that her father is available for communication. She calls her father.*

The Multimedia Telephony communication service can be combined with other IMS communication services to create an even more interesting service offering. Two examples of other IMS communication services, or service enablers, are OMA XML Document

Management (XDM) and OMA Presence (see Section 8.4). The XML Document Management service enabler makes it possible to upload to and manage a user's phone book (contact list) in a server in the network. By doing this the phone book is always available regardless of which device the user is going to use. In the scenario described above the XML Document Management service enabler also kept information about the people Rebecka had on her Presence buddy-list. Presence is another communication service that facilitates communication by presenting other people's availability and willingness to communicate.

> *Rebecka is home again and she is pleasantly surprised about the new phone she has had for a week. It is affordable, has many communication features and provides good quality.*

Why should the operators invest in IMS and Multimedia Telephony? The answer is that they should only invest if the new service can provide added value for the end-users and also have the potential to lower costs for the operator. The main cost driver is of course the need to only maintain a single PS infrastructure instead of two (CS and PS). Classical telecommunication metrics like capacity and coverage are also important to consider. Great capacity and coverage lower cost. In high-density city areas high cell capacity will reduce the number of cells needed and good coverage will enable the operator to have few but large cells in the countryside. It is a fair bet that the operator community will require the Multimedia Telephony voice service to provide similar or even better capacity and coverage than the legacy CS telephony voice service.

Therefore, the Multimedia Telephony communication service should be able to give a capacity that is at least as good as the capacity of the corresponding circuit switched service, given the same quality. The capacity should preferably be significantly better. The Multimedia Telephony service should be designed so that the operators themselves can choose how to optimize their network, i.e. if they want to ensure that the delivered quality is as good as for CS telephony or if they want to increase the capacity by sacrificing some quality.

Another requirement is that the Multimedia Telephony communication service shall be able to establish an emergency voice call. The emergency calls must be routed to the emergency services in accordance with national regulations for where the user is located. The emergency voice call should be prioritized over ordinary calls.

The Multimedia Telephony communication service is an IMS service and is thus by default access agnostic. However, the different access methods have different characteristics and for instance when using GSM/EDGE it may be beneficial to use the lower rate AMR speech codec modes to secure the end-to-end quality. While for WCDMA HSPA, the high peak bit rate enables the use of even the highest AMR wideband modes without any throughput issues. The Multimedia Telephony communication service must handle this and provide a working service regardless of the access methods (e.g. WCDMA, GSM/EDGE, WiMAX, WLAN) that the originating user and the terminating user are using. Basically, this means that it is required that operators should be able to deploy the Multimedia Telephony service on whatever access method that they think is suitable without interoperability issues.

2.4 Summary of the Multimedia Telephony Communication Service

The aim with the Multimedia Telephony standardization effort was to position IMS-based Multimedia Telephony as the standardized, operator-controlled way to realize interoperable IP-based voice and multimedia communication, which is supplemented by telephony-type supplementary services. By that, the Multimedia Telephony should be able to replace the CS telephony for operators that wish to do so, as well as to help to increase the operator's revenue potential by complementing CS telephony, offering a rich person-to-person multimedia experience for the end-users.

Since Multimedia Telephony should be able to replace CS telephony, it is natural that a Multimedia Telephony service should have a usage model similar to traditional CS telephony.

- The Multimedia Telephony communication service will mostly be used for voice and 'speech plus other media' communication. It should be noted that the service is not required to always include speech, but it also caters for other combinations of media (e.g. text and video, or only video).

- The Multimedia Telephony communication service enables a user to connect to any other user, regardless of operator and access technology.

- The Multimedia Telephony communication service supports supplementary services that behave almost identically to the supplementary services supported by CS telephony. The Multimedia Telephony communication service supports conference calls and the creation of user initiated limited conferences (e.g. three-party calls).

But there are things that differentiate Multimedia Telephony from CS telephony. The set of supported media in the specifications include speech, video, text, image sharing, video clip sharing, audio clip sharing and general file sharing. Beside the set of standardized media types, support for other non-standardized media types can also be easily added. A Multimedia Telephony communication may start with only one type of media stream. Additional types of media may be added by the users during the Multimedia Telephony session. Therefore a particular Multimedia Telephony call may consist of only one type of media, e.g. speech.

Multimedia Telephony aims to lower cost for the operator. The use of the IMS platform promises to reduce the long term cost of service deployment when the operators add more services to their network. To help drive cost savings, the Multimedia Telephony communication service must help fast service development and deployments. One requirement is thus that the Multimedia Telephony communication service must be able to be used along several different IMS communication services, at the same time, to allow for the fast development of complete end-user applications.

Further, the design of the Multimedia Telephony communication service is made in such a way that it provides a service performance (e.g. media delay and speech quality), which can be similar to or even better than the performance the users are accustomed to when using the CS telephony service. In the 3GPP community the access methods are being optimized for Multimedia Telephony to secure performance. In the optimization work of the access methods, efforts are spent on further cost savings by enabling a high capacity voice service (higher capacity than for CS telephony) by using the Multimedia Telephony communication service.

Chapter 3

Network Architecture and Service Realization

Gonzalo Camarillo, Shyam Chakraborty, Janne Peisa, Per Synnergren

Multimedia Telephony over cellular systems is a conversational service involving conversational voice, video and text between two- or multiple parties, residing in different access networks across the Internet, and at least one access network is an IP-based cellular network. The architectural concepts of such a system stem from three different network technologies:

- circuit switched telephony with PSTN and ISDN and intelligent networks;

- IP-based packet switched data networks, and the Internet, that serve as the core transport network;

- wireless and mobile systems.

We provide a brief description of these underlying concepts to understand various components of this book that are elaborated further in the following chapters. To summarize this chapter, a presentation of how Multimedia Telephony can be realized over the core transport network and over the wireless access is given.

3.1 Public Switched Telephone Network and Integrated Service Digital Network

The conventional telephony, commonly known as Public Switched Telephone Network (PSTN), provides end-to-end circuit switched full-duplex paths to cater for primarily conversational speech. The PSTN can be end-to-end analog, carrying 4 kHz analog speech signals, or end-to-end digital, carrying 64 kbps digital speech signals, or partly analog and partly digital. The switched circuits may also carry low speed data, for example, G3 fax and computer connection with modems.

The Integrated Service Digital Network (ISDN), on the other hand, provides end-to-end circuit switched digital connection, and, along with speech, it is able to cater for a variety of services and at considerably higher bit rates. The circuit switched ISDN system as shown in

IMS Multimedia Telephony over Cellular Systems   S. Chakraborty, T. Frankkila, J. Peisa and P. Synnergren
© 2007 John Wiley & Sons, Ltd

Figure 3.1 is eventually called narrowband ISDN (N-ISDN[1]). The Global System for Mobile communications (GSM) is primarily structured after the ISDN, with lower access rates, and will be discussed later in this chapter.

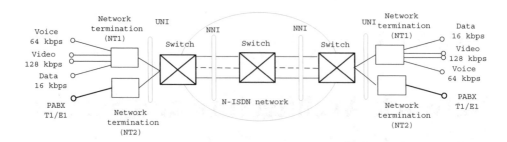

Figure 3.1: Architecture of N-ISDN.

3.2   Data Networks and the Internet

A computer network is a complex organization of computers of various makes, operating systems and application software, communicating over a variety of physical links and through different switches, routers and hubs, which themselves are also dedicated computers. All computers in a network are collectively known as nodes. To make the complex organization of networking comprehensible, the International Standards Organization (ISO) extended the seven-layer Open Systems Interconnection (OSI) reference model in 1981. The layers were defined with the principles that each layer is needed to provide different levels of abstractions and a minimum number of such layers are used. A layer N entity has three responsibilities: it takes services from layer $N-1$, provides services to layer $N+1$, and communicates with its peer process residing at another node, as in Figure 3.2. Each layer operates with a set of rules, known as protocols. A data block generated at a layer N entity is commonly known as layer N Protocol Data Unit (PDU), and passed to the next layer, which receives it as a Service Data Unit (SDU). A layer N PDU is generated by adding control information to the SDU (which is simply the layer $(N+1)$ PDU) in the form of header (and possibly tail bits as well), before passing it to layer $(N-1)$.

In the OSI reference model, the layers are broadly divided into three categories. The lower three layers are responsible for actual transport of information from the source to the destination node. The lowest physical layer deals with physically transmitting and receiving the information between two nodes over a physical link and the data link layer ensures the integrity of information over a single link. While the first two layers operate on a link-by-link basis, the network layer is responsible for routing the data from the source to the destination node, across the network. The transport layer checks end-to-end integrity of information over the network and hides the nuances of the lower layers. The upper three layers, session, presentation and application, are commonly known as application layers, and provide various interactions between application software residing at the communicating nodes.

---

[1]For the sake of brevity, the packet switched broadband ISDN (B-ISDN) will not be discussed.

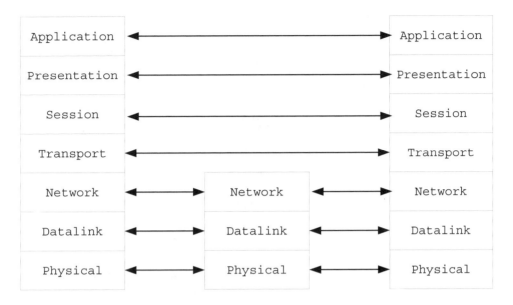

Figure 3.2: ISO/OSI reference model.

3.2.1 Internet Protocol Architecture

The Internet has a four-layer protocol architecture as in Figure 3.3, that is slightly different from the OSI model. At the core of the architecture lies the Transmission Control Protocol (TCP) and Internet Protocol (IP) protocol suite, roughly corresponding to OSI transport and network layers. The IP provides a connectionless service at the network layer, without any flow control (which is done at the transport layer if necessary). The IP layer avails services of the network interface or the link layer, which is a combination of physical and data link layers. The design of IP allows it to match with practically all available link layers. The transport layer sits on top of the network layer and contains primarily two different protocols: the connectionless User Datagram Protocol (UDP) and connection oriented TCP. The transport layer provides services to a variety of applications, such a domain name system, filing system, network management, email, file transfer, security, signaling, WWW, etc.

3.2.2 The Internet

The present Internet is actually a collection of sub-networks, or a 'network of networks', interconnected through 'routers' by a high-speed mesh topology backbone. The networks of the Internet are managed by Network Service Providers (NSP) and Internet Service Providers (ISP). Large NSPs build national or global networks and sell bandwidth to regional NSPs, who in turn resell bandwidth to local ISPs, who in their turn provide services to the users. The sub-networks are connected to each other at Internet exchanges, known as network access points. A network access point is usually an Asynchronous Transfer Mode (ATM) or a gigabit Ethernet switch surrounded by routers. The switches are interconnected in a fully meshed fashion by permanent virtual circuits running over high-speed dense wavelength division multiplexed fiber links. The network access point facilities are hired by the NSPs to install their own routers to exchange traffic coming from the ISPs. Further, many NSPs may establish

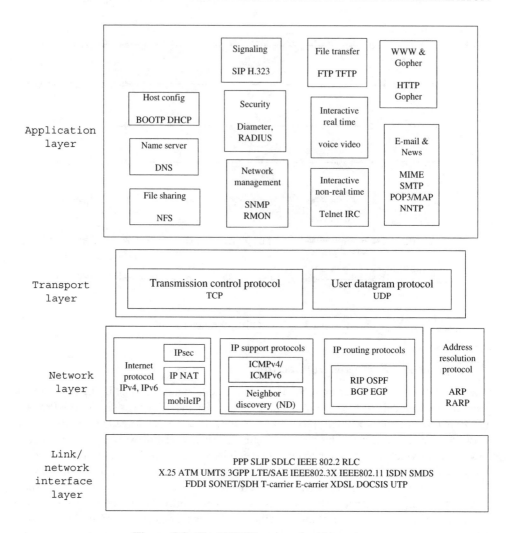

Figure 3.3: The TCP/IP protocol architecture.

private peering arrangements as well, bypassing the switches. The users avail the services of the Internet by connecting through a variety of connections ranging from low bit rate dial-up connection, broadband DSL, LAN, wireless LAN, etc., provided by the ISPs, who also provide such services to the users as Internet transit, domain name, registration and hosting, etc. The ISPs provide the required customer premises equipment and often some additional functionality such as security software, etc. Through the access links, customer premises equipment attaches to the point of presence or to the 'edge of the ISP network', a physical location that houses servers, routers, switches, etc.

3.2.3 Internet Protocol

The purpose of the IP is to provide an unreliable connectionless service (also known as 'best effort') at the network layer, that is, over an interconnected network of sub-networks.

The IP layer receives transport layer PDUs, attaches an IP header to create an IP packet (or IP datagram) and passes it to the link layer to be delivered over the physical network links. In order to fit into the frame format of a link layer, it may also fragment an IP packet into smaller Maximum Transmission Units (MTU). When all the fragments are received, the destination host reassembles them to retrieve the full packet.

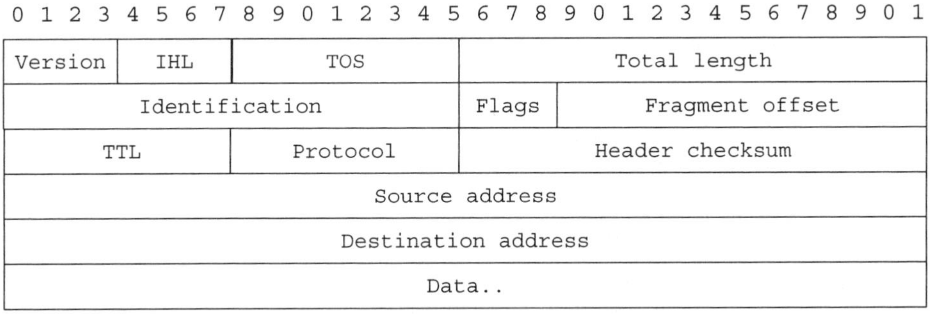

```
0 1 2 3 4 5 6 7 8 9 0 1 2 3 4 5 6 7 8 9 0 1 2 3 4 5 6 7 8 9 0 1
```

Version	IHL	TOS	Total length		
Identification			Flags	Fragment offset	
TTL		Protocol	Header checksum		
Source address					
Destination address					
Data..					

Figure 3.4: IPv4 packet header format.

Currently IP has two major versions, the IPv4 (version 4) and IPv6 (version 6). We will mostly focus our discussion to IPv4, since as of now it is the most dominant version of IP. The IP header format of IPv4 is shown in Figure 3.4, and described briefly below:

- *Version*: defines the specific IP version. This is usually IPv4 (0100) or IPv6 (0110).

- *Internet header length (IHL)*: denotes the IP header length in terms of 4-octet words. The minimum and maximum lengths of the IP header are 20 and 60 octets.

- *Type of service (ToS)*: allows an originating host to inform the requested class of services. While not commonly used earlier, it has found use in Diffserv more recently.

- *Total length*: indicates the length of the entire packet (including the header and the data). Since the size of this field is 16 bits, the maximum length of an IP packet can be 64 kbytes.

- *Identification*: used for fragmentation and reassembly of a large transport layer packet.

- *Flags*: also used for fragmentation and reassembly. The flags consist of More Fragments (MF) and Don't Fragment (DF) bits. MF is used to indicate if the datagram contains additional fragments. DF is used for suppression of fragmentation. The third bit is reserved.

- *Fragment offset*: informs the position of fragmentation in a packet.

- *Time-To-Live (TTL)*: indicates the number of hops a packet is allowed to take before it is discarded. The value of TTL ranges from 0 to 255. Each router decrements the value of TTL by one before transmitting to the next hop and when TTL gets to zero, the packet is discarded.

- *Protocol*: Informs the higher layer protocol carried in the packet. The options include, ICMP, TCP, UDP, OSPF, etc.

- **Header checksum**: provided to ensure correct reception of the header only (not the full packet).

- **Source address**: the 16 bit IPv4 address of the sending host.

- **Destination address**: the 16 bit IPv4 address of the destination host.

- **Options**: a set of options that may be applied to any packet.

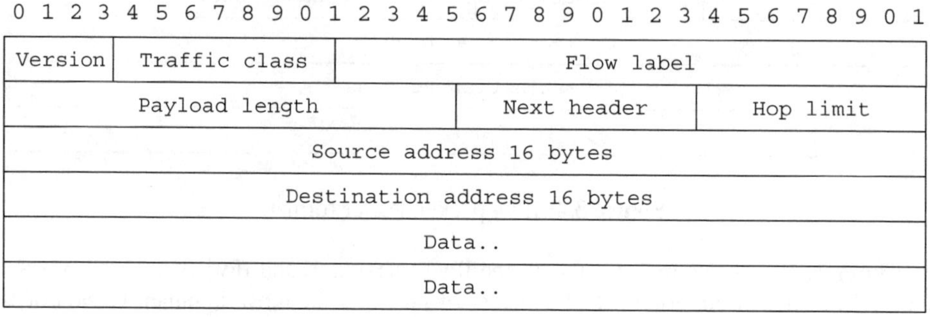

```
0 1 2 3 4 5 6 7 8 9 0 1 2 3 4 5 6 7 8 9 0 1 2 3 4 5 6 7 8 9 0 1
┌─────────┬──────────────┬────────────────────────────────────┐
│ Version │ Traffic class│             Flow label             │
├─────────┴──────────────┼─────────────────────┬──────────────┤
│    Payload length      │    Next header      │  Hop limit   │
├────────────────────────┴─────────────────────┴──────────────┤
│                 Source address 16 bytes                      │
├──────────────────────────────────────────────────────────────┤
│              Destination address 16 bytes                    │
├──────────────────────────────────────────────────────────────┤
│                        Data..                                │
├──────────────────────────────────────────────────────────────┤
│                        Data..                                │
└──────────────────────────────────────────────────────────────┘
```

Figure 3.5: IPv6 packet header format.

The major difference between IPv4 and IPv6 is in the address field. The IPv6 offers a 16 byte address field that is expected to be sufficient to cover everything that needs to have an IP address. The IPv6 header format is shown in Figure 3.5. The fields are described below:

- **Version**: specifies IP version (IPv4 or IPv6).

- **Traffic class**: specifies Internet traffic priority delivery value.

- **Flow label**: used for specifying special router handling for a flow of packets.

- **Payload length**: the 16 bit field as such specifies the length of a datagram of maximum length of 65535 bytes. When cleared to zero, this option denotes a bigger packet known as a jumbogram.

- **Next header**: specifies the next encapsulated protocol.

- **Hop limit**: similar to the IPv4 TTL field. For each router that forwards the packet, the hop limit is decreased by one, and when the value reaches zero, the packet is discarded.

- **Source address**: IPv6 source address.

- **Destination address**: IPv6 destination address.

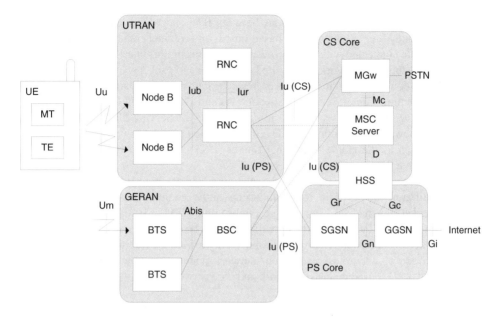

Figure 3.6: Simplified network architecture of modern cellular networks. The network consists of several mobile terminals (also called User Equipment, UE), one or several Radio Access Networks (e.g. UMTS Terrestrial Radio Access Network, UTRAN, and GSM/EDGE Radio Access Network, GERAN) and a core network, possibly divided into circuit and packet switched domains.

3.3 Cellular Systems

The cellular systems are typically composed of a radio access network and a core network. In addition to the core network, which provides the basic connectivity and functions like charging, authentication and mobility, the services are often provided by a service network.

A simplified picture of a typical architecture in modern cellular networks is shown in Figure 3.6. The network consists of mobile terminals, one or several radio access networks and a core network.

The mobile station or more generally the User Equipment can consist of two nodes: the actual mobile terminal, which is typically a hand-held device or a PC-Card and Terminal Equipment, which is then further connected to the mobile terminal. Typically the Terminal Equipment is a laptop.

The two different radio access networks shown in Figure 3.6 are the UMTS Terrestrial Radio Access Network (UTRAN) and GSM/EDGE Radio Access Network (GERAN). UTRAN provides radio access for UMTS terminals, while the GERAN provides access for GSM and EDGE terminals. Both radio access networks provide both circuit switched and packet switched access. The architecture of both radio access networks consists of one or many base stations, which are connected to a control node. The base stations are called Node Bs in UTRAN and Base Transceiver Stations (BTSs) in GERAN. Typically the functionality of base stations is limited, and most of the network intelligence is provided in controllers, called Radio Network Controller (RNC) in UTRAN and Base Station Controller (BSC) in GERAN. The interface between Node B and RNC is based on Asynchronous Transfer Mode

(ATM) and is called Iub, while the interface between BTS and BSC is called Abis. Abis is not a specified interface, but is typically based on circuit switched connections. In UTRAN the controllers are connected with Iur interface. There is no similar interface in the GERAN.

In Figure 3.6 the core network is separated into circuit and packet switched domains. The circuit switched domain shown in Figure 3.6 shows the modern layered architecture, in which the Mobile Switching Center Server (MSC Server) controls the Media GateWays (MGWs) with Mc interface. The packet switched domain consists of Serving GPRS Support Node (SGSN) and Gateway GPRS Support Node (GGSN), which are connected with Gn interface. Some nodes, such as the Home Subscriber Server (HSS), are shared between CS and PS domains (using e.g. D, Gr or Gc interfaces).

The radio access network is connected to the core network with Iu interface. There are two slightly different flavors of the Iu interface, one for the circuit switched core network and one for the packet switched. The Iu interface can be either ATM or IP-based.

The circuit switched domain of the core network can be directly connected to the PSTN, while the packet switched domain provides connectivity toward the Internet with Gi interface.

The functionality and roles of different network nodes are further described in Sections 3.3.1 and 3.3.3 for radio access network and core network respectively.

3.3.1   Radio Access

The role of the radio access network is to provide radio connectivity to as wide an area as desired by the operator. The architecture has traditionally been determined by the need to deploy and maintain several thousands or tens of thousands of radio base stations per network. The large number of base stations has resulted in an architecture in which the base station is a simple node with limited functionality. Most of the functionality has traditionally been located in a controller node (e.g. Base Station Controller in GSM). However, with constant reduction in the hardware cost, the recent updates (such as the High Speed Packet Access) to the 3G standards have resulted in more and more functionality being provided by the base station.

As the UMTS is the most likely candidate for an early introduction of the IMS Multimedia Telephony, we will in the following study the architecture of the UMTS Terrestrial Radio Access Network (UTRAN) in detail.

The initial architecture of the UMTS radio access network was very similar to the architecture of the GSM. However, with the ongoing UMTS evolution, first by introduction of High Speed Packet Access (HSPA) and subsequently by HSPA Evolution and Long Term Evolution an increased amount of functionality is being placed in the base stations.

We will first look at the original architecture of the UMTS network, and then describe the impacts of the ongoing and planned UMTS evolution in Section 3.3.2. The description of the network functionality is intentionally brief. More information on the current UMTS networks can be found in [89].

In Figure 3.7 a simplified UTRAN architecture is shown. UTRAN consists of base stations (called Node Bs) and Radio Network Controller (RNC) nodes. The base stations are connected to the Radio Network Controller via Iub interface, while the Radio Network Controllers are connected to each other with Iur interface. Unlike the GSM Abis interface, both Iub and Iur are standardized interfaces, and (in theory) any base station can be connected to any RNC. The current Iub and Iur interfaces are based on Asynchronous Transfer Mode (ATM). For future releases, it is envisioned that either ATM or IP can be used. The radio access network is connected to the core network with Iu interface. There are two different

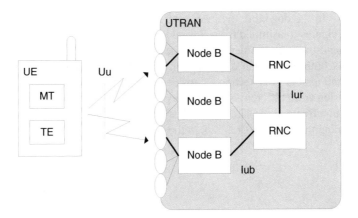

Figure 3.7: The architecture of the UMTS Terrestrial Radio Access Network (UTRAN). A soft hand-over is shown with a bold line.

flavors of the Iu interface, one for circuit switched and one for packet switched core network. The Iu interface can be either ATM or IP-based.

Each base station controls one or more cells. Typically a three sector antenna is used, which results in one base station controlling three cells. The mobiles are connected to cells with radio links (forming the Uu interface).

From the architecture point of view, perhaps the most fundamental property of the UTRAN is the concept of soft hand-over. In soft hand-over the mobile communicates with two or more cells at the same time. This is different from the hard hand-over (used in most other systems, e.g. in GSM), in which the mobile is always connected to a single base station. The set of radio links (to different cells) the mobile terminal is currently communicating with is called the active set. With soft hand-over, the actual hand-over procedure basically consists of first adding a new radio link to the active set, and when the old radio link is no longer usable, removing it from the active set.

The soft hand-over is used for two purposes in UTRAN. First, it provides a method for all base stations to control the transmission power of the mobile. The ability to control the power of the mobile is required, as the transmission of a terminal at the cell border could otherwise cause a high amount of interference to neighboring cells. For systems with frequency reuse higher than one (meaning that the neighboring cells use different frequency) this is not a major problem (as the interference will be on a different frequency), but for UTRAN the possibility for power control of the uplink transmission of the mobile is required for stable system operation. Note that the requirement for power control does not hold for the downlink. Second, the possibility to receive the signal in (or from, in the case of downlink transmission) two or more base stations allows the data transmission to occur only using the power required to reach the best cell. This results in increased capacity for the system. The capacity increase is obtained in both downlink and uplink.

The resulting architectural impacts can be seen in Figure 3.7. In the uplink, the mobile terminal may be communicating with several base stations, and the received uplink data needs to be collected at a central location. In UTRAN this location is RNC, in which transport blocks are received and combined from all base stations. In the special case of a single Node B receiving the signal from multiple cells, the Node B can combine actual received physical signals. In the downlink the mobile terminal will perform the combination of the actual physical signals transmitted from several base stations.

The soft hand-over requirement forces the transmission of the data over Iub to be synchronized in both uplink and downlink. In the uplink the RNC must wait for blocks from all base stations before detecting an erroneous transmission, while in the downlink the actual transmission from different Node Bs must be synchronized[2] so that each base station transmits the same frame at the same time. This leads to the need to buffer the frames in the Node Bs according to the slowest Iub link in the active set.

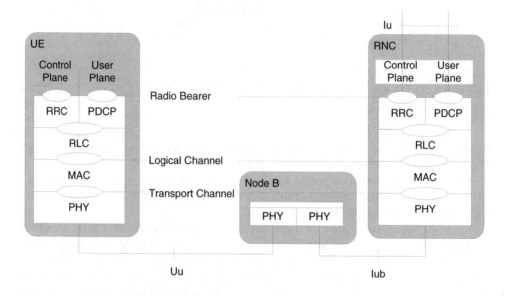

Figure 3.8: The protocol architecture of the UMTS Terrestrial Radio Access Network (UTRAN). Also the names of the logical connections between peer entities are shown in the figure.

The protocol architecture of UTRAN is shown in Figure 3.8. As can be seen, the Node B is only responsible for terminating (parts of) the physical layer, basically consisting of the actual transmission and reception of the physical signal over the air. This includes e.g. modulation of the signal. The remaining part of the physical layer consisting of macro-diversity combining (which is equal to the combining of the received transport blocks in soft hand-over as discussed above) is located in the RNC. Additional Node B functionality includes the following:

- Channel coding, which introduces redundancy into the source data flow, in order to allow the detection or correction of errors introduced by the radio transmission.

- Soft combining of the received signals within different antennas.

- Inner loop power control, which sets the power of the uplink dedicated physical channels. The target quality is received from outer loop power control, and the quality

---

[2]Note that the requirement on the synchronization is not very strict, and some timing difference can be resolved with the rake receiver.

estimates are based on the dedicated physical control channel. The power control commands are sent on the dedicated physical control channel.

The Radio Network Controller is responsible for terminating all user and control plane protocols in the UTRAN, as well as for the radio resource management. The functionality of the protocols include e.g. segmentation, concatenation, error correction using retransmissions, etc. The protocols will be studied in more detail below. Additional functionality of the Radio Network Controller includes the following:

- Radio Resource Management, including functionality such as Congestion and Admission control, allocation of the channelization codes and setting power control targets for the inner loop power control.

- Control of hand-overs, meaning providing the mobility of the radio interface. Hand-overs are based on radio measurements, and may be to/from another system (e.g. UMTS to GSM hand-over).

- Macro Diversity Combining between different Node Bs.

- Ciphering and deciphering, protecting the transmitted data against a non-authorized third party.

- Paging support, consisting of the capability to request a UE to contact the network when the UE is in the Idle mode or in a paging state.

- Broadcast signaling, providing the UE with information necessary to operate within the network.

- Open Loop Power Control, which sets the target quality value for the inner loop power control. The outer loop power control is based on the quality estimates of the transport channel.

Also shown in Figure 3.8 are the logical connections between different protocol entities.

Physical channels provide actual transmission of data over the air interface. There are two kinds of physical channels, dedicated physical channels and common physical channels. The dedicated physical channels, the Dedicated Physical Data CHannel (DPDCH) and the Dedicated Physical Control CHannel (DPCCH), are used to carry the data and physical layer control signaling respectively. The control information is used e.g. for the fast power control. Dedicated physical channels are specified by carrier frequency and scrambling code (and optionally channelization code). In the uplink relative phase is used to separate the data channel from the control channel. The common physical channels, e.g. Physical Random Access CHannel (PRACH) and Primary and Secondary Common Control Physical CHannels (P- and S-CCPCH), are used to carry data and signaling for users that do not have dedicated channels.

The transport channels provide an interface between physical layer and MAC layer. They are characterized by how and with what characteristics data are transferred over the radio interface, not by what is transmitted. The most important transport channels are Dedicated CHannels (DCHs) and common channels such as Random Access CHannel (RACH) and Forward Access CHannel (FACH). The DCH is a channel dedicated to each mobile, while the common channels are common to all mobiles in the cell. The RACH is a contention based uplink channel used for transmission of relatively small amounts of data, e.g. for initial access or non-real-time control or traffic data. FACH is the corresponding downlink channel.

Logical channels provide an interface between RLC and MAC layers. They are characterized by the content of the transmission, not by how the data is transmitted. The most important logical channels are Dedicated Control CHannel (DCCH), which is used to transmit control signaling between the network (both core and access) and the terminal, and Dedicated Traffic CHannel (DTCH), which is used to transmit user plane data between the network and the terminal. Both channels are dedicated to a single terminal. In addition to dedicated logical channel, also common logical channels are available, which are used e.g. by the UEs having no connection with the radio access network.

The user data transmission is mostly performed over the dedicated channels. This solution is very efficient for CS voice, which has a relatively long session time and stable application bit rate. However, for packet data applications, such as web browsing or email, the time to set up a dedicated channel is relatively long, and the application might not use the available bit rate very efficiently. Furthermore, it is somewhat difficult to assign high data rates to a single user. Typically the maximum available data rates are between 64 and 384 kbps over dedicated channels.

The transmission on the dedicated and common channels is interleaved over Transmission Time Interval (TTI). For common channels (with low data rates) typically 10 ms TTI is used, but 20 ms TTI is also possible. For dedicated channels, which are typically used to transmit almost all user data, even longer TTIs are possible in theory, but in practice the 64 kbps bearers use 20 ms TTI and 384 kbps bearers use 10 ms TTI. The 10 ms TTI is the smallest possible TTI length for dedicated channels. The chosen TTI length has an important impact on the latency of the system, and will be discussed in more detail in Chapters 5 and 7.

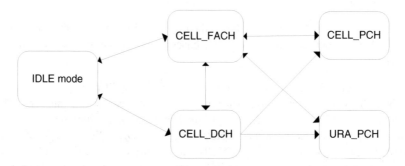

Figure 3.9: The protocol states of the Radio Resource Control (RRC) protocol. Also shown is the Idle mode, in which the UE has no connection to the radio network.

The Radio Resource Control protocol (RRC) is responsible for controlling the radio resources of the UE. The RRC protocol has several states, shown in Figure 3.9.

In Idle mode, the UE is not connected to the radio network. It may still have a connection to the core network, as described in Section 3.3.3. The UE will have to request an RRC connection (with RRC connection request procedure) before any data transmission is possible. The core network may ask the radio access network to contact the UE by means of paging.

In connected mode, there are four RRC protocol states. In the paging states (URA_PCH and CELL_PCH) the UE listens periodically to the paging channel. Similarly to the idle mode, no direct data transmission is possible, but, as the UE maintains a connection to the radio network, it is possible to transmit data directly by moving to CELL_FACH state, without having to request an RRC connection.

In the CELL_FACH state the UE uses Forward Access CHannel (FACH) and Random Access CHannel (RACH) to transmit data in downlink and uplink respectively. Both RACH and FACH are common channels shared between many users and have very limited transmission capability. Typically if the UE has data to transmit, it will be moved to the CELL_DCH state in which the UE has a dedicated channel.

The transitions from the CELL_DCH are most often based on the user activity. If the UE is inactive for a period of time, it can be moved to CELL_FACH, and eventually to either a paging state or to the Idle mode.

The Packet Data Convergence Protocol (PDCP) [36] is mainly responsible for the header compression and decompression of IP data streams (e.g., TCP/IP and RTP/UDP/IP headers for IPv4 and IPv6). Possible methods for header compression include IP Header Compression [62] and RObust Header Compression (ROHC) [51]. The header compression will be described in more detail in Chapter 5.

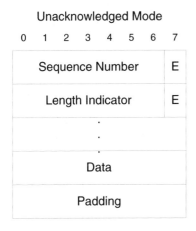
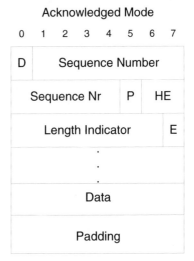

Figure 3.10: The Radio Link Control (RLC) protocol header for Unacknowledged and Acknowledged Modes. Note that the field lengths are in bits, not bytes. The Length Indicator field is optional, and may be repeated one or more times.

The Radio Link Control (RLC) protocol [41] is responsible for the majority of the functionality expected from a link layer protocol, including segmentation and reassembly, concatenation, and padding. It also includes error correction and flow control using a selective repeat ARQ protocol, and in-sequence delivery of upper layer PDUs. The RLC protocol is also responsible for ciphering of data.

The RLC can operate in three different modes:

- Transparent Mode (TM), which does not add any protocol overhead. Segmentation is still possible, but only to predefined PDU sizes. Transparent mode is mostly used for CS traffic.

- Unacknowledged Mode (UM), which provides the possibility for segmentation and concatenation, but no support for retransmissions. The Unacknowledged Mode is most suitable for the transmission of conversational media, which can tolerate moderately high packet loss rates.

- Acknowledged Mode (AM), which provides segmentation, concatenation and retransmissions and flow control using a selective repeat ARQ protocol. The Acknowledged Mode is mostly used for packet data applications as well as for control signaling that require very low packet loss rates.

The RLC header formats for acknowledged and unacknowledged RLC protocols are shown in Figure 3.10. For Unacknowledged Mode, the total header size is 8 bits, consisting of a 7 bit sequence number and a header extension bit (E). If the extension bit is set, the header is followed by a Length Indicator, indicating the end of the SDU and the beginning of the padding. The Length Indicator is only needed if the untransmitted part of the SDU does not fill the RLC PDU completely. For Acknowledged Mode the header size is 16 bits, consisting of a Control/Data indicator (D), a 12 bit sequence number, poll bit (P) and a 2 bit Header Extension (HE) field. The Control/Data indicator is used to indicate if the PDU is a data PDU (consisting of actual data) or a control PDU (consisting a status report for the ARQ operation). The poll bit is used to request a status report.

While the PDCP and RLC protocols are independent of the specific physical or transport channel used, the Medium Access Control (MAC) protocol has been divided into several sub-protocols based on the used transport channel. The MAC-c is an entity that handles the paging channel as well as the forward access channel and the random access channel, while the MAC-d entities handle the dedicated transport channels. With the introduction of the High Speed Packet Access, also additional MAC entities have been specified. These will be described in Section 3.3.2.

The MAC-c functionality includes mapping the logical channels to transport channels, adding a UE identifier on the Random and Forward Access Channels, and selecting the amount of data to be transmitted from each logical channel.

The MAC-d functionality includes mapping and switching different logical channels to transport channels, selecting the amount of data to be transmitted from each logical channel and the ciphering for RLC transparent mode (e.g. for CS services). In the network side, the MAC-d entity also provides flow control towards the MAC-c entity.

3.3.2   Radio Access Evolution

As discussed in Section 3.3.1, using dedicated channels for packet data applications may not be an efficient way to assign very high data rates to individual users. For packet applications higher available peak data rates typically lead to improved performance. In order to improve the packet data performance, the UMTS systems have been enhanced with High Speed Packet Access (HSPA).

The HSPA consists of two components. First, in the downlink a new shared transport channel, the High Speed Downlink Shared CHannel (HS-DSCH), has been added. The HS-DSCH allows the system to assign all available resources to one or more users in a very efficient manner. Second, the uplink dedicated channels have been enhanced to Enhanced Dedicated CHannels (E-DCHs). Even though the uplink channels are still dedicated, the uplink resources can be shared between users in a more efficient manner.

The High Speed Downlink Packet Access (HSDPA) improves the downlink packet data performance in the following ways:

- High Speed Downlink Shared CHannel (HS-DSCH), allowing sharing of resources between many users;

- higher-order modulation, allowing higher peak bit rates;

- 2 ms Transmission Time Interval (TTI), reducing end-to-end delays;

- fast link adaptation, allowing the used data rate to be adjusted to better meet the instantaneous link quality, which will increase the total system capacity;

- fast scheduling, which allows the system to choose the user based on e.g. their instantaneous channel quality, which will also increase the system capacity;

- fast Hybrid Automatic Repeat reQuest (HARQ), which allows the link adaptation to recover from errors, and reduces the time required to provide error recovery by retransmissions.

HS-DSCH is based on shared channel transmission, which means that system resources (channelization codes and the transmission power) are dynamically shared between users. The sharing is done for each 2 ms TTI, and results in more efficient use of available resources in UTRAN compared to the dedicated channels.

HSDPA adds the possibility to use 16-quadrature amplitude modulation (16QAM) in addition to quadrature phase-shift keying (QPSK) modulation for downlink transmission. 16QAM has twice the peak rate capability of QPSK, allowing twice the peak data rate of QPSK and making more efficient use of bandwidth than QPSK. However, in order to use 16QAM, the radio channel conditions need to be better than for QPSK.

The short 2 ms TTI reduces the latency of the system, and also allows fine grained sharing of the resources.

Radio channel conditions experienced by different downlink communication links vary significantly due to fast fading. Fast link adaptation adjusts the transmission parameters (used modulation and channel coding rate) to instantaneous radio conditions. In good radio conditions, high-order modulation and small code rate are used; while in bad radio conditions, robust modulation and coding are used. The current radio conditions are measured by the UE, and reported to the Node B by sending a Channel Quality Indicator (CQI) on a physical control channel especially designed for HSDPA.

As discussed in Section 3.3.1, WCDMA uses power control to compensate for differences and variations in the instantaneous radio channel conditions of the downlink. In principle, power control gives radio links with bad channel conditions a proportionally larger part of the total available cell power. This ensures similar service quality to all communication links, despite differences in radio channel conditions. This strategy has been designed especially for CS voice communication, but may not lead to optimal system capacity for packet data applications, which may tolerate some delay.

HS-DSCH does not adjust to transmission power for each user, but rather adapts the rate to match the current channel conditions. This link adaptation is more efficient than power control for services that tolerate short-term variations in the data rate and delay.

The fast scheduling is based on selecting to which User Equipment (UE) the shared channel transmission should be directed for each TTI. The scheduler determines overall

HSDPA performance. For each TTI, the scheduler decides which users the HS-DSCH should be transmitted to and determines the actual end-user bit rate.

Various scheduling strategies vary from sequentially allocating radio resources among users (round-robin scheduling) to channel-dependent scheduling, such as transmitting always to the user with best channel quality (maximum CQI scheduling). The various possible scheduling strategies are discussed in Chapter 5.

The fast hybrid ARQ is based on storing the received data even when the decoding fails, and requesting a retransmission. The retransmission is then combined with the stored original transmission to enhance the probability of a successful decoding. The effect of the HARQ is to dramatically increase the probability of successful retransmissions. The actual protocol used is N-channel stop-and-wait ARQ.

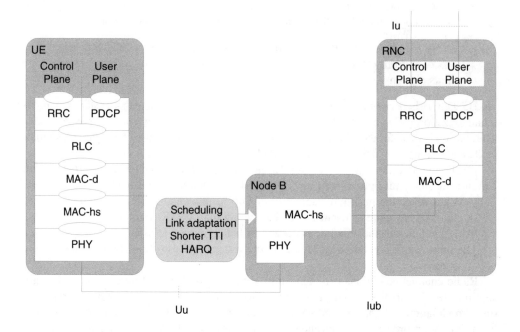

Figure 3.11: The High Speed Downlink Packet Access protocol architecture.

As shown in Figure 3.11, most of the features of HSDPA are implemented in the Node B, largely in a new MAC entity called MAC-hs. The MAC-hs entity is responsible for implementing the HARQ protocol, making scheduling decisions as well as for the fast link adaptation.

The improvements for the Enhanced Dedicated CHannel (E-DCH) in the uplink are similar to HSDPA:

- fast scheduling, allowing Node B to determine when and at what rate individual UEs can transmit;

- fast hybrid ARQ (HARQ) with soft combining, reducing the retransmission delay;

- short 2 ms TTI, which can be used in addition to the normal 10 ms TTI, reducing overall delays.

However, it is important to realize the fundamental differences between the uplink and downlink. The shared resource in the uplink is the interference at the Node B, which depends on the actual transmission power of each UE. In the downlink the shared resource is centralized to the Node B. This difference has significant implication on the scheduler design.

The WCDMA uplink is non-orthogonal and fast power control is essential for the uplink to handle users close to the base station, who might transmit at too high power and interfere with reception of data from users farther out in the cell. The fast power control is also beneficial in order to ensure backwards compatibility of E-DCH and normal dedicated channels. Consequently, fast power control is the primary way to implement fast link adaptation for the uplink. This is in contrast to HSDPA, where a (more or less) constant transmission power with rate adaptation is used.

Soft hand-over is supported in order to limit the amount of interference generated in neighboring cells by allowing uplink power control from multiple cells. It also provides macro-diversity gains. Note that soft hand-over has two aspects: power control by multiple cells (to reduce inter-cell interference) and reception at multiple cells (increasing system capacity by allowing the reception from the best possible cell).

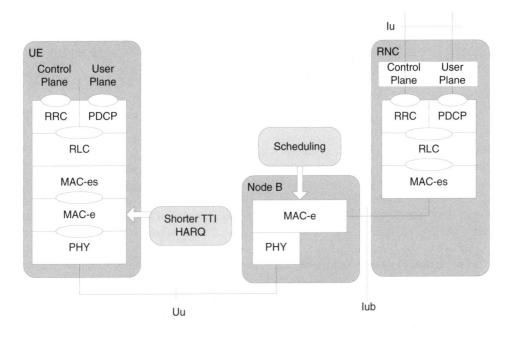

Figure 3.12: The Enhanced Uplink protocol architecture.

The required changes in the UTRAN protocol architecture for E-DCH support are shown in Figure 3.12. Similarly to HSDPA, the hybrid ARQ functionality is located in a new MAC entity in the Node B called MAC-e. However, also the protocol structure in the RNC is updated with a new MAC-es entity in order to support in-sequence delivery, duplicate detection and macro-diversity combining for the enhanced uplink.

In the uplink, the common resource shared among the terminals is the amount of tolerable interference, i.e. the total received power at the Node B. The amount of common uplink resources a terminal is using depends on the data rate used. For packet data applications with bursty traffic pattern, it is possible to use a more relaxed connection admission strategy; a larger number of high rate packet data users can be admitted to the system as the scheduling mechanism can handle the situation when multiple users need to transmit in parallel. However, for conversational services these applications are not expected to be very bursty, and therefore the need for uplink scheduling is not apparent. Especially for voice it is a better strategy to allow transmission of limited amounts of data (typically one or two voice frames) without needing to schedule each individual user. For other multimedia applications with higher data rate requirements, the benefits of the scheduling might be larger.

The scheduling is executed by allocating grants to users. By using the two types of grants, the scheduler can control the transmission behavior of each individual terminal. Absolute grants are used to set an absolute value for the maximum allowed transmission rate for the terminal. Relative grants are used to gradually adjust the allocated rate for a terminal.

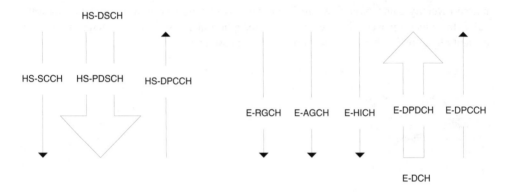

Figure 3.13: The actual physical channels used to carry HSPA traffic channels (HS-DSCH and E-DCH) and related signaling channels.

The actual physical channels used to carry HSPA traffic channels are shown in Figure 3.13. HS-DSCH is carried over a High Speed Physical Downlink Shared CHannel (HS-PDSCH), while E-DCH is carried over an Enhanced Dedicated Physical Data CHannel (E-DPDCH).

The control channels required for HSDPA are the High Speed Shared Control CHannel (HS-SCCH) in the downlink and the High Speed Dedicated Physical Control CHannel (HS-DPCCH) in the uplink. HS-SCCH is used to carry the signaling required for fast scheduling, link adaptation and HARQ. It consists of the link adaptation information (channelization code set, modulation scheme and transport block size), HARQ information (process ID, redundancy version and new data indicator) and UE identity for scheduling. The UE identity is not actually explicitly transmitted, but is coded together with the CRC, and the UE uses the joint decoding of the CRC to detect both transmission errors and transmissions aimed at other UEs. The HS-DPCCH is used to carry link quality information from the UE to the Node B for link adaptation as well as the HARQ feedback signaling (positive and negative acknowledgments).

In addition to the new control channels, an Associated Dedicated Physical CHannel (A-DPCH) is required in the downlink for the uplink power control when using the HSDPA.

For E-DCH four control channels are needed. In downlink, two separate scheduling channels are provided. The Absolute Grant CHannel (E-AGCH) is used to assign absolute grants to individual UEs. The absolute grants specify the exact rate the UE may use to transmit, while the relative grants transmitted on the Relative Grant CHannel (E-RGCH) are used to increase or reduce the current grant of the UE. The third downlink channel, HARQ Indicator CHannel (E-HICH), is used to transmit the HARQ signaling (positive and negative acknowledgments). The uplink Enhanced Dedicated Physical Control CHannel (E-DPCCH) is used to inform the Node B about the exact amount of data transmitted, to provide the HARQ retransmission sequence number (which is used to determine the redundancy version as well as to indicate transmission of new data) and to transmit scheduling requests (indicating that a higher data rate would be required).

In addition to the new physical channels needed to support HSPA, the UTRAN has been improved by introduction of the Fractional Dedicated Physical CHannel (F-DPCH). The fractional DPCH is used to reduce the control signaling of the dedicated physical channels associated with the HSDPA (A-DPCH). It basically consists of a normal physical channel, which is time multiplexed between several users. This allows the system to support significantly more active users in the CELL_DCH state.

There is ongoing work to further evolve the UMTS Radio Access with Long Term Evolution (LTE) and HSPA Evolution.

The LTE will introduce a completely new air interface based on Orthogonal Frequency Division Multiplexing (OFDM) in the downlink and Single Carrier Frequency Division Multiple Access (SC-FDMA) in the uplink. LTE will supporting flexible spectrum allocation from 1.5 MHz up to 20 MHz, and will provide significant performance improvements in application performance and system capacity. The LTE air interface will only support packet access; and in order to support voice and video telephony, IMS Multimedia Telephony service is envisioned to be used. The introduction of the LTE will also change the network architecture significantly for both radio access and core network.

The HSPA Evolution will focus on improving the existing HSPA air interface to support higher data rates and improved capacity. The details of the HSPA evolution in terms of actual performance or architectural changes are still open at the time or writing, but the target capacity for LTE and HSPA evolution are discussed in Chapter 7.

3.3.3   Core Network

The main function of the core network is to provide switching, routing and transit for user traffic. The core network also contains the databases and network management functions and provides support for other additional functions such as charging and lawful interception.

The UMTS Core Network is divided into circuit switched and packet switched domains. The circuit switched domain is used for circuit switched services, such as voice and video telephony applications as well as for circuit switched data transmission. A simplified picture of the architecture of the circuit switched UMTS Core Network is shown in Figure 3.14.

In Figure 3.14 two network nodes are shown, circuit switched Media GateWay (MGW) and the Mobile Switching Center (MSC) Server. These two nodes implement a modern layered architecture Mobile Switching Center.

The Media Gateway is responsible for interconnecting the UMTS network to the PSTN. MGW also interfaces UTRAN with the core network over Iu. Over Iu, the Media Gateway typically supports

- media conversion, such as transcoding from AMR to PCM,

- bearer control,

- payload processing, such as echo cancellation and conferencing,

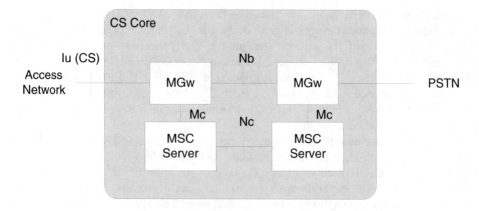

Figure 3.14: UMTS circuit switched Core Network architecture.

in order to provide services typically associated with CS telephony.

Transcoding is used in Media Gateways both to interconnect to the PSTN network and to convert the AMR speech to PCM for links requiring voice sample transmission in PCM format. The actual transcoding capabilities depend on the Media Gateway implementation, but typically include at least transcoding between AMR and PCM.

The Media GateWay controls the used bearers, and may adapt the AMR codec speech rate based on e.g. system load.

Payload processing functionality in the Media Gateway is used to provide circuit switched services such as teleconferencing, echo cancellation (due to acoustic echo from the terminals) and so on.

The Media Gateways are interconnected with Nb interface. The Nb interface can be either IP or ATM based, and provides framing structure for voice frames.

The MSC Server is responsible for the call control and mobility control. The MSC Server is responsible for the control of mobile originated and mobile terminated CS Domain calls. It terminates incoming signaling and translates it to corresponding outgoing signaling.

The MSC Server controls Media Gateways using Mc protocol. The Mc protocol is based on H.248.

The packet switched domain is used for all packet access (e.g. GERAN and UTRAN).

This common PS core network provides the General Packet Radio Service (GPRS), which provides mobility management, session management and transport for Internet Protocol packet services. For more information about GPRS see 3GPP TS23.060 [22].

GPRS can be used for traditional packet switched services such as file downloading or web surfing, but also for conversational services such as Voice over IP. The IMS Multimedia Telephony service will run over the packet switched core network.

The aim of mobility management is to track where the subscribers are, so that mobile phone services can be delivered to them, while the session management is the process of keeping track of the activity of users across sessions.

The packet switched UMTS Core Network consists of Serving GPRS Support Nodes (SGSN), which are the core network edge nodes, and Gateway GPRS Support Nodes (GGSNs), which are the core network gateways, as well as network elements (e.g. EIR, HLR, VLR and AUC) that are shared by both circuit switched and packet switched domains.

A simplified view of the architecture of the packet switched Core Network is shown in Figure 3.15.

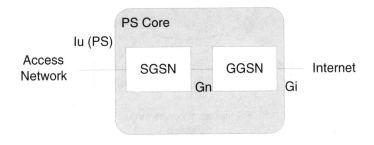

Figure 3.15: UMTS packet switched Core Network architecture.

Both SGSN and GGSN nodes are responsible for routing IP packets from (and to) the external IP networks (e.g. the Internet) to the correct Iu connection (from where it will eventually be delivered to the correct mobile).

In addition to the user plane data transmission, the SGSN stores both subscription and location information needed to handle originating and terminating packet data transfer. The subscription information consists of the International Mobile Subscriber Identity (IMSI), temporary identities and (possibly) Packet Data Protocol (PDP) addresses. The location information consists of either the cell or routing area of the mobile, possibly a VLR number and GGSN address if an active PDP context exists.

Similarly the GGSN stores the IMSI and (possibly) PDP addresses as well as the address of the SGSN where the mobile is registered.

The mobility in the PS core network is handled with GPRS Tunneling Protocol (GTP). GTP allows mobiles to move while maintaining the connection to the Internet as if from one location at the GGSN. It does this by carrying the subscriber's data from the subscriber's current SGSN to the GGSN which is handling the subscriber's session.

The SGSN implements the mobility management functionality for a GPRS user. In the Iu mode three mobility states (PMM states) are present:

- PMM-DETACHED;

- PMM-IDLE;

- PMM-CONNECTED.

A simplified PMM state machine is shown in Figure 3.16. In the PMM-DETACHED state there is no communication between the mobile terminal and the SGSN and no PS connection is available. The mobile terminal is not reachable by the SGSN since the location of the mobile terminal is not known. Hence, IMS services like Multimedia Telephony do not work in PMM-DETACHED.

In order to establish a connection between the mobile terminal and the SGSN, the mobile terminal must perform a GPRS attach procedure. For Multimedia Telephony this should be done at power on. The GPRS attach procedure establishes the PS signaling connection between the mobile terminal and the SGSN, and changes the PMM state to PMM-CONNECTED. In the PMM-CONNECTED state the location of the mobile terminal

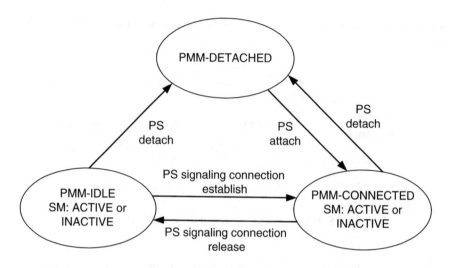

Figure 3.16: PMM state model.

is known in the SGSN with an accuracy of a serving RNC and the location of the mobile terminal is tracked by the serving RNC. After having registered to the IMS the mobile terminal and SGSN should be kept in the PMM-CONNECTED state if fast connection to the IMS is sought after. But, at least, if the RRC states URA_PCH and CELL_PCH are not used (and in some error cases) the mobile terminal and the SGSN will be moved down to the PMM-IDLE state after a period of inactivity. In the SGSN, PS signaling connection release or failed downlink transfer changes the state to PMM-IDLE. The mobile terminal enters the PMM-IDLE state when its PS signaling connection to the SGSN has been released or broken. This release or failure is explicitly indicated by the RNC to the mobile terminal or detected by the mobile terminal. To transfer PS data, the mobile terminal and the SGSN must re-enter the PMM-CONNECTED state by establishing the PS signaling connection again. Therefore, for example, the call setup performance for Multimedia Telephony is better in PMM-CONNECTED state than in PMM-IDLE state.

The PDP context exists in the mobile terminal, the SGSN and the GGSN. The PDP context state machine in these three entities consists of two states, SM: ACTIVE and SM: INACTIVE. The SM: INACTIVE state is characterized by the fact that the data service for a certain PDP address is not activated. If the PDP address is not activated the PDP context contains no routing or mapping information to process the PDUs related to that PDP address and no data can be transferred. In the SM: ACTIVE state, the PDP context contains mapping and routing information for transferring the PDP PDUs for a particular PDP address between the mobile terminal and the GGSN. Thus, the PDP address in use is activated in the mobile terminal, SGSN and GGSN.

In the PMM-IDLE and the PMM-CONNECTED states, GPRS session management may or may not have activated one or more PDP contexts (i.e. the SM state may be ACTIVE or INACTIVE). As a consequence, in the PMM-IDLE state, a PDP context may be established, but no corresponding connection over the Iu interface nor the radio are established. This is the case when camping in RRC states URA_PCH and Cell_PCH.

3.4 Quality of Service

A service like the Multimedia Telephony service is considered end-to-end, which means from a Multimedia Telephony client associated to a mobile terminal to another Multimedia Telephony client associated to another mobile terminal. Therefore, to fulfill the service requirements, the networks supporting Multimedia Telephony service should be able to give end-to-end Quality of Service (QoS) guarantees. The 3GPP have developed a layered QoS architecture. This has to be able to provide QoS for services like the Multimedia Telephony service that uses 3GPP networks (for more information on the 3GPP QoS architecture see 3GPP TS23.107 [40]). Figure 3.17 exemplifies the layered 3GPP QoS architecture.

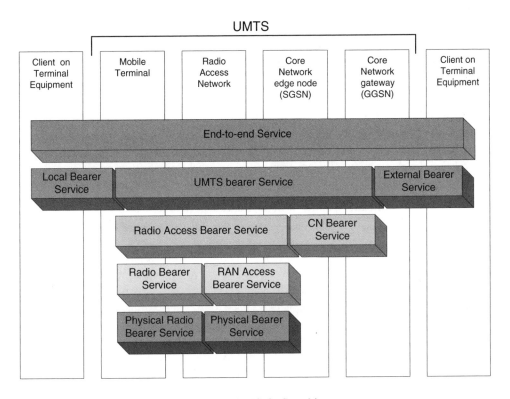

Figure 3.17: UMTS QoS architecture.

IMS services like the Multimedia Telephony service may make use of the UMTS Bearer service when setting up the communication session. In the case of Multimedia Telephony the use of the UMTS Bearer service enables the access network to optimize the radio and core network transport to provide sufficient QoS for the Multimedia Telephony service. It should be noted that the UMTS Bearer service only gives QoS guarantees in the home network of the Multimedia Telephony client (from the mobile terminal to the core network gateway). The UMTS Bearer is set up by either the mobile terminal or the network that requests a suitable QoS for the media flow(s) at Multimedia Telephony session establishment. The key components of the UMTS Bearer are the Radio Access Bearer (RAB) between the mobile terminal and the RAN and the core network bearer between the core network edge node (e.g. the SGSN) and the core network gateway (e.g. the GGSN). The QoS guarantee that defines the RAB and the core network bearer is negotiated at bearer setup by activating a Packet Data Protocol context (PDP context).

The RAB is realized by a Radio Bearer (RB) and an RAN Access Bearer. The Radio Bearer covers all the aspects of the radio interface transport. For an example of a Radio Bearer definition suitable for Multimedia Telephony voice media, see Section 5.6. This bearer service is provided by the UTRAN or the GERAN depending on used radio access type. The RAN Access Bearer together with the Physical Bearer Service provides the transport between RAN and SGSN. A RAN Access Bearer for packet traffic provides different bearer services for a variety of QoS. The RAN Access Bearer is provided by the Iu (UTRAN and possibly GERAN) or the Gb (GERAN) Bearer Service.

It should be noted that to provide an end-to-end service with end-to-end QoS there is a need for an external bearer service (that is outside the scope of 3GPP) and a local bearer service in the interface between the mobile terminal and the entity on which the Multimedia Telephony client is implemented (the terminal equipment) (that also are outside the scope of 3GPP and are not further discussed in this book).

3.4.1   QoS Attributes

GERAN and UTRAN use a common core network for packet transfer. This common core network provides the General Packet Radio Service (GPRS), which is designed to support several QoS levels to allow efficient transfer of non-real-time traffic as well as real-time traffic.

Table 3.1: Quality of Service attributes per traffic class.

Attribute/Traffic class	Conversational class	Streaming class	Interactive class	Background class
Maximum bit rate	X	X	X	X
Delivery order	X	X	X	X
Maximum SDU size	X	X	X	X
SDU format information	X	X		
SDU error ratio	X	X	X	X
Residual bit error ratio	X	X	X	X
Delivery of erroneous SDUs	X	X	X	X
Transfer delay	X	X		
Guaranteed bit rate	X	X		
Traffic handling priority			X	
Allocation/Retention priority	X	X	X	X
Source statistics descriptor	X	X		

In order to send and receive packet data via GPRS, the mobile terminal must activate a PDP context. This operation makes the mobile terminal known in the GGSN and interworking with data network can be started. The mobile terminal, SGSN, GGSN and RAN should support the 3GPP Release 99 Quality of Service. The 3GPP Release 99 Quality of Service provides the means to set up the PDP context that is specified by a set of QoS attributes which are negotiated between the mobile terminal, SGSN, GGSN and RAN. The QoS attributes are negotiated between the mobile terminal and the network at activation or modification of a PDP context. The requested and negotiated attributes are stored in the appropriate Quality of

Service profile that is unique for each PDP context in the mobile terminal, SGSN and GGSN. SGSN negotiates the attributes with RAN using the RAB assignment procedure over the Iu interface and the Packet Flow Context (PFC) procedures over the Gb interface.

One example of QoS attributes are the traffic classes (Table 3.1). Four traffic classes, background, interactive, streaming and conversational, have been introduced in the QoS concept of the 3GPP Release 99 specifications.

Traffic class *conversational* is intended to be used to carry real-time traffic flows that are very delay sensitive. The typical use case is to use this traffic class for the voice stream of a Multimedia Telephony session. This traffic class should also be suitable for video telephony. The real-time data flow is always performed between live (human) end-users. Associated attributes like Guaranteed bit rate and Transfer delay define requirements that should be fulfilled by the GPRS network.

Traffic class *streaming* is mainly intended to be used to carry semi-real-time traffic flows that have less delay sensitivity than conversational flows. The semi-real-time data flow is always aiming at a live (human) destination. A typical use case for this traffic class in a Multimedia Telephony context is the streaming of a video clip from the other end-user or a server during the Multimedia Telephony session.

Traffic class *interactive* is mainly intended to be used by traditional Internet applications (WWW, FTP), messaging applications and SIP signaling. Hence, this interactive traffic class is the 'classical data communication scheme' that on an overall level is characterized by the request response pattern of the end-user. At the message destination there is an entity expecting the message (response) within a certain time. Round-trip delay time is therefore one of the key attributes. In Multimedia Telephony this scheme would be used for the session control signaling, file transfer (e.g. sharing of an image) and messaging.

Traffic class *background* is intended to be used by for instance computers that send and receive data files in the background (e.g. email). The scheme is thus more or less delivery time insensitive but the content of the packets shall be transparently transferred (with low bit error rate).

Table 3.1 shows the QoS attributes per traffic class as defined in 3GPP TS23.107 [40]. Traffic classes together with the associated attributes reflect the typical requirements that should be fulfilled by the RAN and core network in order to provide acceptable quality for the end-user, of a certain service. The only possible way for RAN to make assumptions about the traffic source is the knowledge of the traffic class and the associated QoS attributes. Therefore Release 99 QoS is a helpful tool to be able to optimize the transport for a certain traffic type, like a voice stream in a Multimedia Telephony session.

As mentioned above, the QoS architecture is layered and there are different levels of QoS. The QoS parameters negotiated in the PDP context activation are UMTS Bearer service attributes. However, the RAB attributes are the same set of QoS attributes as for the UMTS Bearer service. Therefore, as an example, a RAB used for packet speech may be referred to as; a 16 kbps Conversational RAB (where 16 kbps is the maximum bit rate attribute and Conversational is the traffic class). Typical examples of RABs and the associated RBs are presented in 3GPP TS25.993 [46], while 3GPP TS34.108 [19] presents RABs and RBs used for testing purposes.

3.5 The IP Multimedia Subsystem

As described in Section 3.3.3, the UMTS core network is divided into the circuit switched and the packet switched domains. The packet switched domain is a data network based on

Internet technologies such as the ones discussed in Section 3.2. Terminals can use the packet switched domain to access services on the public Internet. For example, a UMTS terminal can access its user's email server, which is located somewhere on the Internet, through the packet switched domain. In another example, users can also surf the Web using this domain and the Web browser in their UMTS terminals.

So, the packet switched domain makes all the services that can be found on the Internet available to UMTS users. This way, users can access those services virtually everywhere and at any time.

Given that Internet services include real-time multimedia services such as video conferencing, the packet switched domain can be used to access Internet-based real-time multimedia services as well. For example, a user can use his or her UMTS terminal to establish a Voice over IP session with another UMTS user or with an Internet user using a personal computer with a fixed Internet access.

However, at the time the packet switched domain was being specified, 3GPP noticed that Internet multimedia services were based on several different technologies. Some services used the H.323 [116] protocol suite, some used SIP [172], and many used vendor-specific proprietary protocols. The result was that, unless two terminals used the same technology, they were not able to communicate between them.

This situation was far from ideal from 3GPP's point of view, because 3GPP's goal was that any set of Internet terminals, including UMTS terminals but also regular PCs connected to the Internet, were able to communicate among them. Consequently, 3GPP started working on an architecture to provide multimedia services based on Internet technologies and, in particular, on SIP. This architecture was named the IP Multimedia Subsystem (IMS). For a detailed description of the IMS see [55].

The choice of SIP as the main signaling protocol for the IMS ensured that IMS terminals could communicate with any Internet terminal that used SIP. As time went by, SIP became the main signaling protocol for multimedia services on the Internet. Therefore, IMS terminals are currently able to communicate with a large number of users.

The IMS architecture allows providing services that are horizontally integrated. Service developers move away from the traditional vertically integrated service silos, each of which contains all the functionality needed by the service it provides. Instead, the IMS provides common functionality that may be used by different services (e.g. user authentication). This way, service developers only need to implement the functionality that is specific to a particular service. The rest of the functions needed by the service are provided by the IMS. Functions provided by the IMS that are available to services built on top of it include capability negotiation, authentication, service invocation, addressing, routing, group management, presence, provisioning, session establishment, and charging.

The IMS is often referred to as a service enabling architecture because it enables the creation of services in a rapid fashion. Service developers do not spend time implementing functions that are already provided by the IMS. Instead, they can focus on the service-specific logic each new service requires.

The IMS defines several functions and the interfaces between them. Figure 3.18 shows the IMS architecture. The following sections discuss the IMS functions and the interfaces between them.

Note that 3GPP standardizes logical functions, not nodes. Functions can be mapped to physical nodes in several ways. A single node can implement a single function or several ones. Also, a single logical function can be distributed into several nodes.

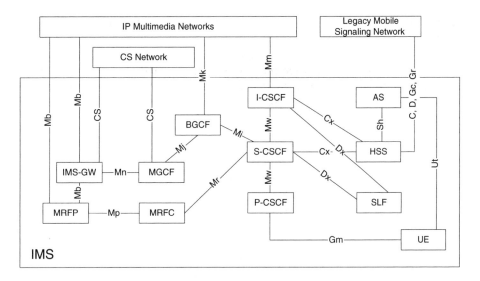

Figure 3.18: The IMS architecture.

3.5.1 The Home Subscriber Server and the Subscription Location Function

The Home Subscriber Server (HSS) is a database that stores the profile of all the users of a domain. A user's profile includes security-related information such as cryptographic keys, service-related information such as to which services the user is subscribed, and other information such as the S-CSCF that has been allocated to the user or the user's location.

An HSS is supposed to be a central database for a domain. However, some networks have more users than a single HSS can handle. These networks contain more than one HSS.

Networks with more than one HSS also contain a Subscription Location Function (SLF). Nodes that need to query an HSS about a particular user first query the SLF, which returns the address of the HSS handling the user.

Both the HSS and the SLF support interfaces based on Diameter [54]. The HSS implements the Cx interface and the SLF implements the Dx interface. The SLF is, effectively, a Diameter redirection agent.

3.5.2 The Call/Session Control Functions

The Call/Session Control Functions (CSCFs) are mainly SIP entities, although they also implement non-SIP interfaces (e.g. Diameter-based interfaces). There are three types of CSCFs in the IMS: Proxy-CSCF (P-CSCF), Serving-CSCF (S-CSCF), and Interrogating-CSCF (I-CSCF).

3.5.3 Proxy-CSCF

From a terminal's perspective, its P-CSCF is a SIP outbound proxy. It is the proxy to which the terminal sends all its outgoing traffic. Additionally, all the terminal's incoming traffic also traverses its P-CSCF.

The P-CSCF's main functions relate to security and signaling compression. A P-CSCF establishes a number of security associations with its terminals in order to exchange SIP traffic with them. These security associations provide integrity protection and may provide confidentiality.

The interface between a terminal and a P-CSCF may traverse a low-bandwidth radio link. Consequently, sending large protocol messages over that interface may take a long time. In order to minimize session establishment times, IMS terminals usually compress the SIP signaling they send to the P-CSCF. The compression mechanism used between them is SigComp [161].

3.5.4   Serving-CSCF

The S-CSCF acts as the SIP registrar for its domain. Additionally, it handles all IMS sessions performing service invocation when needed. When a particular service is applicable to a session being established, the S-CSCF routes the SIP signaling to the Application Server (AS) providing the service.

In order to perform these functions, the S-CSCF needs to have access to the profile of the users it handles. The S-CSCF implements the Diameter-based Cx interface towards the HSS. The S-CSCF obtains user profiles and stores user-related information in the HSS over this interface.

A user's profile includes the user's filter criteria, which are key to service innovation. The S-CSCF obtains a user's filter criteria from the HSS and uses them to decide which services apply to a particular session.

A user's filter criteria are a collection of triggers that are applied to the user's incoming and outgoing SIP messages. If a trigger matches a message the message is relayed to the AS indicated in the trigger. For example, a trigger may match incoming SIP messages received from a particular user. When the S-CSCF, which is the entity that evaluates the filter criteria, receives a message from that particular user, it relays the message to the AS indicated in the trigger.

3.5.5   Interrogating-CSCF

When a SIP entity needs to send a SIP message to a particular domain, it performs a DNS lookup to obtain the address of the SIP server of that domain. A domain's DNS records point to the domain's I-CSCF. Therefore, the I-CSCF handles the domain's incoming traffic.

When an I-CSCF receives an incoming SIP message, it consults the HSS using the Cx interface to discover which S-CSCF should handle this message. Then, the I-CSCF relays the message to that S-CSCF.

3.5.6   The Application Servers

IMS services are implemented in Application Servers. Application Servers communicate with the S-CSCF over the SIP-based ISC interface and with the HSS over the Diameter-based Sh interface.

Although new services are typically implemented in native SIP Application Servers, it is also important to be able to provide legacy services to IMS users. The IMS specifies the Open Service Access – Service Capability Server to interwork with the OSA framework application server and the IP Multimedia Service Switching Function (IM-SSF) to provide Customized Applications for Mobile networks Enhanced Logic (CAMEL) services. The Service Capability Server and the Service Switching Function can access the HSS using the

Diameter-based Sh and the Si interfaces respectively. Both communicate with the S-CSCF using the SIP-based ISC interface.

3.5.7 The Multimedia Resource Function

The Multimedia Resource Function (MRF) performs media-related functions such as media manipulations (e.g. mixing of voice streams in a conference), and playing of tones and announcements. The Multimedia Resource Function is a distributed node. It consists of an MRF Controller (MRFC) and an MRF Processor (MRFP). The MRF Controller controls the MRF processor using the H.248-based [115] Mp interface.

3.5.8 PSTN Interworking Functions

The IMS provides interworking with circuit switched telephony networks through a set of nodes. The Media Gateway Control Function (MGCF) performs the protocol translation between SIP and the signaling protocol used in the circuit switched network (e.g. ISUP). The MGCF also controls a Media GateWay (MGW) using the H.248-based Mn interface. The MGW performs the translation between the RTP-based media used in the IMS and the media format used in the circuit switched network.

The MGCF interworks with the circuit switched network through a signaling gateway. The signaling gateway receives circuit switched signaling over IP (e.g. using SCTP) from the MGCF and sends it to the circuit switched network (e.g. an SS7 network), and vice versa.

The Breakout Gateway Control Function (BGCF) decides which MGCF handles a particular session that will terminate in the circuit switched network. If the BGCF decides that an MGCF in a different domain should handle the session, it can relay the SIP message it received to a BGCF in that domain.

PSTN interworking is further discussed in Section 4.8.

3.5.9 IPv4/IPv6 Interworking Functions

The IMS can be implemented on top of both IPv6 and IPv4. Additionally, IMS terminals may establish sessions with Internet terminals, which may be IPv4-only and IPv6-only terminals. This means that there will be IMS sessions between IPv4 and IPv6 clients.

IPv4–IPv6 conversion is performed at the Interconnect Border Control Function (IBCF), which was originally specified by TISPAN (see Section 3.6). Figure 3.19 shows the architecture of the IBCF.

The IMS-ALG modifies SIP messages in order to perform IPv4–IPv6 conversion. It also controls the Transition GateWay (TrGW), which performs IPv4–IPv6 conversion at the media level.

The IBCF can also implement the so-called Topology Hiding Interwork Gateway functionality. The Topology Hiding Interwork Gateway functionality consists of encrypting the topology-related information contained in outgoing messages. This way, those messages do not disclose any topology-related information to the messages' receivers, which may be a competing operator.

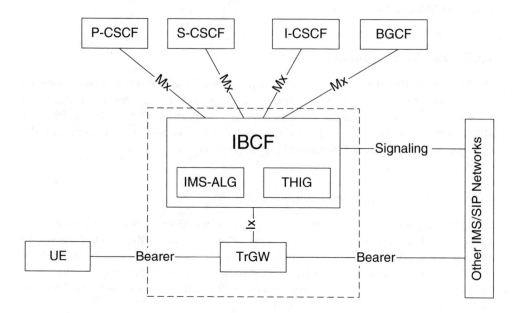

Figure 3.19: The IBCF architecture.

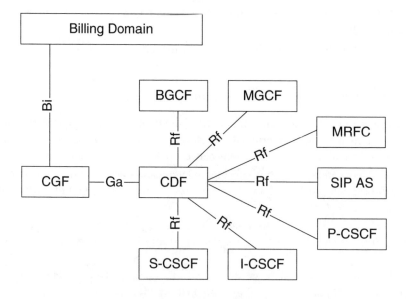

Figure 3.20: The IMS offline charging architecture.

3.5.10 Charging

The IMS provides two types of charging: offline and online. Offline charging is typically used to charge users with permanent subscriptions while online charging is typically used for prepaid users.

3.5.10.1 Offline Charging Architecture

Figure 3.20 shows the offline charging architecture. The SIP elements with information about the session being charged send that information to the Charging Data Function (CDF) over the Diameter-based Rf interface. The CDF generates the Charging Data Records (CDRs) and sends them to the Charging Gateway Function (CGF) over the Ga interface. The CGF uses the Bi interface to send the CDRs to the billing system.

3.5.10.2 Online Charging Architecture

Online charging provides credit control. When a user runs out of credit that user cannot access the service any longer. For example, if a user is involved in a session and runs out of credit, the session is terminated.

Figure 3.21 shows the online charging architecture. The Application Servers and the MRFC involved in a session send information to the Online Charging System (OCS) over the Diameter-based Ro interface.

The IMS GateWay Function (IMS-GWF) also sends information to the OCS over the Ro interface. Additionally, the IMS-GFW can communicate with the S-CSCF using the SIP-based ISC interface. Effectively, the IMS-GWF looks like an AS to the S-CSCF. If a session needs to be terminated due to lack of credit, the IMS-GWF terminates it using the ISC interface.

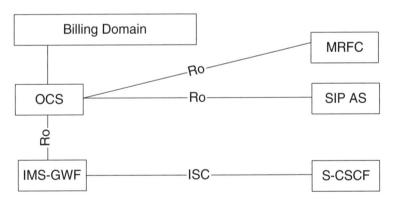

Figure 3.21: The IMS online charging architecture.

3.5.11 Policy and Charging Control

In the IMS, there is a need to control the media plane from the signaling plane. For example, a user may have a voice-only subscription that does not allow the transmission of video. If such a user tries to send a video stream anyway, it should be possible to block that stream. In another example, a user who runs out of credit may ignore the indication to terminate his or her voice session and continue to send voice packets. In this case, it should also be possible to block those packets.

The IMS provides media plane control for policy and charging-related reasons. In the early stages, the IMS provided policy and charging control using separate architectures. Policy control was based on Service Based Local Policy (SBLP) and charging control on Flow Based Charging (FBC). Later, those architectures were combined and, currently, the IMS specifies a single architecture for policy and charging control.

Figure 3.22 shows the IMS Policy and Charging Control (PCC) architecture. Application Functions (AFs) communicate with the Policy and Charging Rules Function (PCRF) over the Diameter-based Rx interface. The PCRF can fetch subscription information from the Subscription Profile Repository (SPR) over the Sp interface and can control the Policy and Charging Enforcement Function (PCEF) in the access gateway through the Diameter-based interface Gx. When a 3GPP cellular access is used, the access gateway is the GGSN.

The OFfline Charging System (OFCS) and the Online Charging System (OCS) can also control the PCEF using the Diameter-based Gz and Gy interfaces respectively.

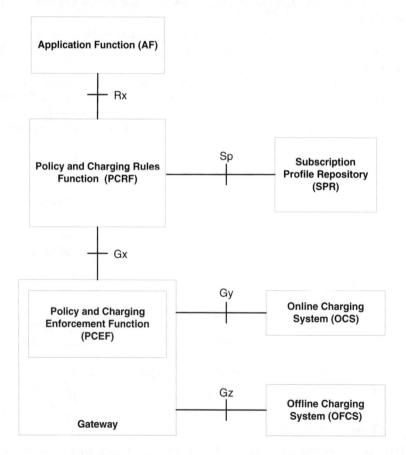

Figure 3.22: The IMS policy and charging control architecture.

Section 4.3.2 discusses policy control in the context of the Multimedia Telephony service.

3.5.12 Home and Visited Domains

Media in sessions between IMS terminals using a 3GPP cellular access flow between their GGSNs. Each GGSN can be located in the user's home or visited domain. When the GGSNs are located in the user's visited domains, media takes the shortest path between the terminals.

However, operators often configure their networks so that the GGSNs are located in the home domains. This gives operators more control over their users' media. Nevertheless, this type of configuration produces non-optimal media routing. In a session between Alice and Bob, media would traverse Alice's visited domain, Alice's home domain, Bob's home domain, and Bob's visited domain.

The location of the GGSN serving a user also impacts the location of the P-CSCF serving the same user. The P-CSCF is always located in the same domain as the GGSN.

3.6 The TISPAN Next Generation Network

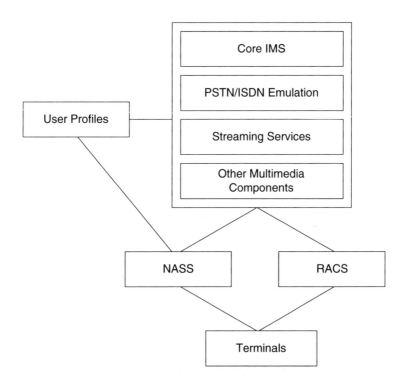

Figure 3.23: The TISPAN NGN architecture.

ETSI TISPAN has developed an architecture to provide multimedia services over fixed accesses. This architecture is called Next Generation Network (NGN) and is based on the IMS.

NGN is an access agnostic architecture able to provide IMS services to users using any type of access. This way, users can enjoy the same set of services regardless of the access they use. Therefore, TISPAN NGN achieves the much expected convergence of mobile and fixed services.

Figure 3.23 shows the TISPAN NGN architecture. The Resource and Admission Control Sub-system (RACS) handles admission control and resource reservation. The Network Attachment Sub-System (NASS) handles network level authentication and IP address provisioning.

The Core IMS is, effectively, the IMS as defined by 3GPP. Figure 3.24 shows the Core IMS architecture. TISPAN defines three nodes that are important from the IMS perspective. They are the Interconnect Border Control Function (IBCF), the Interconnect Border Gateway Function (I-BGF), and the Access Border Gateway Function (A-BGF).

The IBCF controls the I-BGF, which in on the media plane, and can perform policy control towards other networks. The IBCF can also perform the Topology Hiding Interwork Gateway (THIG) functionality, as discussed in Section 3.5.9.

The A-BGF, which is on the media plane, is controlled by the P-CSCF and behaves as an access gateway as discussed in Section 3.5.11.

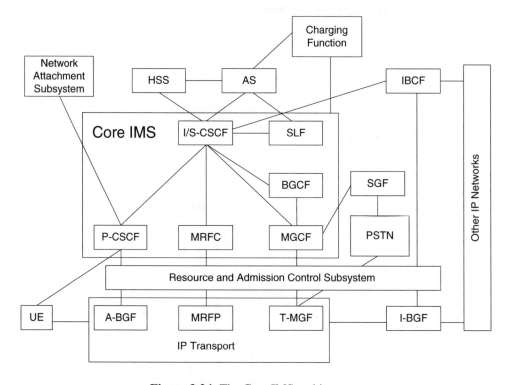

Figure 3.24: The Core IMS architecture.

3.7   Multimedia Telephony Realization

This section presents how Multimedia Telephony could be realized over the core network and over the cellular access to meet the requirements on performance and service flexibility outlined in Chapter 2. The core network realization uses the IMS together with a telephony application server that implements the Multimedia Telephony logic. The realization over the cellular system uses HSPA and includes a number of optimizations of the radio bearers to meet the strict performance requirements of a telephony service.

3.7.1 Core Network and Service Layer Realization

Multimedia Telephony makes use of the IMS functionality in the service layer. In Figure 3.25 one possible realization of Multimedia Telephony is shown. The system presented includes the IMS as outlined in Section 3.5 and information about the functions and interfaces can be found there. The text below highlights some of the functions that are of special interest for Multimedia Telephony and are part of this particular realization of the Multimedia Telephony communication service. The functions of interest are marked in the figure with numbers that correspond to the numbers in the text below.

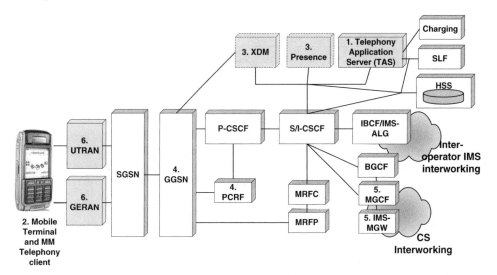

Figure 3.25: Realization of Multimedia Telephony from an architectural view.

The application server (1), which here is called the Telephony Application Server (TAS), is a crucial part of the Multimedia Telephony infrastructure. The TAS contains service-specific logic for the Multimedia Telephony communication service. This service-specific logic includes mechanisms that handle setup, teardown and modification of Multimedia Telephony sessions. Logic to handle supplementary services like conferencing and communication diversion is also included. For more information about Multimedia Telephony session handling and supplementary services see Sections 4.4 to 4.7. The TAS also does charging and subscriber handling with the help of the HSS.

The Multimedia Telephony client (2) is another crucial part of the system. The Multimedia Telephony client includes the Multimedia Telephony communication service, potentially other IMS communication services like Presence and the Multimedia Telephony application. The Multimedia Telephony communication service is the part that originates or terminates Multimedia Telephony session signaling and media while the Multimedia Telephony application implements amongst other things the end-user to client interfaces (see Figure 3.26). When producing media streams, the Multimedia Telephony client accesses shared resources like codec and protocol stacks implemented in the mobile terminal. For more information about codecs and protocols used in Multimedia Telephony see Sections 5.1 and 5.2. The Multimedia Telephony client implements media transport processing mechanisms to handle jitter and packet losses and it also handles media adaptation if needed. These mechanisms are presented in Sections 5.3 and 5.4.

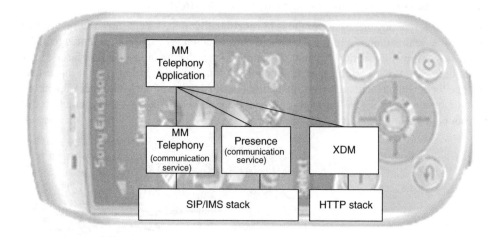

Figure 3.26: Terminal architecture for a Multimedia Telephony client on the application level. On top is shown the Multimedia Telephony application, which may utilize different IMS communication services and XDM. A common SIP stack is used for the different IMS communication services and XDM that use XCAP over HTTP.

XML Document Management (XDM) and Presence (3) are service enablers that can be used together with Multimedia Telephony to ease communication. XDM helps users to store and manage their contact lists in the network rather than in the mobile terminal as of today (for more information about XDM see Section 8.4.2). Presence is a service enabler that allows a user to subscribe to other users' Presence information. Usually the Presence information shows the availability and willingness of a user to communicate, but other information like geographical position and current mood can be included as part of the Presence information (for more information about the Presence communication service, see Section 8.4.1).

The different media types supported by the involved IMS communication services need different QoS. To help control the QoS, 3GPP has developed the Policy and Charging Control (PCC) architecture. The PCRF (4), which is part of the Policy and Charging Control (PCC) architecture, will most probably be a key component in many Multimedia Telephony realizations. The PCRF controls the media plane by the information provided in the control plane via SIP and SDP (for more information about SDP usage in Multimedia Telephony, see Section 5.2.5) and by that PCRF helps operators to provide users with radio bearers that have suitable QoS for the media types they intend to use. PCRF can also help operators to block users who do not have valid subscriptions. The PCRF is an integral part of the system when using network initiated QoS, which is a new feature in 3GPP release 7. For more information about network initiated QoS see Sections 4.3.1.1 and 4.3.2.1. The GGSN (4) is the end-point of the UMTS bearer service, which means that the end-points for the QoS negotiation are the GGSN and the mobile terminal. Therefore, in the PCC architecture the GGSN acts as the policy enforcement point enforcing policies regarding QoS for Multimedia Telephony and other IMS communication services.

CS interworking is needed for fast uptake of Multimedia Telephony. The MGCF and the IMS-MGW (5) provide the functionality of translating CS signaling to/from SIP and perform media conversion. For more information about IMS interworking with CS networks and media considerations of the interworking, see Sections 4.8 and 5.7.

A 3GPP compliant mobile terminal is connected to IMS via either GERAN or UTRAN (6). Both access methods can provide a number of different radio bearer realizations that can be used for Multimedia Telephony. However, UMTS is the most likely candidate for early introduction of Multimedia Telephony. Therefore, in Section 3.7.2 one possible UMTS radio bearer realization for Multimedia Telephony is outlined. The realization is further described, in more detail, in Section 5.6. This radio bearer realization was used for the performance evaluations for Multimedia Telephony in Chapter 7.

3.7.2 Outline of a Radio Bearer Realization

The Multimedia Telephony communication service is an IMS service that can be realized over any access technology. However, since Multimedia Telephony aims to replace CS telephony, the realization of the Multimedia Telephony must meet strict performance requirements such as low end-to-end delay and low packet loss rates. And at the same time, Multimedia Telephony should be a flexible communication tool that allows the end-users to choose between a set of different communication methods, including voice, video and text communication. This flexibility also puts requirements on the underlying access. As a result of the different requirements, certain access methods are more suitable for Multimedia Telephony than others. In Figure 3.27 one possible radio realization suitable for Multimedia Telephony is outlined. Various features used in the realization are marked in the figure with numbers that correspond to the numbers in the text below.

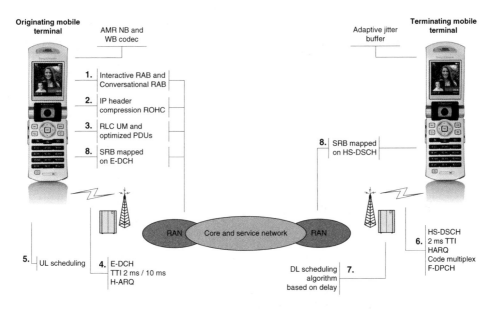

Figure 3.27: Radio bearer realization for Multimedia Telephony.

Multimedia Telephony use combinations of RABs (1) to provide differentiated QoS for different types of media. In the simplest case, Multimedia Telephony basic voice call, two RABs are used. One RAB is used for IMS related SIP signaling, and that RAB uses the

Interactive traffic class. The speech data is sent over a Conversational RAB. In the case of more complex media combinations, more advanced RAB combinations may be used. For instance, in the case of Multimedia Telephony video calls, two Conversational RABs can be used (one RAB is used for speech transfer and another RAB for video transfer). The use of RABs for Multimedia Telephony is depicted in Figure 3.28. For more information about the different traffic classes and the QoS architecture see Section 3.4.

IP technology has a number of benefits, but one apparent drawback is quite large overhead of real-time media due to generous IP header sizes. To be able to compete with CS telephony in terms of capacity, these headers need to be compressed. A number of header compression algorithms have been included in the 3GPP specifications but the preferred technology is called ROHC (2). ROHC can compress the IP, UDP and RTP headers from 60 bytes (IPv6) or 40 bytes (IPv4) down to 3 bytes. For more information about header compression and ROHC see Section 5.5.

Real-time media, like speech and video, are quite tolerant to packet losses but are intolerant to delay variations. Therefore, to avoid time-consuming retransmissions, UDP is used rather than TCP as the IP transport protocol. A radio realization suitable for real-time media should also avoid slow retransmissions on the radio link layer. Therefore, the proposed radio realization uses RLC UM (3) as the radio link control protocol mode for speech and video. For non-real-time data such as the SIP signaling, the radio bearer uses RLC AM. Optimized RLC PDU size is beneficial for capacity. When Multimedia Telephony is implemented on a 3GPP compliant mobile terminal, the speech will be encoded by AMR-NB or AMR-WB. This knowledge, plus the knowledge of how ROHC works, is used to optimize the RLC PDU size so that the padding on the RLC layer is minimized. For more information about the optimized RLC PDUs see Section 5.6.

The media flexibility aspects of Multimedia Telephony (i.e. the instant adding and dropping of media) and the variation of packet sizes due to ROHC and other 'bursty' IMS service enablers like Presence that may be used in parallel with Multimedia Telephony make it beneficial to use radio bearers that can transfer packets of varying sizes during e.g. speech transfer without causing significant delay variations in the media transfer. In the uplink, the preferred radio technology is the E-DCH (4). For more information about E-DCH see Section 3.3.2. E-DCH can be realized using 2 ms or 10 ms TTIs. Both options perform well for Multimedia Telephony. The addition of a Hybrid ARQ enables fast retransmissions and soft combination of multiple attempts, which is beneficial for capacity as well as media quality. The E-DCH can be scheduled to increase control over the interference level. However, for low bandwidth media like speech, the uplink scheduling of E-DCH (5) may not increase capacity significantly; rather the overhead associated with the uplink scheduling may reduce capacity. Therefore, in the uplink voice capacity evaluations, non-scheduled mode is used (see Section 7.3.4).

In the downlink, HSDPA is the preferred radio technology. HS-DSCH (6) is the shared channel used for data transfer. It dynamically shares the resources between multiple mobile terminals. The number of mobile terminals that can share the HS-DSCH may in some cases be limited by the codes used by the associated power control channel (the A-DPCH or the F-DPCH). The F-DPCH is preferred since it is economical in terms of code and power usage. It can be viewed as a shared power control channel for which one code is shared among a set of mobile terminals (up to 10) to carry power control commands. Using F-DPCH for Multimedia Telephony, it is highly unlikely that the system becomes code limited in downlink. The HS-DSCH channel uses 2 ms TTI and HARQ for fast retransmissions and soft combining of multiple attempts. The data rate is continuously adapted to radio conditions

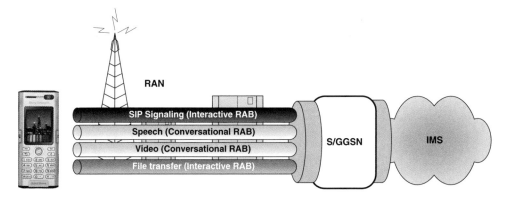

Figure 3.28: Multimedia Telephony using 'multi-RABs', N Interactive RABs for SIP signaling and non-real-time media and one or two Conversational RABs for speech and video.

using link adaptation algorithms that change maximum possible transport block size. A good downlink scheduler is the key to getting good downlink capacity for Multimedia Telephony over HSDPA. Scheduling algorithms for HSDPA are not standardized; instead it is up to every vendor to implement suitable scheduling algorithms. The radio bearer realization proposed in this book uses a delay scheduler (7) for scheduling of real-time media streams like voice and video. The delay scheduler increases priority of each packet based on the delay of the packet in the scheduling queue. Multimedia Telephony downlink voice capacity for the delay scheduler and other scheduling strategies are presented in Section 7.3.3.

The signaling radio bearer (SRB) (8), which carries e.g. the RRC and GPRS SM signaling, is mapped onto the E-DCH and the HS-DSCH. This reduces the delay of various RRC and GPRS SM procedures and thus helps to provide efficient cell reselections (hard hand-over) in downlink and lower Multimedia Telephony session setup delay. Calculations of Multimedia Telephony session setup delays, having the signaling radio bearers mapped on the HSPA channels, may be found in Section 7.7.

Chapter 4

Session Control

Gonzalo Camarillo, Per Synnergren

This chapter explains how the control plane of the Multimedia Telephony service works. The first part of the chapter explains the Session Initiation Protocol (SIP) that is used by the Multimedia Telephony clients and IMS to set up, modify and release the Multimedia Telephony sessions. SIP, together with the Session Description Protocol (SDP), is also used to realize supplementary services like communication diversion, conferencing and communication barring that are present in ISDN/PSTN. To exemplify the Multimedia Telephony session management procedures and the supported supplementary services a number of example flows are provided that describe how Multimedia Telephony sessions are set up, modified and released and how supplementary services are realized using SIP/SDP.

Quality of Service (QoS) is becoming increasingly important with larger volumes of traffic in the packet switched domain. The use of IMS as the service layer provides the possibility to negotiate QoS for the individual media components of the Multimedia Telephony sessions at session establishment as well as during the session. The IMS uses the UMTS bearer service provided by the 3GPP QoS architecture. This chapter describes the GPRS session management signaling and the policy control and charging architecture, which all are key components to enable QoS negotiation in IMS sessions.

4.1   SIP

SIP (Session Initiation Protocol) [172] is the main signaling protocol of the IMS and thus used by the Multimedia Telephony communication service. SIP is a text-based rendezvous protocol that provides user mobility and session establishment.

SIP provides user mobility through registrations. A user is always reachable under the same identifier regardless of the user's current location. Every time a user becomes available at a user agent, the user agent registers its location with the user's registrar, which is located in the user's home domain. This way, the home domain is informed about how to route incoming session requests to the user agent where the user is currently available.

In SIP, users are identified by URIs. SIP URIs have a similar format to email addresses (e.g. <sip:alice@example.com>).

SIP provides session establishment through a two-way session description exchange called the offer/answer model [170]. A user agent generates a session description that contains

IMS Multimedia Telephony over Cellular Systems   S. Chakraborty, T. Frankkila, J. Peisa and P. Synnergren
© 2007 John Wiley & Sons, Ltd

the information needed to establish the session (e.g. IP addresses to be used to transfer the media) and sends it to the remote user agent. This session description is referred to as the *offer*.

On receiving the offer, the remote user agent generates its own session description, which is referred to as the *answer*.

Both the offer and the answer are written in a session description format which needs to be understood by both user agents. The default session description format is SDP (Session Description Protocol) [84].

Once the offer/answer exchange completes, the user agents can start exchanging media between them. At that point, the session is considered to be established.

4.1.1 Logical Entities

In addition to user agents, which are the terminals or software that users use to connect to the SIP network, SIP defines the following logical entities: proxy servers, redirect servers, and registrars.

Proxy and redirect servers help route SIP messages towards their destination. A proxy servers receives a message, makes a routing decision based on the message's contents and on any extra information available to the proxy (e.g. registration information), and forwards the message to an appropriate next SIP hop. A redirect server, on the other hand, makes a routing decision and informs the entity that sent the message to the server so that this entity forwards the message to the appropriate next SIP hop.

Registrars are the SIP entities that handle SIP registration. When a user registers at a registrar, a binding is created between an AoR (Address of Records) and a contact URI. A user's AoR is the URI under which the user is always available (i.e. the URI users would write in their business cards). A contact URI is the URI where the user is currently available. For example, a registration can bind the AoR <sip:Alice.Smith@example.com> to the contact URI <sip:alice@192.0.2.1>.

4.1.2 IMS Registration

As discussed in Section 4.1.1, registrations create a binding between an AoR and a contact URI. In the IMS, registrations create an extra state in the network, in addition to that type of binding.

IMS users may have several AoRs, which in the IMS are referred to as public user identities. For example, a user may have a public user identity that consists of a SIP URI and another public user identity that consists of a tel URI (e.g. <tel:+4681234567>). A single IMS registration may bind several public user identities to a contact URI.

When a user registers to the IMS, an S-CSCF is assigned to the user. This S-CSCF acts as the user's registrar. Once a user registers, the user's S-CSCF obtains the user's profile from the HSS. This way, when the user establishes a session, the S-CSCF can invoke the appropriate Application Servers (ASs) based on the user's filter criteria, as discussed in Section 3.5.4.

Additionally, IMS registrations are used to choose a security mechanism to be used between the IMS terminals and the network and to set it up. When IPsec is used, the registration is used to set up the IPsec security associations that will protect the SIP signaling between the IMS terminal and the P-CSCF.

IMS registrations are also used to set up the signaling compression state between the IMS terminal and the P-CSCF. This compression state will be used throughout the duration of the registration to compress all the SIP messages exchanged between the IMS terminals and the P-CSCF.

Figure 4.1 shows an IMS registration. The P-CSCF routes the SIP REGISTER request (2) to the I-CSCF of the user's home domain. The I-CSCF consults the HSS (3) and is informed (4) that there is no S-CSCF assigned to the user. The I-CSCF assigns an S-CSCF to the user and relays the SIP REGISTER request (5) to it. The S-CSCF informs the HSS (6) that it will be handling the user and downloads (7) the authentication vectors needed to authenticate the user. The S-CSCF generates a challenge based on the authentication vectors and challenges the IMS terminal (8).

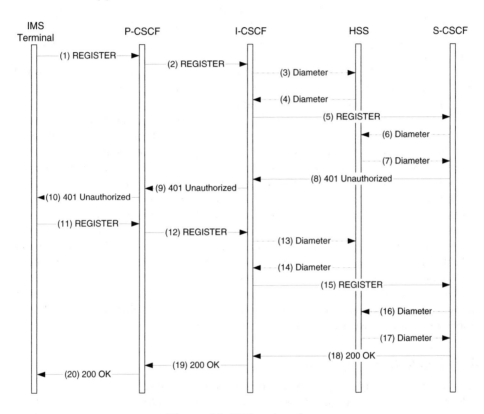

Figure 4.1: IMS registration.

On receiving the challenge (10), the IMS terminal generates a new SIP REGISTER request (11) with its credentials. The I-CSCF consults the HSS (13) and is informed (14) about the S-CSCF that has been assigned to the user (the I-CSCF does not store this type of information). The I-CSCF relays the SIP REGISTER request (15) to the S-CSCF assigned to the user. The S-CSCF checks that the credentials provided by the IMS terminal are valid and informs the HSS (16) that the user has been successfully registered.

4.1.3 IMS Session Establishment

The remaining sections of this chapter contain a few Multimedia Telephony signaling flows, including flows showing Multimedia Telephony session establishments. However, all those flows abstract out the IMS nodes and represent them by a black box called the IMS core. This section discusses what happens in that black box at session establishment time.

Figure 4.2 shows the establishment of an IMS session. An IMS terminal generates a SIP INVITE request (1) and sends it to its P-CSCF, which forwards it (3) to the S-CSCF assigned to the calling user. On receiving the SIP INVITE request, the S-CSCF evaluates the calling user's filter criteria to decide whether or not it needs to invoke any AS. In this example, no AS is invoked. However, Figure 4.2 clearly shows the points where they could be invoked, as a result of an S-CSCF evaluating filter criteria. For examples of flows where ASs are invoked at both the originating and the terminating sides see Figures 4.9 and 4.10.

After evaluating the filter criteria, the S-CSCF relays the SIP INVITE request (5) to the I-CSCF of the called user's domain. The I-CSCF consults the HSS (7) and is informed (8) about the S-CSCF handling the called user. The I-CSCF relays the SIP INVITE request (9) to that S-CSCF. The S-CSCF evaluates the called user's filter criteria to discover that no AS needs to be invoked for this session. Consequently, the S-CSCF relays the SIP INVITE request (11) to the P-CSCF assigned to the called user. Finally, the P-CSCF relays the SIP INVITE request to the called user's IMS terminal.

The SIP INVITE request received by the called user's terminal uses the QoS precondition extension. This means that the terminal should not alert the called user until QoS is available for the session being established. Consequently, the terminal returns a provisional response (15) and starts performing resource reservation on its access network. The calling user's terminal starts doing resource reservation on its access network on reception of that provisional response (20). Once the calling user's terminal is done with its resource reservation, it sends a SIP UPDATE request (31) indicating so. Once the called user's terminal receives this SIP UPDATE request (35) and is ready with its own resource reservation, it starts alerting the called user. The called user's terminal generates a provisional response (41) indicating that alerting has started. When the called user accepts the invitation to the session, the terminal generates a successful final response (57).

The SIP PRACK requests in the flow and their responses are used to provide a reliable delivery of provisional responses (here the SIP 180 Ringing and SIP 183 Session Progress responses). They are needed when the preconditions extension is used.

4.2 Signaling Compression

As an option, signaling compression can be applied to the SIP signaling for e.g. Multimedia Telephony. The use of SIP for IMS session establishment in cellular systems such as UMTS may lead to unnecessarily long setup times if the available bandwidth for SIP signaling is low. In order to reduce the setup times, the Signaling Compression scheme SigComp (see RFC 3320 [161] and RFC 3321 [85]) has been developed by the IETF to compress text-based protocols such as SIP. In an IMS system, SigComp is implemented in the IMS terminal and in the P-CSCF. The SIP messages are thus compressed over the air interface and in the mobile core network. The use of compression must be negotiated between the terminal and the P-CSCF. This is in general done during the IMS registration but SigComp can be invoked at any time.

The heart of SigComp is the Universal Decompressor Virtual Machine (UDVM). The UDVM is especially designed for decompression. It receives the compressed SIP message, loads the decompressor code plus a dictionary, and then starts executing. The compression algorithm is not specified. The choice of compression algorithm is left to the implementer using the specified UDVM instructions. Thus, SigComp will be implemented using many

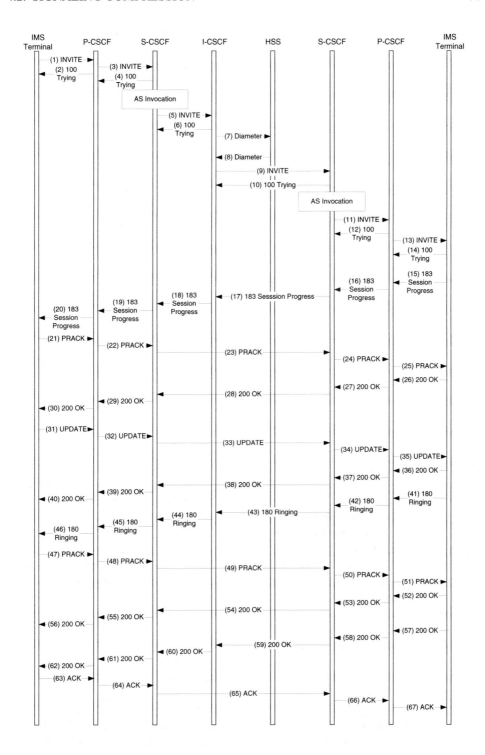

Figure 4.2: IMS session establishment.

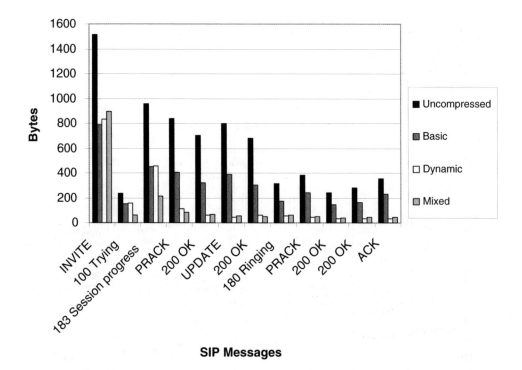

SIP Messages

Figure 4.3: Message sizes for a session setup sequence, subjected to Basic, Dynamic and Mixed compression.

different compression algorithms. However, it is reasonable to assume that the majority of algorithms will belong to the general class of dictionary compression algorithms. Common for these algorithms is that during the processing of the text, all repeated occurrences or matches of symbol strings are replaced by a code, typically giving the offset and length of the match. The performance of the compression algorithm depends on the number and lengths of the matches as well as on the form of the code used to represent the match. Using this class of compression algorithms, the compression efficiency is increased by using a SIP static dictionary (see RFC 3485 [80]), which contains typical SIP key words and phrases.

To further increase SigComp performance, SigComp extended operation can be used. SigComp extended operations are specified in RFC 3321 [85]. Figure 4.3 exemplifies the compression efficiency of SigComp for three cases. The cases are basic compression (as in RFC 3320 [161], including the use of the SIP static dictionary), dynamic compression and mixed compression. In dynamic compression the SIP dictionary together with sent acknowledged messages are used as the dictionary for compression. In mixed compression the static dictionary together with both sent acknowledged messages and received messages are used as the dictionary for compression.

The IMS signaling flow used in the evaluation corresponds to the SIP messages sent/received by the originating terminal in Figure 4.2. This flow and the SIP messages used (including the size of the messages) in the flow are described in 3GPP TS24.228 [13]. The compression factor (size uncompressed/size compressed) for the entire sequence is 1.9:1 for Basic, 3.8:1 and 4.4:1 for Dynamic and Mixed compression respectively. Based on this, in later calculations of Multimedia Session setup latencies a SigComp compression factor of 3:1 is assumed (see Section 7.7).

4.3 Controlling QoS

This section describes the GPRS session management procedures and policy control procedures needed to control QoS when establishing and modifying IMS sessions such as Multimedia Telephony sessions.

4.3.1 GPRS Session Management Signaling

If the Multimedia Telephony service is to be deployed in a network compatible with 3GPP release 6 or earlier 3GPP releases, the PDP context activation will be mobile terminal originated. Thus, the mobile terminals start the QoS negotiation. The mobile terminal includes a set of QoS attributes with values suitable for the media stream(s) of the Multimedia Telephony session in the PDP context activation request. Figure 4.4 shows the GPRS session management flow of activating a PDP context using mobile terminal originated PDP context activation. The following list explains the numbered events in the flow represented in the figure.

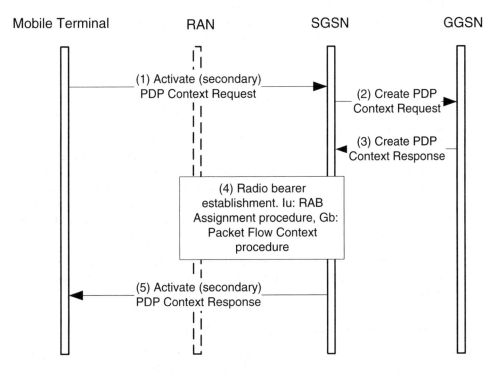

Figure 4.4: GPRS session management flow for a mobile terminal originated establishment of PDP context.

1. The mobile terminal generates an Active PDP Context Request message (when activating a so-called primary PDP context) or an Activate Secondary PDP Context Request message (when activating a so-called secondary PDP context). The choice of activating a primary or secondary PDP context depends on whether the PDP context should have a new PDP (IP) address or if it should reuse the PDP address from an already activated PDP context. Typically a media stream in a Multimedia Telephony session should use a secondary PDP context. The primary PDP context will in most cases be used to transfer SIP signaling. The mobile terminal includes the desired QoS by adding the set of values for the QoS attributes in the request message. In case of secondary PDP context activation a Traffic Flow Template (TFT) is in most cases provided. The TFT contains attributes that specify an IP header filter that is used to direct data packets received from the interconnected packet data network to the newly activated PDP context. The Activate PDP (Secondary) Context Request is sent to the SGSN.

2. The SGSN validates the Activate (Secondary) PDP Context Request by using information from the mobile terminal like the PDP type (typically IPv4 or IPv6), the PDP address (no information is provided by the mobile terminal in case a dynamic PDP address is sought) and the Access Point Name (APN, that is a reference point to a certain packet data network). If the validation is successful, the SGSN sends a Create PDP Context Request to the GGSN. If there is a request for a primary PDP context and there is no static PDP address of the mobile terminal, the GGSN allocates a PDP address. Otherwise the PDP address information is already provided by the mobile terminal.

3. The GGSN generates a new entry in its PDP context table, stores and activates the TFT (if provided) and checks if the requested QoS is compatible with operator policies. Thereafter the GGSN responds to SGSN with a Create PDP Context Response including the negotiated QoS profile.

4. The SGSN interoperates with the radio access network using different procedures depending whether the SGSN is operating in the Gb mode or the Iu mode. In the case of Iu mode it starts the RAB assignment procedure to set up a RAB with suitable QoS that is derived from the negotiated QoS profile of the PDP context. In the case of Gb the PFC procedure is started. In that case a QoS profile is sent to the base station subsystem giving the base station subsystem information about the negotiated QoS.

5. The SGSN returns an Activate (Secondary) PDP Context Response. The SGSN is now able to route PDP PDUs between the GGSN and the mobile terminal, and to start related charging.

4.3.1.1   The Network Requested Secondary PDP Contexts Activation

One drawback with the current 3GPP QoS architecture is that it does not allow the operator to control the QoS to a degree that is desirable and possible. For services like Multimedia Telephony, where the subscriber is paying for a service experience rather than for just a bit pipe, it is the responsibility of the operator to deliver the service to the subscriber with sufficient quality, and to determine the adequate QoS mechanisms. Therefore, in 3GPP release 7 it has been agreed to add network initiated QoS for secondary PDP context activation, giving the operator full control of the QoS negotiation procedure. With this addition, the core network becomes the entity that decides the suitable QoS for a certain media stream associated to a service request. However, for network initiated QoS to make sense, the core network must get the service/media information from the service layer. This information is provided by the policy control and charging architecture; see Section 4.3.2. The addition of network initiated QoS also allows usage of 3GPP QoS for IMS clients that are unaware of what kind of underlying access network it uses (e.g. a PC client using a WCDMA modem).

To support network initiated QoS a number of new additions must be done to the 3GPP protocols:

- Network Requested Secondary PDP Context Activation (NRSPCA): A new procedure to let GGSN request a new secondary PDP context. The parameters passed down to SGSN must include a QoS profile and an uplink filter to be installed in the mobile terminal. Further, a new procedure is needed to let the SGSN request a new PDP context from the mobile terminal. The parameters passed down to the mobile terminal must include the QoS profile and the uplink filter.

- Network controlled modification of uplink filters: A new parameter in the network initiated PDP modification procedure must be added as well as the associated UpLink filter (i.e. a UL TFT).

- Capability negotiation in bearer session establishment and modification: Information that the UE can handle NRSPCA must be transferred from the mobile terminal to SGSN and GGSN when establishing the primary PDP context. The GGSN must also pass down information to the mobile terminal whether mobile initiated or network initiated secondary PDP context activation should take place. The network must also be able to trigger the change of QoS initiation mode at any time, for instance in order to handle inter-SGSN routing area updates (and the SGSNs having different capabilities).

- Support for network initiated QoS in the policy control and charging architecture.

Figure 4.5 shows the GPRS session management flow of activating a PDP context using NRSPCA. The following list explains the numbers in the figure.

1. The GGSN gets information from the policy control that a Multimedia Telephony session is being started and that a secondary PDP context suitable for carrying voice is needed. This triggers the GGSN to request a secondary PDP context with a suitable QoS that is preconfigured by the operator.

2. A new NRSPCA message is sent from GGSN to SGSN. This message carries the requested QoS profile and the downlink filter (DL TFT), which later will be used by the mobile terminal to request the secondary PDP context in step 4. The UpLink filter (the UL TFT) is also included.

3. A new NRSPCA message is sent from SGSN to the mobile terminal with the same content as in step 2.

4. The mobile terminal installs the UL TFT and starts an Activate Secondary PDP Context procedure by sending the Activate Secondary PDP Context Request message to SGSN. It includes the DL TFT and the QoS profile the mobile terminal received from the network.

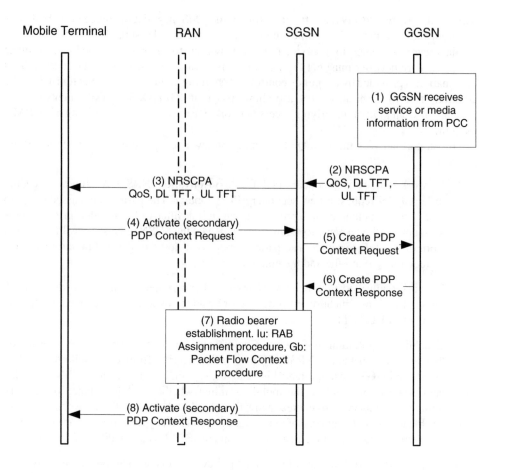

Figure 4.5: Network requested secondary PDP context activation.

5. The SGSN validates the Activate (Secondary) PDP Context Request. If the validation is successful, the SGSN sends a Create PDP Context Request to the GGSN.

6. The GGSN activates the TFT. Thereafter the GGSN responds to SGSN with a Create PDP Context Response including the negotiated QoS profile.

7. The SGSN interoperates with the radio access network using different procedures depending if the SGSN is operating in the Gb mode or the Iu mode. In the case of Iu mode it starts the RAB assignment procedure to set up a RAB with suitable QoS derived from the negotiated QoS profile from the PDP context. In the case of Gb the PFC procedure is started, and a QoS profile is sent to the base station subsystem giving the base station subsystem information about the negotiated QoS.

8. The SGSN returns an Activate Secondary PDP Context Response. The SGSN is now able to route PDP PDUs between the GGSN and the mobile terminal using the new RAB.

4.3.2 Policy Control Signaling

Policy control is an important mechanism for a Multimedia Telephony service since it is the mechanism that is used to coordinate events such as Multimedia Telephony session initiation/modification in the application layer and resource management events such as bearer establishment/modification in the IP bearer layer. Policy control is the mechanism that enables IMS to enforce suitable QoS for a given service or media flow.

In 3GPP release 6, two different architectures were used for policy control and charging. These two architectures were the following:

- Service Based Local Policy (SBLP): A Policy Decision Function (PDF) was used to authorize the QoS for flows and a Policy Enforcement Point (PEP) enforces the QoS policy and accordingly accepts or rejects the bearer setup. For more information on SBLP, see 3GPP TS23.207 [9].

- Flow Based Charging (FBC): Charging rules were applied on a per flow basis. The Traffic Plane Function (TPF) requests charging rules from the charging rules function (CRF) which gets the flow information from the Application Function (AF). The AF in Multimedia Telephony would be the IMS core (or in more detail the P-CSCF). For more information on FBC, see 3GPP TS23.125 [35].

One part of the IMS evolution in 3GPP release 7 was to merge SBLP and FBC into one Policy Control and Charging (PCC) architecture. This was done to harmonize the policy control and charging procedures. For more information on PCC, see 3GPP TS23.203 [39]. In this book it is assumed that Multimedia Telephony is deployed using PCC. The logical architecture of PCC is shown in Figure 4.6.

The interfaces (or reference points) between the logical entities are specified by 3GPP and have the following functionality:

- The Rx interface enables transport of application level session information from the application function (the IMS core) to the Policy Charging Rules Function (PCRF). Such information may include IP filter information to identify the media flow for policy control and differentiated charging and media bandwidth requirements.

- The Gx interface enable the PCRF to have dynamic control over the Policy Charging Enforcement Function (PCEF). Basically the Gx interface enables the signaling of policy decisions and indication of bearer termination to and from the GGSN.

- The Sp interface allows the PCRF to request subscription information related to the IP transport level policies from the Subscription Profile Repository (SPR) based on a subscriber ID.

- The Gy interface allows online credit control for service data flow based charging.

- The Gz interface enables transport of service data flow based offline charging information.

The policy control functionalities over the Rx and Gx interfaces are of great importance for the session control of Multimedia Telephony sessions. The policy control works on a service data flow level and performs QoS control by authorization and enforcement of the maximum QoS that is authorized for a service data flow. It can also perform gating control to allow

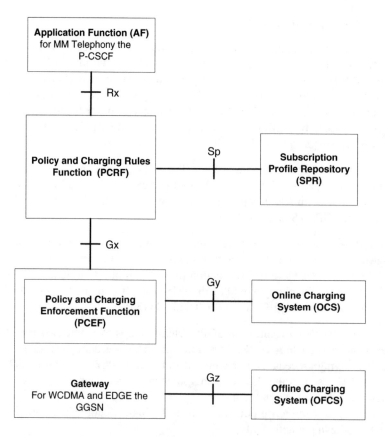

Figure 4.6: Overall PCC logical architecture.

or block packets belonging to a service data flow and event reporting to notify and react to application events and events related to the resource situation in the GGSN. From a pure Multimedia Telephony session initiation/modification point of view it is the QoS control that is of most interest. Some of the key features of the QoS control are as follows:

- The QoS control has the ability to handle aggregation of multiple service data flows on one PDP context, i.e. it is the combination of the authorized QoS information for the different flows that is the authorized QoS for the PDP context. This enables multiple media streams to share one PDP context and thus also one RAB.

- The enforcement of authorized QoS can force a downgrading of the requested bearer QoS by the GGSN as part of the activation/modification of the PDP context and initiation/modification of the radio bearer. This may happen due to operator policies and/or network capabilities.

- It has also been agreed that the PCRF should be able to provide network initiated QoS and thus be able to use the new NRSPCA functionality.

Figure 4.7 shows an example of QoS control over the Rx and Gx interfaces triggered by a Multimedia Telephony session initiation. The following list explains the numbered events in the flow.

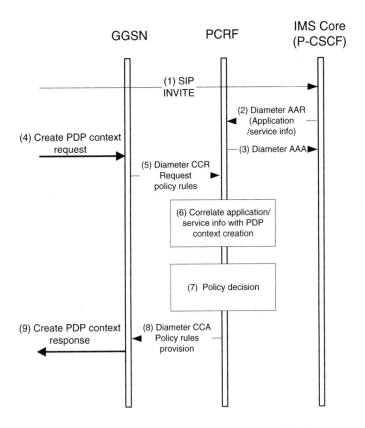

Figure 4.7: PCC flow using mobile initiated QoS.

1. A SIP request, in this case a SIP INVITE request, carrying SDP to either initiate or modify the Multimedia Telephony session is received by the IMS core (or more in detail the P-CSCF).

2. SDP information such as media bandwidth is passed down to the PCRF using a Diameter AAR message.

3. The PCRF acknowledges the Diameter AAR with a Diameter AAA message.

4. At some point in time the Multimedia Telephony client decides to activate the PDP context used for the media stream. Therefore, the GGSN will receive a Create PDP context request message.

5. The GGSN that is the policy charging enforcement point needs policy rules to decide if it should accept, modify or reject the request of creating a PDP context and sends a request of policy rules to the PCRF by using a Diameter CCR message.

6. The PCRF correlates the incoming request for policy rules with the service/application service information it has stored.

7. The PCRF makes the authorization and policy decision.

8. The PCRF sends the policy decision to the GGSN using a Diameter CCA message. After having received the policy decision the GGSN enforces the decision that may result in acceptance, modification or rejection of the PDP context activation procedure.

9. In this case the PDP context activation was accepted and a Create PDP context response is sent back to the SGSN (not shown in the figure).

4.3.2.1 Network Initiated QoS

The addition of NRSPCA has implications on the PCRF. To make NRSPCA work the PCRF must be able to map service information to bearer QoS and indicate this to GGSN in PCC rules. The mapping should be based on a combination of service information received over the Rx interface, policies provisioned by the operator and potentially subscriber information. The service information may be the media flow description, IMS communication service identifier (feature tag) or address information for network provided services like streaming. The PCRF should also be able to make a policy-based decision on when a PDP context should be activated (early or late in an IMS session setup) to be able to have different strategies for different services with different service requirements.

Figure 4.8 shows what the QoS control over the Rx and Gx interfaces may look like when using network initiated QoS. The two different flows show two different options that could be called Early-Rx and Late-Rx. The difference is when the Diameter AAA message is sent from PCRF to IMS core.

In Early-Rx the Diameter AAA message is sent back after the PCRF has made its policy decision but before the activation of the secondary PDP context is finalized. This means that the hold time of the SIP INVITE message that triggered the policy decision and resource reservation is smaller than for the case of Late-Rx. This method could be used if the delay of the Multimedia Telephony session establishment is to be optimized.

When using Late-Rx the Diameter AAA message is sent back after the PCRF has got information over the Gx interface that the activation of the secondary PDP context is done. Hence, the SIP INVITE will be held in the IMS core until the resource reservation is finalized. This method guarantees that resources are available when the SIP message is sent to the next entity. The advantage with this scheme is that the Multimedia Telephony client that sends the SIP INVITE does not need any lower layer interface to the mobile terminal in order to figure out when the resources are set up. The reception of a SIP response to the SIP INVITE request indicating that the Multimedia Telephony session establishment progresses is implicit information that resources are available for the Multimedia Telephony session.

4.4 Establishment of Multimedia Telephony Sessions

The SIP INVITE method is used to establish Multimedia Telephony sessions. The SIP method in its simplest form uses a three-way handshake to establish a SIP session. In that case a SIP INVITE message is sent from the originating client to the terminating client. When the user answers the call, the client responds with a SIP 200 OK message. When receiving the SIP 200 OK message the originating client acknowledges the message with a SIP ACK.

However, to meet the requirements for establishing the Multimedia Telephony sessions, more than the three SIP messages mentioned earlier are needed in the initiation phase. One general requirement for a Multimedia Telephony session is to provide a ringing indication on the originating side when the terminating Multimedia Telephony client starts to ring.

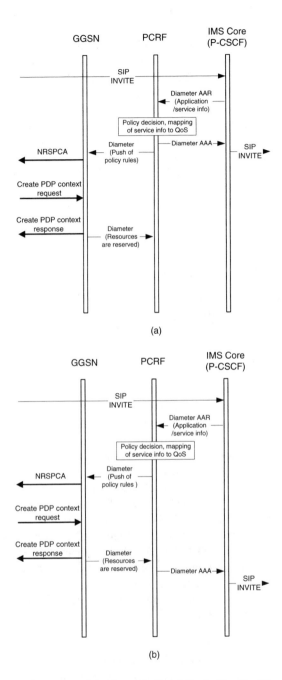

Figure 4.8: PCC flow using network initiated QoS. (a) Early-Rx; the Diameter AAA message is sent back to the IMS core before the PDP context activation is finalized. (b) Late-Rx; the Diameter AAA message is sent back to the IMS core after the PDP context activation is finalized.

This is done by adding an informal SIP response called SIP 180 Ringing, which is sent before the final response (the SIP 200 OK response) is sent. There is also a requirement that the Multimedia Telephony client should be able to cope with reservation of resources, i.e. establishment of PDP contexts and radio bearers for the purpose of carrying the media in the Multimedia Telephony session. The signaling flow must be able to handle both successful and unsuccessful establishment of radio bearers without causing a condition that is called 'ghost ringing'. The term 'ghost ringing' means that the terminating Multimedia Telephony client starts to ring even though the radio bearer for media was rejected. Thus, in the case of ghost ringing, when the terminating Multimedia Telephony user answers the call, he/she will only hear silence since no media can pass through.

The signaling flow to establish a Multimedia Telephony session including resource reservation is slightly different depending on whether mobile terminal initiated QoS or network initiated QoS is used on the originating side. Therefore, there are two sections below describing Multimedia Telephony session establishment for mobile terminal initiated QoS and network initiated QoS.

4.4.1   Using Mobile Terminal Initiated QoS

When using mobile terminal initiated QoS, it is mandatory for the Multimedia Telephony client to be able to get knowledge about the status of the resource reservation. The basic idea is that when the Multimedia Telephony client knows that it does not have resources reserved for the media stream it indicates using the SDP attribute 'inactive' (see RFC 3108 [131]) and using the SIP precondition method (see RFC 3312 [57] and RFC 4032 [56]) that it has no resources available. The SIP precondition method makes it possible for a Multimedia Telephony client to suspend the establishment of the Multimedia Telephony session until the resource reservation is finalized. Basically, the SIP precondition method triggers additional exchange of SIP messages within the Multimedia Telephony session establishment procedure. This gives the possibility to include a second SDP exchange in order to signal when the Multimedia Telephony client gets information that the resources are reserved and it is ready to send and receive media. The signaling flow in Figure 4.9 shows the case when both Multimedia Telephony clients need to perform resource reservation and establish RABs for the media. The signaling flow in Figure 4.2 also uses SIP preconditioning. It can be noted that the Multimedia Telephony session setup flow uses fewer SIP messages than the general IMS session establishment in Figure 4.2. This is a result of a number of optimizations, such as the use of SIP PRACK only for provisional responses that carry SDP and the start of resource reservation on the originating side before the reception of the SIP response. To describe the different events in more detail a numbered list is added below. The numbers correspond to specific messages and events marked in the figure. The following text explains the specific message or event in more detail.

1. The Multimedia Telephony client A has received a request to establish a Multimedia Telephony session to Multimedia Telephony user B that is using Multimedia Telephony client B. A SIP INVITE request is created. The request-URI of that SIP INVITE is either a SIP URI (e.g. sip:userB@operatorB.com) or a tel URI (e.g. tel:+1-212-555-2222, see RFC 2806 [182]) that addresses Multimedia Telephony user B. The SIP INVITE request should also contain the public user identity of the calling Multimedia Telephony user A to allow Originating Identification Presentation (see Section 4.7.4). Further, the SIP INVITE request also provides the IP address of the

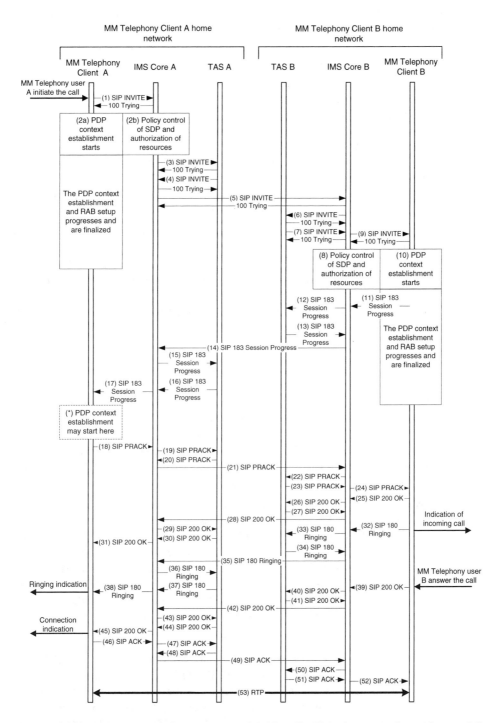

Figure 4.9: Establishment of a two-party Multimedia Telephony session using mobile initiated quality of service.

originating Multimedia Telephony client, information if the Multimedia Telephony client supports SigComp and if privacy is requested (Originating Identification Restriction). The SIP INVITE request must also indicate the level of support for reliable provisional responses and the precondition mechanism.

The SIP INVITE request must contain a description of the user plane session. In Multimedia Telephony sessions, SDP is used to describe the user plane part of the session. For more information about the SDP offer and the media capability of the Multimedia Telephony client, see Sections 5.1 and 5.2. Important attributes in the SDP are the 'inactive' attribute and the precondition attributes. When sending the SIP INVITE request it is assumed that the Multimedia Telephony client knows that the PDP context and radio bearer for media are not set up. Therefore it indicates in the SDP that the session is 'inactive' and that the preconditions to start the session are not met and hence local resource reservation must occur.

The SIP INVITE request is sent to IMS core A, which acknowledges the request with a SIP 100 Trying response. The SIP 100 Trying response is an informal response that informs the sending entity that the receiving entity has received the SIP INVITE request. When the originating Multimedia Telephony client gets information that the SIP INVITE request was successfully delivered it turns off the SIP retransmission timer (for the SIP INVITE request) to prevent multiple transmissions of the SIP INVITE, as it may take a while for the terminating side to respond to the invitation request.

2. (a) In 3GPP release 6 it is allowed to start resource reservation directly after having sent the SIP INVITE. This is the case shown here. The other possibility, which may be more often used, is to start resource reservation after the reception of the first SDP answer. This happens at step 17 and is marked in the figure with a (*). Based on the session description in the SIP INVITE sent in step 1, the Multimedia Telephony client starts to establish a PDP context with suitable QoS. The PDP context activation results in the establishment of a radio bearer with suitable radio bearer parameters for the service (for a description of radio bearers for media, see Section 5.6).

(b) When the IMS core A receives the SIP INVITE request it sends the session information to the PCRF to verify that the session description does not violate any operator controlled rules for the Multimedia Telephony session and to authorize resources for the Multimedia Telephony session being initiated. The PCRF also gets information from GGSN that a PDP context is being established for the Multimedia Telephony session. Given that resources are authorized by the PCRF for the Multimedia Telephony session, the PDP context activation and the Multimedia Telephony session establishment can progress, otherwise both processes are terminated by the network.

3. The SIP INVITE request is routed to the Telephony Application Server (TAS A). The TAS acknowledges the reception of the SIP INVITE request with a SIP 100 Trying response. It is proposed in 3GPP, but not decided at the time of writing this book, that a feature tag (for more information on feature tag see RFC 3840 [173]) is included in the SIP INVITE request to identify that the request is a Multimedia Telephony session invitation that should be routed to a TAS. This methodology is already used for 3GPP CSICS (for more information about 3GPP CSICS, see Section 8.1) and OMA PoC

(for more information about OMA PoC, see Section 8.2). As an example of the format of feature tags, 3GPP CSICS service enabler uses the feature tags +g.3gpp-cs-voice and +g.3gpp-cs-video.

4–7. The SIP INVITE request is routed from TAS A to IMS core B via IMS core A and TAS B. If the original SIP INVITE request contained a request-URI that was a tel-URI, it was translated to a globally routable SIP URI before the SIP INVITE request leaves IMS core A and is sent to IMS core B. Usually the ENUM-DNS protocol (see RFC 2916 [76]) is used for this process.

8. Before the SIP INVITE request is sent to Multimedia Telephony client B, a policy control of the SDP offer is done to verify that the session description does not violate any operator controlled rules for the Multimedia Telephony session. If the Multimedia Telephony client B is allowed to reserve resources for a Multimedia Telephony session, the GGSN is notified about the positive policy decision.

9. The SIP INVITE request is sent from IMS core B to Multimedia Telephony client B.

10. When the Multimedia Telephony client has received the SDP offer in the SIP INVITE request it determines the set of codecs it is capable of supporting for this Multimedia Telephony session and calculates the probable session bandwidth. Based on this information the PDP context activation procedure is started. Given that resources are authorized by the PCRF in step 8 for the Multimedia Telephony session, the PDP context activation progresses, otherwise it is terminated by the network. In the successful case, the PDP context activation results in a resource reservation, i.e. the establishment of a radio bearer with suitable radio bearer parameters for the service (for a description of radio bearers for media, see Section 5.6).

11. Since Multimedia Telephony client A marked that the preconditions of the session are not met and it thus needs to do a local resource reservation, Multimedia Telephony client B needs to answer the SIP INVITE request with a SIP 183 Session Progress response. Multimedia Telephony client B is QoS aware and indicates this using the precondition attribute. It indicates that it has no resources available and that the SDP and thus the media are still inactive. The SDP answer will also contain possible codec alternatives for the Multimedia Telephony session. The SDP answer will mark the intersection of the codecs that Multimedia Telephony client A indicated in the SDP offer and the set of codecs Multimedia Telephony client B is capable of supporting as possible codec alternatives. It will also contain the IP address and port numbers to be used for the media stream(s).

12–17. The SIP 183 Session Progress is sent back to Multimedia Telephony client A following the same path as the SIP INVITE request. When Multimedia Telephony client A receives the SIP 183 Session Progress response it also gets the SDP answer indicating which codec(s) can be used in the Multimedia Telephony session. In the earlier 3GPP releases it was mandatory for Multimedia Telephony client A to start resource reservation here; however, most implementations of Multimedia Telephony clients following the 3GPP release 6 and 7 specifications for IMS will start the resource reservation after step 17, though it is possible to do otherwise. However, if the PDP context activation was started in step 2a there is a quite high possibility that it has finalized at this point in time. Thus, if there is a need to modify the

PDP context because the initially requested QoS has a mismatch against the possible media codecs in the SDP answer, it could be done after step 17 without interfering with the activation procedure started at step 2a.

18. In most cases the initial SIP INVITE will trigger paging and procedures to set up the radio connection on the terminating side before the SIP INVITE request can be sent to Multimedia Telephony client B. Therefore, in most cases the PDP context activation and radio bearer setup initiated in step 2a will be finalized before step 18. In the initial SIP INVITE request, Multimedia Telephony client A indicated support for reliable provisional responses. Therefore, Multimedia Telephony client A can use a provisional response to SIP 183 Session Progress to indicate that the resource reservation is done and its SDP is now active (using the 'sendrecv' attribute) and that the preconditions for the Multimedia Telephony session are met by including a second SDP offer. This provisional response is a SIP PRACK response (see RFC 3262 [171]). It could be noted that even if the resource reservation was not finalized because it may have been started after step 17, the SIP 183 Session Progress should be acknowledged by a provisional response since it carried vital SDP information.

19–24. The SIP PRACK is sent to Multimedia Telephony client B.

25–31. The SIP PRACK is acknowledged by a SIP 200 OK response sent from Multimedia Telephony client A to Multimedia Telephony client B. In this case it is most probable that the resource reservation that started at step 10 is finalized. Therefore, the terminating Multimedia Telephony client B can answer back with a final SDP answer indicating that it is ready with the resource reservation.

32–38. When the following two conditions are fulfilled the terminating Multimedia Telephony client B starts to ring: (1) PDP context activation and radio bearer setup started at step 10 must be finalized in order to have a radio bearer ready for media and (2) the SIP PRACK request must have been received and the final SDP answer must have been sent back to Multimedia Telephony client A. When the Multimedia Telephony client B starts to ring it sends a SIP 180 Ringing response back to Multimedia Telephony client A, which can then give its user an indication that is has started to ring on the terminating side. Given that the SIP 200 OK response sent in step 25 included the final SDP answer, the SIP 180 Ringing does not have to carry any SDP.

39–45. At some point in time Multimedia Telephony user B answers the call. A final SIP 200 OK response is sent back to Multimedia Telephony client B. When the SIP 200 OK response is received, the Multimedia Telephony user A gets a connection indication.

46–52. The SIP 200 OK response is acknowledged by a SIP ACK message that is sent from Multimedia Telephony client A to Multimedia Telephony client B.

53. Media starts to flow between the two Multimedia Telephony clients.

More information about basic communication procedures in IMS can be found in 3GPP TS24.229 [25] and the basic Multimedia Telephony session procedures are to be written in 3GPP TS24.173 [24].

4.4.2   Using Network Initiated QoS

Network initiated QoS, which enables the use of QoS even if the Multimedia Telephony clients have no QoS knowledge, is proposed as a possible enhancement in 3GPP release 7. This would enable the use of QoS in Multimedia Telephony also for instance for a PC client that has its PS connection via a switch connected to an HSPA modem. Since the Multimedia Telephony client needs no knowledge about QoS and resource reservations, the signaling flow for a Multimedia Telephony session establishment becomes slightly different than for the mobile originated QoS case. It should however be noted that at the time of writing the network initiated QoS method and the resulting signaling flows were not yet part of the 3GPP release 7 IMS and Multimedia Telephony specifications. One possible realization of a signaling flow used for the realization of an establishment of a Multimedia Telephony session is shown in Figure 4.10. The signaling flow is described in more detail in the numbered list below.

1. The Multimedia Telephony client A has received a request to establish a Multimedia Telephony session to Multimedia Telephony user B that is using Multimedia Telephony client B. A SIP INVITE request is constructed. The main difference between this case and the case described in step 1 of Section 4.4.1 is that the Multimedia Telephony client has no knowledge of the QoS and resource reservation status. Therefore, the SIP INVITE request must either omit the precondition attributes and indicate that the SDP is active or indicate that the preconditions are met and the SDP is active. The SIP INVITE request is sent to IMS core A and is acknowledged by a SIP 100 Trying from IMS core A.

2. (a) Upon reception of the SIP INVITE request, the IMS core A interacts with the PCRF in order to do a policy control of the SDP (i.e. verify that the session description does not violate any operator controlled rules for the Multimedia Telephony session), to authorize resources and to initiate a network requested PDP context for the media. It should be noted that in this case it is purely up to the network to decide what kind of QoS is suitable for the requested service.

    (b) The policy decision is enforced in the GGSN that initiates a network requested secondary PDP context for the media resulting in the establishment of a radio bearer with suitable radio bearer parameters for the service (for a description of radio bearers for media, see Section 5.6).

3–7. The SIP INVITE request is kept in the IMS core A until the IMS core A gets a final response from the PCRF. If the policy control uses so-called Late-Rx (for more information about Late-Rx, see Section 4.3.2), the final response from the PCRF to the IMS core A will be sent after the radio bearer establishment and the PDP context activation are finalized. Therefore, after the PDP context activation and radio bearer setup are finalized the Multimedia Telephony session establishment will progress and the SIP INVITE is sent to IMS core B via TAS A and TAS B.

8. (a) Upon reception of the SIP INVITE request, the IMS core B interacts with the PCRF in order to do a policy control of the SDP (i.e. verify that the session description does not violate any operator controlled rules for the Multimedia Telephony session), to authorize resources and to initiate a network requested PDP context for the media.

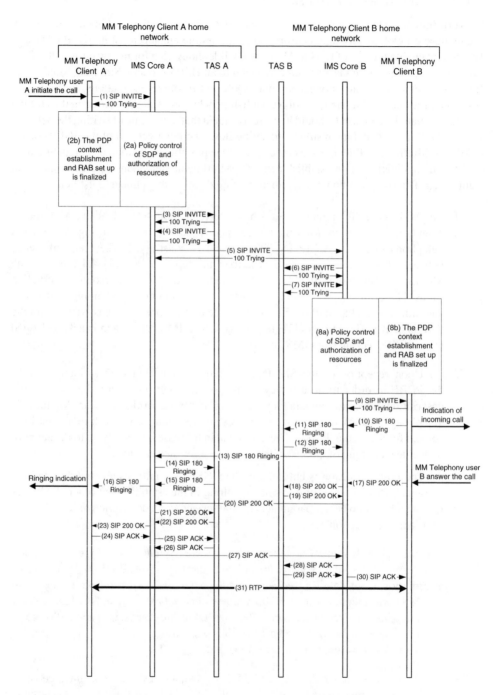

Figure 4.10: Establishment of a two-party Multimedia Telephony session using network initiated quality of service.

(b) The policy decision is enforced in the GGSN that initiates a network requested secondary PDP context for the media, which results in the establishment of a radio bearer with suitable radio bearer parameters for the service (for a description of radio bearers for media, see Section 5.6).

9. The SIP INVITE request is sent from IMS core B to Multimedia Telephony client B.

10–16. The terminating Multimedia Telephony client B starts to ring. When the Multimedia Telephony client B starts to ring it sends a SIP 180 Ringing response back to Multimedia Telephony client A that can give its user an indication that is has started to ring on the terminating side.

17–23. At some point in time Multimedia Telephony user B answers the call. A final SIP 200 OK response is sent back to Multimedia Telephony client B carrying the final SDP answer. When the SIP 200 OK response is received, the Multimedia Telephony user A gets a connection indication. The final SDP answer will mark the intersection of the codecs that Multimedia Telephony client A indicated in the SDP offer and the set of codecs Multimedia Telephony client B is capable of supporting as possible codec alternatives. It will also contain the address and port information to be used by the media stream(s).

24–30. The SIP 200 OK response is acknowledged by a SIP ACK message that is sent from Multimedia Telephony client A to Multimedia Telephony client B.

31. Media starts to flow between the two Multimedia Telephony clients.

4.5 Modification of Multimedia Telephony Sessions

The base specification for SIP (RFC 3261 [172]) defines the SIP INVITE method for both the initiation and the modification of SIP sessions. However, in RFC 3311 [164] another method to modify SIP sessions is specified. The second method uses the SIP UPDATE method to update session description parameters. The difference between the methods is that the SIP INVITE method impacts the dialog state while the SIP UPDATE method does not. This means that the SIP UPDATE method can be used to modify the session before the initial SIP INVITE has been answered. For instance, the SIP UPDATE method enables the caller or the callee to put the call on hold before the call is actually answered. In Multimedia Telephony session both methods are supported. The choice of method is up to implementation but there are some rules of thumb to follow. The SIP UPDATE method is suitable for updates of the Multimedia Telephony session that the receiving Multimedia Telephony client accepts without any user interaction using automatic consent. If the update of the Multimedia Telephony session needs manual consent from the user, the SIP INVITE method is a better choice. The reason for this is that the SIP INVITE method stops the SIP retransmission timers and no SIP messages are retransmitted until the user has accepted/rejected the modification of the Multimedia Telephony session. If the SIP UPDATE method is used, the SIP UPDATE message will be periodically retransmitted until the user has accepted/rejected the modification of the Multimedia Telephony session and SIP 200 OK answer is received, which wastes resources. The property that the timers are stopped may also be useful when the modification of the Multimedia Telephony session leads to updates of lower layers. For instance, the modification of the Multimedia Telephony session may lead to the update of

PDP contexts and radio bearers. The procedures to modify the PDP contexts and radio bearers are time consuming and the final response to the modification request should not be sent until the lower layers have been modified. Using the SIP INVITE method would then remove the need for SIP retransmissions. This section presents both the SIP INVITE method and the SIP UPDATE method covering two different use cases.

4.5.1   The SIP INVITE Method

In this case, the user of Multimedia Telephony client A wants to turn on e.g. the video clip feature to show a funny clip she has stored on the terminal. However, before the user of Multimedia Telephony client A is allowed to share the content, the user of Multimedia Telephony client B must give her consent that she accepts the reception of the video clip. Further, the addition of the MSRP connection to transfer the video clip calls for the setup of a new PDP context and RAB. In this example it is assumed that the Multimedia Telephony clients are QoS aware and that the originating Multimedia Telephony client has already set up a best effort radio bearer that can be used to convey the MSRP media before it starts the update of the Multimedia Telephony session. The resulting SIP INVITE flow is described in Figure 4.11 and the list below.

1. The Multimedia Telephony client A has received a request to turn on the video clip feature in the ongoing Multimedia Telephony session between Multimedia Telephony client A and Multimedia Telephony client B. Multimedia Telephony client A has already a suitable PDP context and radio bearer established for the video clip feature while Multimedia Telephony client B has not. A SIP INVITE request is constructed. This kind of invitation message for a service change in an already ongoing session is often called a SIP reINVITE to differentiate it from the initial SIP INVITE request. But for this scenario the main difference between this SIP reINVITE request and the initial SIP INVITE request described in step 1 of Section 4.4.1 is that the Multimedia Telephony client A already has resources reserved. Therefore, the SIP INVITE request indicates that the SDP is active and that the preconditions are met. Further, the video clip feature uses MSRP as the application layer protocol. Therefore, the Multimedia Telephony client A creates a local MSRP URL, which can be used for the communication between the two Multimedia Telephony clients. The SDP offer must contain the generated MSRP URL and a local port number for the MSRP communication. The SIP INVITE request is sent to IMS core A and is acknowledged by a SIP 100 Trying from IMS core A.

2–6. The SIP INVITE request is sent to IMS core B.

7. Before the SIP (re)INVITE request is sent to Multimedia Telephony client B, a policy control of the SDP offer is done to verify that the new session description does not violate any operator controlled rules for the Multimedia Telephony session. The GGSN is notified about the positive policy decision.

8. The SIP INVITE request is sent to the Multimedia Telephony client B.

9. When the Multimedia Telephony client has received the new SDP offer in the SIP INVITE request it determines whether it needs to reserve new resources or not. In this case it is decided that a new PDP context and radio bearer are needed for

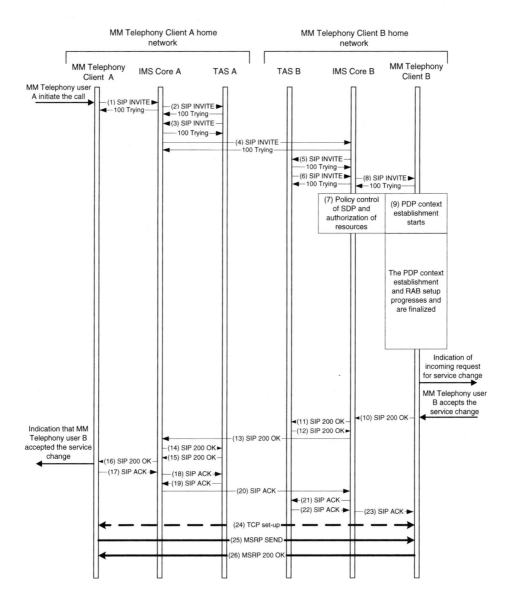

Figure 4.11: Example of a modification of a two-party Multimedia Telephony session using the SIP (re-)INVITE method.

the video clip feature to be used in the Multimedia Telephony session. The PDP context activation procedure is started. Given that the resources were authorized by the PCRF in step 7 for the Multimedia Telephony session, the PDP context activation progresses, otherwise it is terminated by the network. In the successful case, the PDP context activation results in the establishment of a radio bearer with suitable radio bearer parameters for the video clip feature. When the PDP context activation and radio bearer setup are finalized, the Multimedia Telephony client B indicates to its user that there is an incoming request for using the video clip feature in the Multimedia Telephony session.

10–16. The Multimedia Telephony user B accepts the incoming request for enhancing the media content of the Multimedia Telephony session. Multimedia Telephony client B constructs a SIP 200 OK response carrying the SDP answer that indicates that the terminating Multimedia Telephony client B listens on the MSRP TCP from the originating Multimedia Telephony client A. The Multimedia Telephony client B generates an MSRP URL and a local port number for the MSRP communication and includes it in the SDP answer. Further the SDP answer indicates that Multimedia Telephony client B's local resource reservation is finalized by marking the SDP as active and setting the preconditions as met. The SIP 200 OK response is sent to Multimedia Telephony client A.

17–23. The SIP 200 OK response is acknowledged by a SIP ACK message that is sent from Multimedia Telephony client A to Multimedia Telephony client B.

24. The originating Multimedia Telephony client A establishes a TCP connection to the host address and the port as specified in the MSRP URL received in the SDP answer from the terminating Multimedia Telephony client B.

25. The originating Multimedia Telephony client A sends the first message over the MSRP session with an MSRP SEND request using the established TCP connection.

26. The terminating Multimedia Telephony client B acknowledges the reception of the MSRP SEND request with an MSRP 200 (OK) response using the established TCP connection.

4.5.2   The SIP UPDATE Method

A use case suitable for the SIP UPDATE method is: the Multimedia Telephony client gets an indication from the radio layers of the mobile terminal that it needs to change one or more media related parameters. One such example is a sudden change of the radio access, which triggers the need to lower the maximum possible session bandwidth. The resulting SIP UPDATE flow is described in Figure 4.12 and below.

1. The Multimedia Telephony client A has received information that it needs to renegotiate media related parameters such as the session bandwidth. A SIP UPDATE request, which identifies the session being modified, is created, including SDP with an updated session description. The SIP UPDATE request is sent from Multimedia Telephony client A to IMS core A.

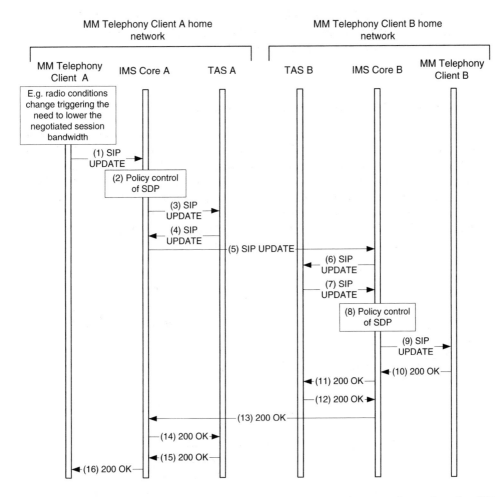

Figure 4.12: Modification of a two-party Multimedia Telephony session using the SIP UPDATE method.

2. Upon reception of the SIP UPDATE request the IMS core A needs to interact with the policy control to verify that the new session description does not violate any operator controlled rules for the Multimedia Telephony session.

3–7. The SIP UPDATE request is sent from IMS core A to IMS core B via the involved TASs.

8. When the IMS core B receives the SIP UPDATE request, it signals the modification of the Multimedia Telephony session to the policy control. The policy control to verify that the new session description does not violate any operator controlled rules for the Multimedia Telephony session for user B.

9. The SIP UPDATE request is sent to the Multimedia Telephony client B from IMS core B.

10–16. The Multimedia Telephony client B acknowledges the SIP UPDATE request by sending a SIP 200 OK answer back to Multimedia Telephony client A. The SIP 200 OK answer passes through all involved nodes.

4.6 Release of Multimedia Telephony Sessions

The SIP BYE request is used to abandon SIP sessions. In a two-party Multimedia Telephony session, abandonment of the SIP session by one of the users implies that the Multimedia Telephony session is terminated. The transmission of the SIP BYE request will then terminate the Multimedia Telephony related states for the two users in the IMS cores and the TASs. It should be noted that, in a Multimedia Telephony conference call, a SIP BYE request from one user just means that the particular user leaves the conference. The Multimedia Telephony conference session itself is not affected. The flow in Figure 4.13 and below describes a user initiated termination of a two-party Multimedia Telephony session.

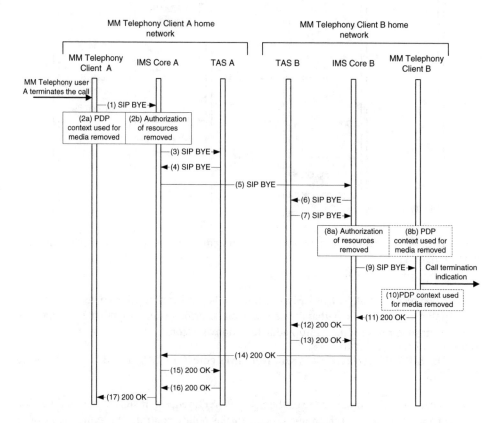

Figure 4.13: The termination of a two-party Multimedia Telephony session. The dashed line used for the boxes in steps 8b and 10 indicates that the PDP context deactivation may occur at different times depending on the mobile terminal and the network support of network initiated deactivation of PDP contexts.

1. The Multimedia Telephony client A has received a request to terminate the Multimedia Telephony session from its user. A SIP BYE request is created that identifies the session being terminated. The SIP BYE request is sent from Multimedia Telephony client A to IMS core A.

2. Upon reception of the SIP BYE request the IMS core A removes the authorization for the resources that had been issued for the Multimedia Telephony session. Depending on the mobile terminal and whether the network supports network initiated PDP context deactivation or not, the PDP context used for the media in the Multimedia Telephony session is either terminated by the mobile terminal after the transmission of the SIP BYE request or by the network after the policy control has removed the media authorization.

3–7. The SIP BYE request is sent from IMS core A to IMS core B via the involved Telephony Application Servers. The Telephony Application Servers involved remove the states associated to the Multimedia Telephony session.

8. When the IMS core B receives the SIP BYE request, it signals the termination of the Multimedia Telephony session to the policy control that removes the authorization for the previously assigned media resources. If the mobile terminal and the network support network initiated PDP context deactivation, then the policy control initiates the deactivation of the PDP context used for the media in the Multimedia Telephony session.

9. The SIP BYE request is sent to the Multimedia Telephony client B from IMS core B.

10. If the mobile terminal and/or the network doesn't support network initiated deactivation of PDP contexts, then the Multimedia Telephony client initiates the deactivation of the PDP context used for the media in the Multimedia Telephony session.

11–17. The Multimedia Telephony client B acknowledges the SIP BYE request by sending a SIP 200 OK answer back to Multimedia Telephony client A. The SIP 200 OK answer passes through all involved nodes.

4.7 Supplementary Services

Besides defining the basic communication features of the Multimedia Telephony service, a set of supplementary services has been defined to make the Multimedia Telephony service offering complete. The set of supplementary services used by Multimedia Telephony is based on IMS and have similar functionally like the existing ISDN/PSTN supplementary services. It should be noted that the functionality of the supplementary services developed for IMS are not 100% identical to the functionality of their ISDN/PSTN equivalents. To indicate this difference, the set of IMS supplementary services are often referred to as the simulation services, only simulating the set of PSTN/ISDN supplementary services. The IMS supplementary services were first specified by TISPAN. The following simulation services are supported by Multimedia Telephony:

- Communication DIVersion (CDIV). This simulation service allows the diversion of communications and the corresponding service interworking with the PSTN/ISDN network.

- CONFerence (CONF). This simulation service provides the possibility to hold conferences with three or more users.

- Message Waiting Indication (MWI). This simulation service supports an indication sent to the user to provide him with information about the status of a voice/video/multimedia mail box.

- Originating Indication Presentation (OIP)/Originating Indication Restriction (OIR). These simulation services support the presentation or restriction of an identity to the terminating user. They are the simulation of the ISDN/PSTN CLIP/CLIR services.

- Terminating Indication Presentation (TIP)/Terminating Indication Restriction (TIR). These simulation services support the presentation or restriction of an identity of the terminating user to the originating user. They are the simulation of the ISDN/PSTN COLP/COLR services.

- Communication Hold (HOLD). This simulation service supports the possibility of suspending the communication (on hold) while for example another communication with another user is to be done.

- Communication Barring (CB). This simulation service allows communications to be rejected regardless whether the originating identity is anonymous or not.

- Explicit Communication Transfer (ECT). This simulation service provides a party involved in a communication to transfer that communication to a third party.

- Communication Diversion: Communication Forwarding on Mobile Subscriber Not Reachable (CFNRc). This simulation service is a new addition to the CDIV family of simulation services and it supports the ability to complete a requested communication to a user without having to make a new communication attempt when the destination B is not reachable by attempting to reach the user via a second registered destination.

In the sections below, the simulation services are presented in more detail with an informative signaling flow that exemplifies how the services work.

4.7.1   Communication Diversion

The CDIV services are a set of services that enable a diverting user to divert the communication to another destination. This set of services has a general requirement that it shall be possible for the user or the network to identify an alternative destination for a Multimedia Telephony session or individual media of a Multimedia Telephony session. It shall also be possible for the redirection of the Multimedia Telephony session or individual media to be initiated at various stages of the Multimedia Telephony session. For example:

- during the initial request of a Multimedia Telephony session (called Communication Forwarding Unconditional (CFU));

- during the establishment of a Multimedia Telephony session (called Communication Deflection (CD)).

Redirection can be applied for all Multimedia Telephony sessions unconditionally or it can be caused by any set of list of events or conditions. Typical causes may be:

- if the destination party is already in another session (called Communication Forwarding on Busy user (CFB));

- if the destination party is unreachable or unavailable in some other way (called Communication Forwarding on Not Logged-in (CFNL) or Communication Forwarding on No Reply (CFNR)).

To exemplify the CDIV services two examples are given below.

4.7.1.1 Communication Forwarding Unconditional

The first example in Figure 4.14 shows an example of call forwarding using the Communication Forwarding Unconditional method. The Multimedia Telephony user B has activated the call forwarding service indicating to the network that she is not available at Multimedia Telephony client B1; rather she can be reached on Multimedia Telephony client B2 instead. Multimedia Telephony user A calls Multimedia Telephony user B on her usual number that is linked to Multimedia Telephony client B1 but is redirected to Multimedia Telephony client B2 instead. It should be noted that the Multimedia Telephony session initiation flow on both originating and terminating side is simplified and in this example the signaling flow does not contain signaling needed for resource reservation or the ringing mechanisms. To get an understanding of which SIP messages are omitted see the more detailed Multimedia Telephony session initiation flows in Sections 4.4.1 and 4.4.2. The numbered list below briefly describes the messages sent and the actions taken by various nodes when diverting the Multimedia Telephony session to another destination.

1–3. The Multimedia Telephony client A has received a request to initiate a Multimedia Telephony session with Multimedia Telephony client B1 from its user. A SIP INVITE request is created and it is sent using a SIP URI that addresses Multimedia Telephony client B1. In this case the SIP INVITE is sent from the Multimedia Telephony client A to TAS B via the IMS core A and the IMS core B.

4. Upon reception of the SIP INVITE request the TAS B recognizes that Multimedia Telephony user B has registered that she is not available at Multimedia Telephony client B1. Instead the Multimedia Telephony user B is available at Multimedia Telephony client B2. Therefore, TAS B executes the Communication Forwarding Unconditional logic.

5–7. The SIP INVITE request is acknowledged by a SIP 181 Call Is Being Forwarded response that is sent to Multimedia Telephony client A via the IMS core B and the IMS core A. On reception of the SIP 181 Call Is Being Forwarded, the Multimedia Telephony client A may indicate that the call is being forwarded to its user.

8–9. When the TAS B has executed the Communication Forwarding Unconditional logic it creates a SIP INVITE request that contains the SIP URI that addresses Multimedia Telephony client B2. This SIP INVITE is sent to Multimedia client B2 via IMS core B.

10–14. The user receives an indication that she is invited to a Multimedia Telephony call and when the user answers the call, the Multimedia Telephony client B2 acknowledges the SIP INVITE request by sending a SIP 200 OK answer back to the Multimedia Telephony client A. The SIP 200 OK answer passes through all involved nodes.

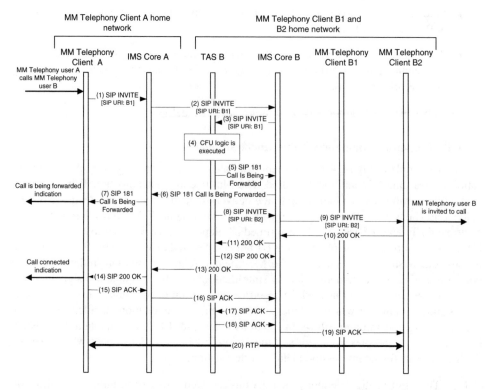

Figure 4.14: Signaling flow for call forwarding using the Communication Forwarding Unconditional method.

15–19. Multimedia client A acknowledges the SIP 200 OK by sending a SIP ACK message back to Multimedia client B2.

   20. The media starts to flow between the Multimedia Telephony client A and the Multimedia Telephony client B2.

4.7.1.2   Communication Deflection

The second example presented in Figure 4.15 shows an example of call forwarding using the Communication Deflection method. In this case the Multimedia Telephony user B has also activated the call forwarding service. The difference from the Communication Forwarding Unconditional service is that in this case it is the Multimedia Telephony client B1 that has the knowledge that the user is now available at Multimedia Telephony client B2 rather than the network. It should be noted that the Multimedia Telephony session initiation flow on both originating and terminating side is simplified and in this example the signaling flow does not contain signaling needed for resource reservation or the ringing mechanisms. To get an understanding of which SIP messages are omitted see the more detailed Multimedia Telephony session initiation flows in Sections 4.4.1 and 4.4.2.

1–5. The Multimedia Telephony client A has received a request to initiate a Multimedia Telephony session with Multimedia Telephony client B1 from its user. A SIP INVITE

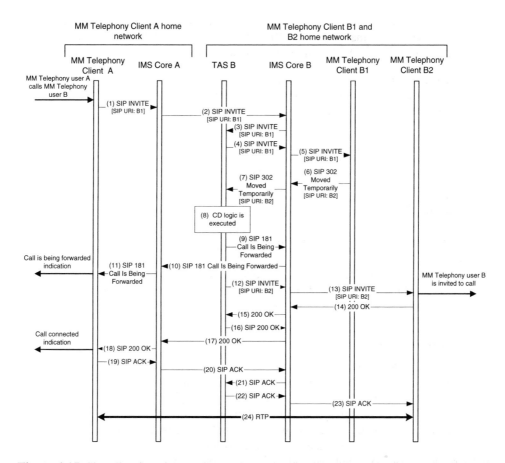

Figure 4.15: Signaling flow for call forwarding using the Communication Deflection method.

request is created and it is sent using a SIP URI that addresses Multimedia Telephony client B1. In this case the SIP INVITE is sent from the Multimedia Telephony client A to Multimedia Telephony client B1 via the IMS core A, the IMS core B and TAS B.

6–7. Upon reception of the SIP INVITE request the Multimedia Telephony client B1 recognizes that Multimedia Telephony user B has registered that she is not available at Multimedia Telephony client B1. Instead the Multimedia Telephony user B is available at Multimedia Telephony client B2. Therefore, Multimedia Telephony client B1 responds with a SIP 302 Moved Temporarily response that contains the SIP URI of Multimedia Telephony client B2 to TAS B.

8. The TAS B executes the Communication Deflection logic.

9–11. TAS B sends a SIP 181 Call Is Being Forwarded response to Multimedia Telephony client A via the IMS core B and the IMS core A. On reception of the SIP 181 Call Is Being Forwarded, the Multimedia Telephony client A may indicate that the call is being forwarded to its user.

12–13. When the TAS B has executed the Communication Deflection logic it creates a SIP INVITE request that contain the SIP URI that addresses Multimedia Telephony client B2. This SIP INVITE is sent to Multimedia Telephony client B2 via IMS core B.

14–18. The user receives an indication that she is invited to a Multimedia Telephony call and when the user answers the call, the Multimedia Telephony client B2 acknowledges the SIP INVITE request by sending a SIP 200 OK answer back to the Multimedia Telephony client A. The SIP 200 OK answer passes through all involved nodes.

19–23. Multimedia Telephony client A acknowledges the SIP 200 OK by sending a SIP ACK message back to Multimedia Telephony client B2.

24. The media starts to flow between the Multimedia Telephony client A and the Multimedia Telephony client B2.

The two other methods Communication Forwarding on Busy user and Communication Forwarding on Not Logged-in have related and quite similar signaling flows to the simulation service Communication Forwarding on Mobile Subscriber Not Reachable (CFNRc); see Section 4.7.9. More information about the different CDIV services can be found in ETSI TS 183 004 [67] and 3GPP TS24.173 [24].

4.7.2   Conference

The CONF simulation service enables a user to participate and control a simultaneous communication involving a number of users. The CONF simulation service provides means to create conferences to which users have to join by calling in to the conference bridge or the creation of ad-hoc multiparty calls to which users are invited. In this section the flow in Figure 4.16 demonstrates the creation of a conference and the invitation of a second user is shown. It should be noted that the conference session initiation flow on both originating and terminating side is simplified and in this example signaling needed for the resource reservation or the ringing mechanisms is not shown. To get an understanding of which SIP messages are omitted see the more detailed Multimedia Telephony session initiation flows in Sections 4.4.1 and 4.4.2. The numbered list below briefly describes the messages sent and the actions taken by various nodes in the Multimedia Telephony Conferencing session establishment.

1–2. The Multimedia Telephony client A has received a request to initiate a Multimedia Telephony conference session from its user. A SIP INVITE request is created and it is sent to the Conference URI. In this case the SIP INVITE is sent from the Multimedia Telephony client A to TAS A that acts as the conference server via IMS core A.

3. Upon reception of the SIP INVITE request the TAS A creates the conference and starts to maintain states for it.

4–7. The SIP INVITE request is acknowledged by a SIP 200 OK response that is sent to Multimedia Telephony client A via the IMS core A. On reception of the SIP 200 OK, the Multimedia Telephony client A sends a SIP ACK to the TAS A via IMS core A.

8. The media starts to flow between Multimedia Telephony client A and the conference server (TAS A).

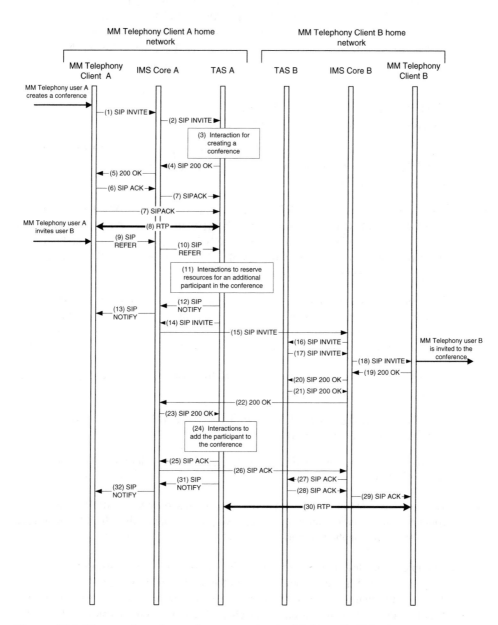

Figure 4.16: The creation of a conference session by Multimedia Telephony user A and the invitation of Multimedia Telephony user B to the conference.

9–10. The Multimedia Telephony client A receives a request to invite a user using Multimedia Telephony client B to the Multimedia Telephony conference session. A SIP REFER request is created containing a request URI indicating the conference URI and a Refer-To header indicating the address (SIP URI or tel URI) of the user who is being invited to the conference. It should be noted that the user can be invited by sending the SIP REFER request directly to the Multimedia Telephony client B instead of sending it to the conference focus that is hosting the conference.

11. The TAS A receives the SIP REFER request and it checks if the conference URI is allocated. If the conference URI is allocated, it authorizes the request and allocates resources for an additional participant in the conference.

12–13. The TAS A creates and sends a SIP NOTIFY response to the SIP REFER request containing information about the progress of the SIP REFER request processing to Multimedia Telephony client A.

14–18. The TAS A creates a SIP INVITE request that is sent to the SIP URI or tel URI found in the SIP REFER request. Since the SIP URI or tel URI is the address for Multimedia Telephony client B, the SIP INVITE is sent to Multimedia Telephony client B via a set of nodes. In this example, the TAS B is in the SIP message route but it doesn't have to be in the route.

19–23. The user receives an indication that she is invited to the conference and when the user answers the call, the Multimedia Telephony client B acknowledges the SIP INVITE request by sending a SIP 200 OK answer back to the conference server TAS A. The SIP 200 OK answer passes through all involved nodes.

24. The TAS A interacts with the media resource function to add another RTP flow to the conference.

25–29. TAS A acknowledges the SIP 200 OK by sending a SIP ACK message back to Multimedia Telephony client B.

30. The media starts to flow between Multimedia Telephony client B and the conference server (TAS A).

31–32. The TAS A creates and sends a final SIP NOTIFY response to the SIP REFER request containing information about the successful progressing of the SIP REFER request to Multimedia Telephony client A.

The example above describes the situation when a user creates a Multimedia Telephony conferencing session and invites another user via the conferencing server. Besides this use case, additional participants to the conference can either join the conference by themselves or be invited to the conference directly via client-to-client communication. The CONF simulation service also supports the joining of two or more active ongoing Multimedia Telephony sessions to a so-called three-way session.

It should be noted that the conferencing architecture used for the CONF simulation service applies to any kind of media stream by which users may want to communicate, this includes e.g. audio and video media streams as well as instant message based conferences or gaming. The specifications also include the possibility to use floor control, as part of the conferencing service to offer control of shared conference resources using the Binary Floor Control Protocol (BFCP) being specified by the IETF, see [53]. More information about the CONF service can be found in ETSI TS 183 005 [69], 3GPP TS24.173 [24] and 3GPP TS24.147 [20].

4.7.3   Message Waiting Indication

The MWI simulation service is used to notify a Multimedia Telephony client that subscribes to the MWI service about status change of a so-called message account. The message account is an entity that retains multimedia messages (e.g. voice, video, and fax) intended for a particular Multimedia Telephony user. In this section the flow in Figure 4.17 demonstrates the subscription of the MWI service and a notification of a status change of the message account.

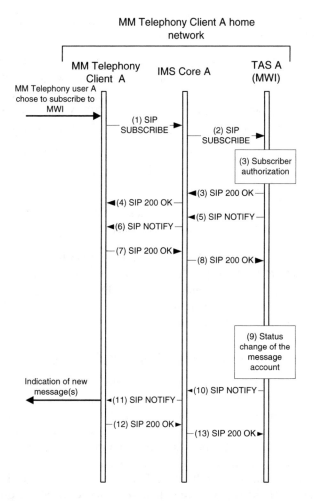

Figure 4.17: Signaling flow for the subscription of the MWI service and a status change of the message account.

1–2. The Multimedia Telephony client A subscribes to the Message Waiting Indication service by sending a SIP SUBSCRIBE request to the MWI server that here is assumed to be co-located with TAS A.

3. When receiving the SIP SUBSCRIBE, the TAS A authorizes the subscriber and identifies the requested message account.

4–5. If the authorization is successful (as shown here), the TAS sends a SIP 200 OK response back to the Multimedia Telephony client A.

6–7. Immediately after successful subscription, the TAS A sends a SIP NOTIFY request to the Multimedia Telephony client A to synchronize the current state of the message account. This initial SIP NOTIFY, however, does not contain any extended information about available message.

8–9. The Multimedia Telephony client A acknowledges the SIP NOTIFY by sending a SIP 200 OK response back to TAS A.

10. After the TAS has generated a SIP NOTIFY request to inform the subscriber's Multimedia Telephony client about the subscription state, the TAS waits for the change of the message account status. As soon as new message(s) are deposited into the message account the TAS generates a SIP NOTIFY request to indicate the change in the message account status to the Multimedia Telephony client.

11–12. When a new message has arrived, the TAS A sends a SIP NOTIFY request to the Multimedia Telephony client A. This notification sent by the TAS A contains an extended set of message waiting information about the newly deposited message(s).

13–14. The Multimedia Telephony client A acknowledges the SIP NOTIFY request by sending a SIP 200 OK response message back to TAS A.

More information about the MWI service can be found in ETSI TS 183 006 [71], 3GPP TS24.173 [24] and RFC 3842 [136]. It should be noted that there is a possibility to implement MWI without using the Message Waiting Indication Event Package specified in RFC 3846 [125]. In that case the message waiting indication is sent using the SIP MESSAGE method instead of the SIP NOTIFY method.

4.7.4   Originating Indication Presentation/Restriction

The OIP service provides the terminating Multimedia Telephony user with the possibility of receiving trusted (i.e. network provided) identity information in order to identify the user that originates the call. In addition to the trusted identity information, the identity information from the originating Multimedia Telephony user can also include additional identity information generated by the originating Multimedia Telephony user.

The OIR service is a service offered to the originating Mobile Telephony user. It restricts presentation of the originating Multimedia Telephony user's identity information to the terminating user. When the OIR service is applicable and activated, the originating network provides the destination network with the indication that the originating user's identity information is not allowed to be presented to the terminating user. Therefore, no originating Multimedia Telephony user's identity information is included in the SIP requests sent to the terminating user. But the presentation restriction function does not influence the forwarding of the originating Multimedia Telephony user's identity information within the network as part of the simulation service procedures.

The identity information is sent in the basic Multimedia Telephony session signaling and hence there is no additional OIP and OIR signaling flows. The OIP and OIR services make use of the following SIP headers and conform to the mentioned specifications:

- P-Asserted-Identity, see RFC 3325 [124] and RFC 3966 [175];

- P-Preferred-Identity, see RFC 3325 [124] and RFC 3966 [175];

- Privacy, see RFC 3323 [160] and RFC 3325 [124].

The basic idea is that the originating Multimedia Telephony client should insert its public user identity in the P-Preferred-Identity header in any initial SIP request. This information is sent to the originating Multimedia Telephony user's network. Following the rules of RFC 3325 [124], in the originating user's network, and if the P-Preferred-Identity header matches one of the registered users, the identity information in the P-Preferred-Identity header is moved to a P-Asserted-Identity header. Thereafter the public user identity of the originating Multimedia Telephony user is conveyed by the P-Asserted-Identity header field. Therefore, to support the OIP service, a terminating Multimedia Telephony client must support the receipt of one or more P-Asserted-Identity header fields in SIP requests initiating a dialog or stand-alone transactions. The information in the P-Asserted-Identity header field(s) is presented to the terminating user as part of the OIP service. In the case of OIR, either the TAS (in the case of so-called permanent mode OIR) or the originating Multimedia Telephony client (in the case of so-called temporary mode OIR) includes a Privacy header indicating that the identity needs to be restricted. In the case of OIR, the public user identity information in the From header field and in the P-Asserted-Identity header fields are removed by the network before the SIP message is received by the terminating Multimedia Telephony client. Therefore, if no P-Asserted-Identity header fields are present but a Privacy header field is present, then the terminating Multimedia Telephony client gets an indication that the identity has been withheld due to OIR.

The service activation/deactivation of the OIP and OIR services may be done using XCAP, web-based provisioning or by the operator. There are defined XML documents for the purpose of activating and deactivating originating identity services for use with XCAP in ETSI TS 183 023 [65].

More information about the OIP/OIR services can be found in ETSI TS 183 007 [72] and 3GPP TS24.173 [24].

4.7.5 Terminating Indication Presentation/Restriction

The TIP service provides the originating party with the possibility of receiving trusted identity information in order to identify the terminating party. The TIR service is a service offered to the connected party which enables the connected party to prevent presentation of the terminating identity information to the originating party.

The TIP and TIR services are closely related to the OIP and OIR services. TIP and TIR work in a similar way as the OIP and OIR service with the only difference being that the public user identity in the P-Asserted-Identity header field and/or the privacy request in the Privacy header field are carried in a SIP response that is sent from the terminating Multimedia

Telephony client to the originating Multimedia Telephony client instead of the other way around when using a SIP request.

More information about the TIP/TIR services can be found in ETSI TS 183 008 [73] and 3GPP TS24.173 [24].

4.7.6   Communication Hold

The HOLD supplementary service enables a Multimedia Telephony user to suspend the media stream(s) of an established Multimedia Telephony session, and resume the media stream(s) at a later time. The HOLD service utilizes a set of procedures in accordance with RFC 3264 [170]. Thus, these are the SDP parameters to use when putting media streams on hold:

- 'inactive' if the media stream was previously set to 'recvonly';

- 'sendonly' if the media stream was previously set to 'sendrecv'.

And the following parameters are used to resume media stream(s):

- 'recvonly' if the media stream was previously set to 'inactive';

- 'sendrecv' if the media stream was previously set to 'sendonly'.

The Multimedia Telephony client generates the SDP offer and sends it either in a SIP (re-)INVITE request or a SIP UPDATE request as shown in Sections 4.5.1 and 4.5.2. The example in Figure 4.18 highlights the possibility of having announcements for the Multimedia Telephony user that is put on hold.

1–7. The Multimedia Telephony client A has received information that it should set the media stream(s) on hold from its user. A SIP UPDATE request is created that identifies the session being modified including a new SDP offer indicating that it contains the 'sendonly' attribute for the media stream(s). The SIP UPDATE request is sent from Multimedia Telephony client A to Multimedia Telephony client B via all SIP entities in the path.

8–10. The Multimedia Telephony client B acknowledges the SIP UPDATE request by sending a SIP 200 OK response with a SDP answer that contains the 'recvonly' attribute for the media stream(s).

11. When the media resource function has received the SIP 200 OK response, it may be considered as an event to direct the media resource function to start playing the announcement. The media resource function may receive the SIP 200 OK either via the TAS B or IMS core B.

12–15. The SIP 200 OK response is sent to Multimedia Telephony client A.

More information about the HOLD service can be found in ETSI TS 183 010 [68] and 3GPP TS24.173 [24].

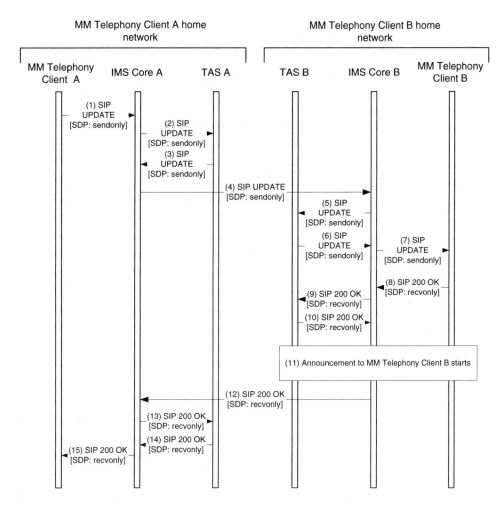

Figure 4.18: Signaling flow for setting a Multimedia Telephony user on hold and to play announcements to that user.

4.7.7 Communication Barring

The CB services are a set of services that is used to reject communications. The following services are supported:

- Incoming Communication Barring (ICB) rejects incoming communications that fulfill provisioned or configured conditions on behalf of the terminating Multimedia Telephony user.

- Anonymous Communication Rejection (ACR) is a special case of ICB, which allows rejection of incoming communications from an anonymous source on behalf of the terminating Multimedia Telephony user.

- Outgoing Communication Barring (OCB) is a service that rejects outgoing communications that fulfill provisioned or configured conditions on behalf of the originating Multimedia Telephony user.

The ICB service makes it possible for a Multimedia Telephony user to apply barring of certain categories of incoming communications by the use of a configured or provisioned barring program in the network. The barring program is a set of rules that may or may not apply barring for the incoming communication depending on whether the asserted originating public identity matches a specific public user identity.

The ACR service is a special case of ICB that allows the barring of a communication from a restricted originating public user identity. Basically the rule applied for ACB is if 'condition = anonymous' then 'allow = false'. The possibility of rejection of anonymous calls is a regulatory requirement in many countries. Therefore the ACR service is highlighted as an own service differentiated from the ICB service.

The OCB service makes it possible for a Multimedia Telephony user to apply barring of certain categories of outgoing communications by the use of a configured or provisioned barring program in the network. An example of a rule that may be provisioned/configured in the barring program is to apply OCB when the request URI matches a specific public user identity.

To exemplify the CB service, a signaling flow of the ACR service is given in Figure 4.19 and listed below.

1–4. The Multimedia Telephony user A calls Multimedia Telephony user B. But Multimedia Telephony user A has requested that his public user identity shall be restricted (anonymous). Therefore a SIP INVITE request is sent from Multimedia Telephony client A to IMS core B with a Privacy header field set to 'id' (or 'header').

5. The IMS core B evaluates the request URI and identifies that Multimedia Telephony user B subscribes to the ACR service. Therefore the SIP INVITE request must be processed by the 'ACR part' of the application server.

6. The SIP INVITE request is forwarded to the ACR server which here is assumed to be co-located with the TAS B.

7–11. The TAS B identifies that the call is anonymous and thus rejects the call by answering with a SIP 433 Anonymity Disallowed response.

More information about the CB service can be found in ETSI TS 183 011 [66] and 3GPP TS24.173 [24].

4.7.8 Explicit Communication Transfer

The ECT service provides the possibility for a party involved in a Multimedia Telephony session to transfer the communication to a third party. Thus, there are three actors active in a transfer, acting in the following roles: The transferor is the party that initiates the transfer of the active communication that it has with the transferee. The transferee is the party that stays in the communication which is transferred. The transfer target is the party to which the communication is transferred and which replaces the transferor in the communication. There are two types of transfer that can be done: blind transfer or consultative transfer.

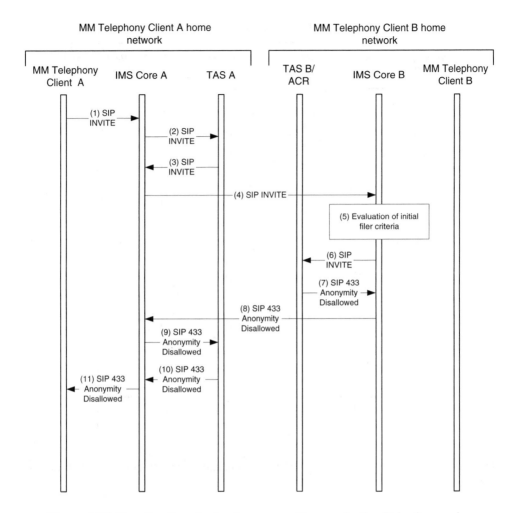

Figure 4.19: Signaling flow for the Anonymous Communication Rejection service.

The main difference is that in the blind transfer case the session between the transferor and the transferee is terminated before the new session between the transferee and the transfer target is established. While for a consultative transfer the session between the transferee and the transfer target is set up before the session between the transferor and the transferee is terminated. In the example in Figure 4.20 a blind transfer is shown. It should be noted that the Multimedia Telephony session initiation flow on both the originating and terminating side is simplified and in this example signaling needed for the resource reservation or the ringing mechanisms is not shown. To get an understanding of which SIP messages are omitted see the more detailed Multimedia Telephony session initiation flows in Sections 4.4.1 and 4.4.2.

1. A communication is ongoing and media is flowing between Multimedia Telephony client A and Multimedia Telephony client B. The Multimedia Telephony user B decides to transfer the call to create a Multimedia Telephony session between Multimedia Telephony client A and the Multimedia Telephony client C.

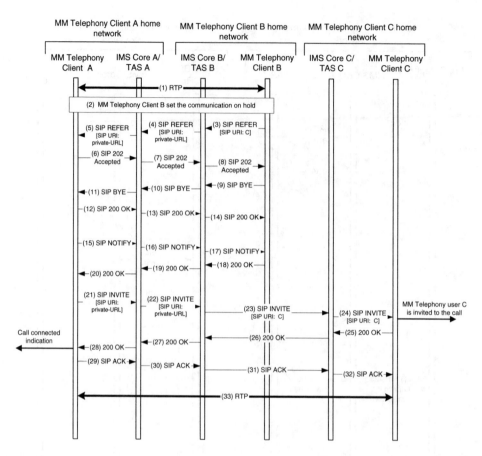

Figure 4.20: Signaling flow for Explicit Communication Transfer using the blind transfer option.

2. Multimedia Telephony user B sets the communication between Multimedia Telephony client A and the Multimedia Telephony client B on hold. For signaling flows of setting a call on hold see Section 4.7.6.

3. The Multimedia Telephony client B creates a SIP REFER request that has a Referred-To header indicating the SIP URI of Multimedia Telephony client C. This SIP REFER request is sent to TAS B in the SIP dialog between Multimedia client A and Multimedia Telephony client B.

4–5. TAS B checks whether Multimedia Telephony client B is allowed to transfer calls. If it is allowed to transfer the call then TAS B generates an ECT Session Identifier URI, addressed to itself, with the new destination information that will be needed for the new session. It replaces the Refer-To value with the ECT Session Identifier URI. This ensures that TAS B will remain in the loop even though the call will be transferred from Multimedia Telephony client B. The SIP REFER request with the ECT Session Identifier URI is sent to Multimedia Telephony client A via IMS core A/TAS A.

6–8. The SIP REFER request is accepted by Multimedia Telephony client A. A SIP 202 Accepted response is sent to TAS B.

9–11. Since the SIP REFER request was accepted, the Multimedia Telephony client B terminates the existing Multimedia Telephony session by sending a SIP BYE request to Multimedia Telephony client A.

12–14. The Multimedia Telephony client A acknowledges the SIP BYE request with a SIP 200 OK response.

15–20. The usage of the SIP REFER method causes the transmission of a notification (SIP NOTIFY).

21–22. The Multimedia Telephony client A initiates a new session by sending an INVITE request to TAS B's ECT Session Identifier URI (which represents Multimedia Telephony client C). IMS core A/TAS A routes the INVITE request to TAS B using the ECT Session Identifier URI using normal SIP routing procedures.

23–24. Upon receiving the SIP INVITE request to the ECT Session Identifier URI that was inserted by TAS B, the TAS B replaces it with the Request URI of the user of Multimedia Telephony client C and creates a SIP INVITE sent towards Multimedia Telephony client C.

25–28. The normal terminating method applies for Multimedia Telephony client C. The call shall be treated as a call from Multimedia Telephony client A to Multimedia Telephony client C. Therefore, the SIP 200 OK response is sent to Multimedia Telephony client A via IMS core C/TAS C, IMS core B/TAS B and IMS core A/TAS A.

29–32. The SIP 200 OK message is acknowledged by a SIP ACK.

33. The media starts to flow between the Multimedia Telephony client A and the Multimedia Telephony client C.

More information about the ECT service can be found in ETSI TS 183 029 [70] and 3GPP TS24.173 [24].

4.7.9 Communication Diversion: Communication Forwarding on Mobile Subscriber Not Reachable

The CFNRc service is a new addition to the set of CDIV services that already includes CFB, CFNL and CFNR which enable the communication to be diverted when the user is busy, not logged in or there is no reply. The difference is that the CFNRc service targets a mobile terminal specific scenario, namely the case when the user is not reachable due to for instance being out of coverage. Thus, the CFNRc service was first discussed in 3GPP and later it was agreed to include CFNRc also in the TISPAN specifications that originally aimed to specify simulation services for a fixed IP telephony service. It should be noted that this service was not specified at the time of producing this book. Therefore, the signaling flow is a best guess of how it may look like.

Figure 4.21 shows an example of call forwarding using the Communication Forwarding on Mobile Subscriber Not Reachable method. In this case the Multimedia Telephony user B has activated the call forwarding service but she expects to be reached via Multimedia Telephony client B1. However, at the moment the user is out of coverage and when Multimedia Telephony user A calls, she will be forwarded to Multimedia Telephony client B2. It should be noted that the Multimedia Telephony session initiation flow on both originating and terminating side is simplified and in this example signaling needed for the resource reservation or the ringing mechanisms is not shown. To get an understanding of which SIP messages are omitted see the more detailed Multimedia Telephony session initiation flows in Sections 4.4.1 and 4.4.2.

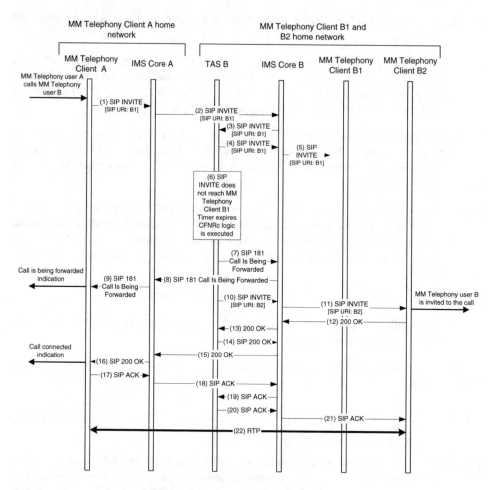

Figure 4.21: Signaling flow for call forwarding using the Communication Forwarding on Mobile Subscriber Not Reachable method.

1–4. The Multimedia Telephony client A has received a request to initiate a Multimedia Telephony session with Multimedia Telephony client B1 from its user. A SIP INVITE request is created and it is sent using a SIP URI that addresses Multimedia Telephony client B1. In this case the SIP INVITE is sent from the Multimedia Telephony client A to IMS core B via the IMS core A and TAS B.

5. Due to poor coverage the SIP INVITE cannot be successfully delivered to Multimedia Telephony client B1.

6. After a while without any response to the SIP INVITE request a timer expires in IMS core B. Then the TAS B decides that Multimedia Telephony client B1 is not reachable and executes the Communication Forwarding on Mobile Subscriber Not Reachable logic.

7–9. TAS B sends a SIP 181 Call Is Being Forwarded response to Multimedia Telephony client A via the IMS core B and the IMS core A. On reception of the SIP 181 Call Is Being Forwarded, the Multimedia Telephony client A may indicate that the call is being forwarded to its user.

10–11. When the TAS B has executed the Communication Deflection logic it creates a SIP INVITE request that contains the SIP URI that addresses Multimedia Telephony client B2. This SIP INVITE is sent to Multimedia Telephony client B2 via IMS core B.

12–16. The user receives an indication that she is invited to a Multimedia Telephony call and when the user answers the call, the Multimedia Telephony client B2 acknowledges the SIP INVITE request by sending a SIP 200 OK answer back to the Multimedia Telephony client A. The SIP 200 OK answer passes through all involved nodes.

17–21. Multimedia Telephony client A acknowledges the SIP 200 OK by sending a SIP ACK message back to Multimedia Telephony client B2.

22. The media starts to flow between the Multimedia Telephony client A and the Multimedia Telephony client B2.

More information about the CFNRc service can be found in 3GPP TS24.173 [24].

4.8 Interworking with CS Networks

It is required that IMS shall be able to interwork with Bearer Independent Call Control (BICC) and ISDN User Part (ISUP) protocol based legacy CS networks, e.g. PSTN, ISDN, CS PLMNs, in order to provide the ability to support basic voice calls (see 3GPP TS22.228 [43]) and supplementary services, between e.g. a Multimedia Telephony client using the IMS network and user equipment located in a CS network. To be able to support the delivery of basic voice calls between the IMS and CS networks, basic protocol interworking between SIP and BICC or ISUP (as specified in ITU-T Recommendation Q.1902.1 [107] and ITU-T Recommendation [105] respectively) has to occur at a control plane level. Therefore, the Media Gateway Control Function (MGCF) was developed. It is the MGCF that provides the protocol mapping functionality within the IMS. The user plane interworking between the media delivered over the IP bearers used by IMS and the CS network bearers is done by the functions within the IP Multimedia Media GateWay function (IM-MGW). Figure 4.22 shows the IMS to CS network logical interworking reference model. The Signaling GateWay (SGW) performs the call related signaling conversion to or from BICC/ISUP based MTP transport networks to BICC/ISUP based SCTP/IP transport networks. It should therefore be noted that the SGW is not required where SCTP/IP is used by the CS network.

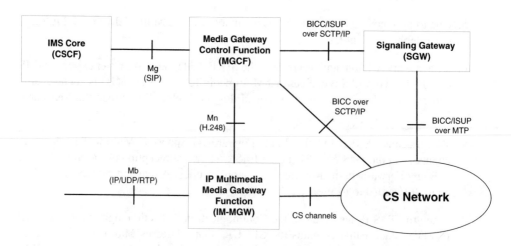

Figure 4.22: The IMS to CS interworking architecture.

The interface Mg between the IMS core and the MGCF uses SIP. The MGCF acts as an originating or terminating Multimedia Telephony client in the SIP domain when it translates incoming or outgoing call requests to/from the CS network. The Multimedia Telephony session initiation flows over the Mg interface for the interworking case will not differ substantially from the Multimedia Telephony session initiation flows between two Multimedia Telephony clients. The MGCF is aware about its resource reservation status and will act as a Multimedia Telephony client that uses mobile terminal initiated QoS and preconditions; see Section 4.4.1. For more information about the principles of the interworking mechanisms and the BICC/ISUP to SIP conversion see 3GPP TS29.163 [26]. The Mn interface between MGCF and the IM-MGW uses H.248 and more information can be found in 3GPP TS29.332 [30]. The Mb interface uses RTP/UDP/IP to convey real-time media like voice in the IMS network. It should be noted that the user plane interworking may need transcoding between e.g. PCM encoded data to/from the CS network to AMR encoded data from/to the IMS network. For more information about the user plane interworking see 3GPP TS29.163 [26].

Chapter 5

Media Flow

Daniel Enström, Tomas Frankkila, Per Fröjdh, Janne Peisa, Krister Svanbro

Voice coding has been an essential feature in digital cellular networks since the introduction of the GSM system. Significant developments have been made during the last 20 years in order both to increase the perceived quality and to reduce the required bit rate.

Video coding became an essential feature of mobile phones with the introduction of video telephony services in third generation systems. Video coding is also used in multimedia messaging and streaming services.

Text coding was introduced in order to support hearing disabled persons. Older systems send text over the telephone system by encoding each character into a series of bits and using an in-band modem to generate tones for each bit, for example ITU-T V.18 [106]. The tones are then encoded with the normal voice codec. For IP systems, the text can now be encoded and transmitted directly, thereby minimizing the required bandwidth while still providing greater flexibility than the previous systems.

With the introduction of IMS Multimedia Telephony over IP, these media types are now merged into one single service that is transported with IP. This chapter describes how the service is realized in cellular systems.

Section 5.1 describes the codecs that are used for speech, video and text. Section 5.2 describes the protocols that are used for transporting the encoded media and the procedure for defining the media types, the codecs and the codec configurations to be used in the session.

Transporting real-time media over IP networks means that the transport will be affected by delay jitter and packet loss. Handling of these transport impairments is described in Section 5.3. The clients must also handle changing operating conditions and when the users want to add or remove media components in an ongoing session. This is described in Section 5.4.

Section 5.5 describes the compression of IP, UDP and RTP headers that is required to reduce the overhead that these protocols introduce. Section 5.6 describes how the radio bearers have been designed in order to maximize the capacity of the system.

Interworking with legacy systems is an important aspect of the Multimedia Telephony service since one can expect that legacy products will be available for some time. Section 5.7 describes how the quality can be maximized when interworking with GERAN and UTRAN by avoiding tandem coding.

IMS Multimedia Telephony over Cellular Systems   S. Chakraborty, T. Frankkila, J. Peisa and P. Synnergren
© 2007 John Wiley & Sons, Ltd

Section 5.8 outlines the special requirements that the Multimedia Telephony, Media Handling and Interaction specification includes. The level of detail included in this section is quite different for different media types. The reason is that for some media it is possible to mainly reference another specification while other media require much more detail of specification.

5.1   Media Coding

Digital cellular systems require that the analog media is encoded to a digital form. There are several contradictory requirements for the encoding process, the following for example:

- High bit rates are usually required for high quality. On the other hand, high bit rates require significant transmission resources, which gives an expensive service. It is therefore essential to select a codec that delivers high quality speech but still has a sufficiently low bit rate.

- Using long blocks in the encoding process usually gives a higher compression ratio but also introduces delay, which is disadvantageous for real-time services where short delays are required for good conversational quality. When the delay is really long, then the users will believe that the level of interactivity is low and will regard the service as mainly a half-duplex service.

- Advanced algorithms, such as transforms and vector quantization, give an improved quality but also increase the complexity. High computational requirements mean that more advanced processors need to be used. This has a negative impact on the power consumption and gives shorter battery lifetimes. The alternative is to use larger batteries, which increases the physical size of the phone, the weight of the phone and usually also gives more expensive phones.

Media codecs need to provide a suitable compromise between all these aspects in order to be useful for cellular services. Another useful feature is the capability to adapt to different and varying radio conditions and network load levels.

This section reviews several different codecs for the voice, video and text media types respectively and outlines their respective usefulness in multimedia telephony.

5.1.1   Speech

5.1.1.1   Narrowband and Wideband Speech Codecs

There are two main classes of speech codecs, narrowband and wideband. Narrowband speech codecs use 8 kHz sampling frequency when converting the analog signal to digital form. This allows for encoding sound frequencies up to 4 kHz. Furthermore, the input signal is often filtered with an IRS [97] or a ModIRS filter [101] to make the sound signal compatible with traditional land-line circuit switched telephone systems (PSTN). This reduces the bandwidth to 300–3400 Hz.

Wideband speech codecs use 16 kHz sampling frequency in the analog to digital conversion process. The signal is further filtered to an audio bandwidth of 100–7000 Hz [101]. The advantages with wideband speech codecs, compared with narrowband speech codecs, are improved sound quality, intelligibility, naturalness and presence [123].

5.1.1.2 Waveform Coding

Waveform coding of speech has been used for a very long time. It is the simplest form of encoding of an analog signal that has been converted to digital form. Each sample is encoded with a predetermined number of bits. Prediction may be used in order to explore the inter-sample dependences and thereby reduce the bit rate.

The most commonplace waveform codec is Pulse Code Modulation (PCM) [96], which is used in the land-line telephone network. The PCM codec has a very low complexity, since it does not use any inter-sample prediction, and gives also a very good speech quality. The main drawback is that the bit rate is quite high, 64 kbps, which makes this codec unsuitable for wireless networks since the capacity requirement would be too high.

Another waveform codec is the Adaptive Differential PCM (ADPCM) codec [98], which uses prediction in order to reduce the bit rate. ADPCM also gives a very good quality while still having low complexity requirements. It has a lower bit rate than PCM, 32 kbps being the most commonly used mode, but the bit rate is still too high to be realistic for wireless systems.

The ADPCM codec can operate with a few different bit rates, 16, 24, 32 or 40 kbps. The 32 kbps mode is the most commonly used mode.

The PCM and ADPCM codecs are both narrowband codecs. One example of a wideband waveform codec is the Sub-Band ADPCM (SB-ADPCM) codec [95]. The SB-ADPCM codec can operate with three different bit rates, 48, 56 and 64 kbps, and the 64 kbps rate is the most commonly used mode.

The main drawback with all these waveform codecs is the relatively high bit rate, which is undesirable for cellular systems.

5.1.1.3 CELP Speech Codecs

Most modern speech codecs are based on the Codebook Excited Linear Predictive (CELP) codec model [49], where speech is synthesized according to a model of how the speech is produced by humans. The synthesis, in the decoder, follows the following steps:

1. A fixed code book generates an innovation. This simulates the air flow created by the lungs.

2. The innovation is passed through a pitch predictor, which is also often called adaptive code book. This creates pitch pulses in a similar way as the vocal cords create pulses in the air flow.

3. The excitation after the pitch predictor is passed through an LPC predictor. The LPC predictor shapes the signal in a similar way as the vocal tract would do for the air flow.

CELP codecs typically use block-based encoding where the sound is usually encoded in frames of 20 ms. Most CELP codecs also use the analysis-by-synthesis method in the encoder to determine innovation and excitation parameters [174]. Analysis-by-synthesis, in its purest form, is very complex since one basically executes the decoder with every possible parameter combination and selects the parameter combination that minimizes the error between the original signal and the synthesized signal. Preselection and structured search methods are therefore often used to reduce the complexity to a more manageable level.

A multitude of CELP speech codecs have been developed during the recent decades. Several codecs have been standardized by ITU-T, for example the Conjugate Structure

Algebraic CELP (CS-ACELP) codec [99]. Several variants of the G.729 code are also available in annexes. Other speech codecs have been especially designed for cellular systems and have been standardized by ETSI, 3GPP and TIA/EIA, for example GSM-EFR [10], AMR [4], AMR-WB [14] and TDMA-EFR [181].

Out of these codecs, AMR and AMR-WB are of especial interest for mobile IMS voice services since they were originally designed for cellular systems and were also designed to work well for a large set of different and varying channel conditions. These codecs have the following advantages over all the other alternatives:

- The quality is very good. The highest codec modes of AMR and AMR-WB give a quality for narrowband and wideband services respectively that is not exceeded by any other codec under the same operating conditions and given the same bit rate requirements.

- Both AMR and AMR-WB have several codec modes. The AMR codec includes eight codec modes ranging from 4.75 kbps to 12.2 kbps (see Table 5.1) and the AMR-WB codec has nine codec modes ranging from 6.60 kbps to 23.85 kbps (see Table 5.2). This allows for adapting the codec rate to different network loads and channel conditions. It also allows for developing several service variants that can be differentiated by quality; see Chapter 2.

- The complexity is reasonable. One of the fundamental requirements in the selection of these codecs was that the complexity must be manageable and should not increase the processing requirements in the mobile phones too much.

- These codecs, especially AMR, are today available in most GSM phones. Since there are over 2 billion GSM customers in the world in more than 210 countries [83], this gives excellent opportunities to maximize the quality by using Tandem-Free Operation (TFO) between IMS services and traditional circuit switched services; see also Section 5.7.1.

As described above, the AMR and AMR-WB codecs can operate at a number of codec modes. The possible codec modes and bit rates are shown in Tables 5.1 and 5.2.

Table 5.1: Codec modes for AMR.

Codec mode [kbps]	Comment
AMR 12.2	Same as GSM-EFR
AMR 10.2	
AMR 7.95	
AMR 7.4	Same as TDMA-EFR
AMR 6.7	Same as PDC-EFR
AMR 5.9	
AMR 5.15	Default codec mode for PoC
AMR 4.75	

For the circuit switched GERAN and UTRAN systems, the adaptive feature of the AMR and AMR-WB codecs allows for adapting the source coding and channel coding bit rates so that the service quality can be optimized for a variety of operating conditions. For good

Table 5.2: Codec modes for AMR-WB.

Codec mode [kbps]
AMR-WB 23.85
AMR-WB 23.05
AMR-WB 19.85
AMR-WB 18.25
AMR-WB 15.85
AMR-WB 14.25
AMR-WB 12.65
AMR-WB 8.85
AMR-WB 6.60

channel conditions, the best possible quality can be delivered by using the AMR 12.2 and a high rate AMR-WB mode respectively. For degraded channels, the bit rate used for source coding can be reduced, giving room for more channel coding.

For HSPA, the adaptive feature can be used in a similar way. Lower codec modes allow for using smaller transport blocks, which gives room for using more channel coding for the transport blocks. For high system loads, it is also advantageous to reduce the bit rate. This can be done in two ways:

1. A lower codec mode bit rate gives smaller packet sizes that can be transmitted with smaller transport blocks. Since the required transmission power is proportional to the transport block size, this means less interference, which allows more users into the cell.

2. Since a lower codec mode bit rate gives smaller packets, this can be used to encapsulate more packets into one transport block. This reduces the packet rate, which means that more TTIs will be available for other users.

The adaptive feature of AMR and AMR-WB is therefore important also for packet switched systems since it allows for making different trade-offs between capacity and quality for different system load levels.

5.1.1.4   Unequal Error Protection vs. Equal Error Protection

In the legacy circuit switched systems, the transport impairment that was the main focus was handling the large amount of bit errors that arise when the channel conditions get worse. Bit errors arise when the noise, or interference, is so high that the modulation can no longer reliably regenerate the data bits from the transmitted symbols. The common solution to this problem has been to apply channel coding to the bitstream from the source encoder. There is however almost always a limit for how much channel coding one can add in any given system, which means that there is also a limit for how many bit errors the channel codec can correct.

The radio channel in wireless systems often encounters fading, which gives bursts of bit errors. Most received frames have very few bit errors or are even error-free while there are some frames that have a very large amount of bit errors. Given the limited room for channel coding, it is impracticable to design the channel coding to handle the short bursts of high bit error rates. Most wireless systems instead use interleaving to distribute the encoded bit over several frames and scrambling to distribute the encoded bits within a frame. If the channel then introduces a short burst of bit errors, then the interleaving and scrambling will distribute the bit errors over several frames when sorting the data bits back into the original order. The hope is that the channel decoder will then be able to correct the now reduced peak error rates for a large amount of frames.

Unequal Error Protection (UEP) [162] is often used in circuit switched services since the encoded bits from the speech encoder have very different error sensitivity. The encoded bits are often divided into three or more classes, class 1a, class 1b and class 2. With UEP, the more sensitive bits, class 1 bits, are protected with a stronger channel code than the less sensitive bits, the class 2 bits. In some cases, the class 2 bits may even be left unprotected. In addition, a Cyclic Redundancy Code (CRC) [82] is often also applied on the most sensitive bits, class 1a bits, to allow for detecting any residual errors after the channel decoder. The error concealment (see Section 5.1.1.5) is then activated when the CRC check fails. The UEP thus allows for having a few bit errors in the class 2 bits, even after channel decoding. For best possible performance, the bit errors should be distributed fairly well so that the occurrences when the channel codec and the CRC fail are also well distributed over time.

For cellular VoIP services, the possibilities to use UEP are small, since the whole IP packet is typically transmitted in one transport block, which is protected with the same channel coding rate, so-called Equal Error Protection (EEP). Especially for cellular VoIP systems using ARQ, such as HSPA (see Section 5.6), the transport blocks that carry the speech frames are retransmitted until it is properly decoded. There is however an upper limit for how many times a transport block may be transmitted, and the complete transport block is discarded if this limit is exceeded.

The equal error protection is also a design choice because one desirable feature is to separate source coding and transport in order to achieve service agnostic transport networks (or transport agnostic services). This allows for using the exact same transport functions regardless of the service, which gives a faster development, simplified integration and testing, and also faster roll-out of new systems and services. Due to the service agnostic transport functions, a packet received with error will be rejected and will not be forwarded up to the application. Since the design of the MAC layer protocols are unaware of the type of data that is being transmitted, it is not possible to take advantage of UEP in these kinds of systems. For VoIP in HSPA, one can thus expect that the speech frames are either properly received or not received at all.

5.1.1.5   Error Concealment

The Error Concealment Unit (ECU) is a function that tries to cover short interruptions in the media flow arising from packet losses. When a frame is received with errors, the ECU tries to smoothen the sound so that strong distortions become much less audible. The ECU plays an important role for real-time speech services. Even though channel coding, in the form of Forward Error Correction (FEC) or Automatic Repeat Request (ARQ) [133], is used to protect the data, the receiver must always be prepared to handle frames with bit errors or even lost frames.

The speech decoder needs to synthesize a speech frame at every speech frame interval in order to produce a continuous audio stream to the listener even if the received frame is severely damaged. The receiving application must therefore be prepared to detect errors in the received frame and apply error concealment when required. The error concealment function thus has the obligation to generate a speech frame that will minimize the distortions that will be presented to the listener.

When errors in the received speech data frame are detected, traditional error concealment methods for circuit switched services work by copying some of the bits (LPC, pitch predictor and gain bits) from the previous speech frame, applying smoothing or muting to the decoded speech parameters and synthesizing the frame. The innovation bits are often taken from the received frame even if they probably contain some bit errors. Using the innovation bits from the received frame is often better than repeating the innovation sequence from the previous frame. This is because repeated signals often tend to give tonal distortions that are quite audible and using the received bits gives some randomization and the repetitions can then be avoided.

As described in Section 5.1.1.4, one cannot expect to use any received innovation bits because the whole frame will be lost.[3] It is therefore essential that the ECU for VoIP is developed to handle this situation. A common solution for this problem is to randomize the innovation bits.

There are however also occurrences in CS where the complete frame is lost and where the innovation bits are not available. One example is when the frame was stolen for system signaling. Another example is in mobile-to-mobile calls where there are bit errors in the first hop. Depending on the system design, the received erroneous frame may be transmitted over the second air interface but it may also be rejected and a dummy frame may be transmitted instead. Error concealment units designed for cellular systems thus already handle both cases. When the frame is available, some speech codec bits may be used, even if there are bit errors. When the frame is not available, then the ECU will have to generate the innovation itself.

The ECU for AMR, specified in [3], has been designed to handle both the case of receiving a frame with bit errors and not receiving the frame at all. The error concealment uses the repeat-and-mute concept for both cases. When the frame is lost, it also randomizes the innovation bits. This latter case, when the frame is not available, is identical to the packet losses that occur in VoIP services. A properly designed ECU for CS, which handles both the case of erroneous frames and the case of lost frames, will work equally well for VoIP.

5.1.1.6 Error Resilient Speech Service

One of the most challenging aspects of creating a real-time service in any system is how to make the service resilient to the various types of impairments that may occur. In the literature, the most commonly discussed problem is handling of packet losses. The typical reason for packet losses is channel interference due to either poor coverage or high system load. For dedicated channels in UTRAN, the fast and precise power control typically manages to maintain a BLock Error Rate (BLER) that is very close to the defined target value. For shared channels, one can expect that the packet loss rate varies more over time and also depends more on the current cell load. This is especially true for shared channels that also use retransmissions. At low load levels, there is a sufficient amount of resources available to allow for retransmitting packets until the packet can be properly decoded. In this case, the packet loss rate should be close to zero, even for users that have quite poor channel conditions. For high load levels, this will not be the case and one can expect higher packet loss rates.

The scheduler also plays a very important role in shared channels. If the scheduler has been optimized for voice services, then it is likely that packet losses will be distributed fairly well. If, on the other hand, the scheduler has been optimized for, for example, data

---

[3] An exception to this rule is discussed in Section 5.7.

transmission, then one can expect both high packet loss rates and packet loss bursts of considerable length. This is further discussed in Chapter 7.

When using AMR or AMR-WB, it is possible to make the speech service quite resilient towards high packet loss rates and also towards poor scheduler designs. In [127], the advantages of using application layer redundancy are shown. When combining application layer redundancy with the adaptive feature of AMR and AMR-WB, it is possible to design a solution that is virtually bit rate neutral. One such solution is to use the AMR 12.2 kbps mode when no redundancy is used, then adapt to the AMR 6.7 kbps mode when single (or 100%) redundancy is used and then adapt further down to the AMR 4.75 kbps mode when double (or 200%) redundancy is applied. The bit rates for the respective redundancy modes for such a solution are shown in Table 5.3.

Table 5.3: Bit rates for the compared configurations.

Codec	Source coding [kbps]	Redundancy [kbps]	Total codec bit rate [kbps]
AMR 12.2, no redundancy	12.2	0	12.2
AMR 6.7, 100% redundancy	6.7	6.7	13.4
AMR 4.75, 200% redundancy	4.75	9.5	14.25

The performance of this scheme has been evaluated by applying random losses to a VoIP stream. The quality was evaluated with PESQ [108]. The results for packet loss rates up to 30% are shown in Figure 5.1. As can be seen in this figure, the AMR 12.2 kbps mode offers the highest quality when the packet loss rate is low. The redundancy scheme gives a substantial speech quality improvement for high packet loss rates, even when switching to a lower rate codec mode.

This example shows that, for VoIP, a multi-rate codec that adapts the source coding bit rate and the application layer redundancy has a clear advantage over fixed-rate codecs. One should however expect that the packet loss rate will be low for most normal operating conditions and will only be severe when the network load is high or when the user is in a spot of poor coverage. Using a low rate codec mode together with a significant amount of redundancy all the time would therefore not maximize the performance. It is important that the scheme is adaptive, so that maximum quality is delivered when the packet loss rate is low and the level of redundancy is increased when the packet loss rate increases.

It should however be noted that packet loss is not the only problem that real-time services on wireless links will experience. Other important problems are:

- delay;

- delay jitter;

- late loss;

- interruptions in the media flow due to for example hand-over;

- acoustic echo;

- speech level or gain control; and

- clock drift.

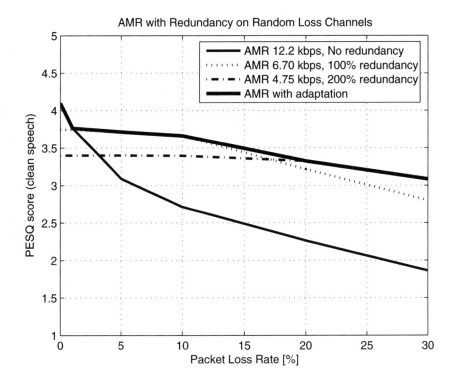

Figure 5.1: Performance of AMR for various packet loss rates when different redundancy levels are used.

Some of these problems are the same for packet switched and circuit switched systems, for example delay, interruptions due to hand-over, echo and gain control. Other problems, like delay jitter and late loss, are unique for packet switched systems.

The most well-known transport impairments are delay, delay jitter and late loss. The effects of delay and delay jitter are discussed quite extensively in Section 5.3 and are thus not further addressed here.

An often occurring impairment in cellular systems is interruptions in the media flow caused by either hard hand-over or frame stealing caused by the need for signaling. Hard hand-over causes interruptions of several consecutive frames. It is therefore essential that the system is designed in such a way that the hand-over time is minimized. Frame stealing can also be designed so that the stolen frames are distributed over a short period of time and consecutive stolen frames are avoided. These impairments can most of the time be solved with a proper system design. Given that the interruptions are infrequent enough, the degradations should be small enough to be acceptable. Interruptions may be either a larger or a smaller problem in VoIP systems. For UTRAN DCH bearers that use soft hand-over, the interruptions should be non-existing as for circuit switched services. In HSPA, one can however expect that the interruptions due to hard hand-over are similar to the interruptions caused by hard hand-over in GERAN.

Another commonly occurring problem is echoes. The system design, and especially the interleaving used to improve the performance of the channel codec, introduces a fairly large amount of delay. Long delays have been proven to be quite annoying since it is hard to maintain a fluent conversation when the delay is long. Another problem with long delays is that the sound produced in the other end echoes back to the person that is currently talking. In for example ITU-T Recommendation G.108 [104], it is shown how the quality is degraded depending on one-way delay and the level of the echo. The problem with maintaining a fluent conversation is solved by designing the system to have a short enough delay. The problems with echoes are solved by introducing Echo Cancellers (EC), which are sometimes also referred to as Acoustic Echo Cancellers (AEC). Echo cancellers are implemented in most mobile terminals. Network echo cancellers are often also used if the echo canceller in the mobile terminal is not working sufficiently well or if there are plenty of echoes from the other participant in the session.

The problem with echoes can be expected to be worse for VoIP than for circuit switched systems. This is because it is desirable to send the VoIP packets end-to-end, without decoding them in any network node since this reduces the complexity in the network nodes and gives a more scalable solution. One should therefore expect that there will be no network echo cancellers in the media path and that the VoIP clients, even VoIP clients for the fixed IP system, must handle echo properly.

Closely related to the acoustic echo problem is the problem with maintaining a correct sound level of the speech. Also in this case, circuit switched systems typically have network nodes that can adjust the level, if it is found that the mobile terminals do a poor job. Similarly as for echoes, one cannot expect that there will be a network node that corrects the speech level in end-to-end VoIP sessions. When designing VoIP clients, one should therefore pay especial attention to the Automatic Gain Control (AGC) function.

Clock drift may become a larger problem for VoIP than for CS terminals. The sampling frequency in terminals can be assumed to be quite exact. It is however possible to implement the Multimedia Telephony client in, for example, a laptop and use the mobile phone as a modem. In this case, the difference in sampling frequency between the end-points can be quite considerable, which the VoIP applications must be prepared to handle.

5.1.2   Video

Video is synonymous with moving pictures and is used here as a term for describing various storage or transmission formats for a sequence of frames. The frames represent snapshots in time and when viewed in sequence at a high enough frame rate by the human eye, they are perceived as a picture in motion. Depending on frame rate, there is a sliding cross-over between 'slide show' and 'video'. In contrast to (narrowband) speech where the sampling rate is 8 kHz, video only requires a very low temporal sampling rate. Ten frames per second is enough for a sequence of frames to be perceived as motion, although higher rates are usually needed to represent smooth motion. Movies in movie theaters use 24 frames per second and on TV we see 25 or 30 frames per second. Considering that screen size and viewing distance also have an impact on the user's experience, it is often enough to use a smaller frame rate for mobile devices with their small screen dimensions. Typical frame rates for 3G systems vary from 10 frames per second for video telephony up to 25 or 30 for mobile TV.

As uncompressed natural video contains a lot of data it requires substantial channel bandwidth for transport. It is fortunate however that video has many characteristics that make it suitable for compression. A high degree of temporal redundancy between frames as well as spatial redundancy within frames are typical. This allows for inter-frame and intra-frame compression to be successfully deployed. A crucial factor in video compression is to utilize redundancies with lossy rather than lossless compression techniques. The fact that finer distortions are not perceived by the human eye makes it possible to compress video signals to sizes several thousand times smaller than the original signals. Comparing compression ratios of several thousands for lossy compression with typical compression ratios around 2 to 4 for lossless compression, it is easy to understand that lossy compression is ubiquitous in video compression and is used for all mobile applications.

A well-known example of lossy video compression is interlace in standard TV signals where every second line in the picture is skipped in an alternating pattern. Interlaced video was designed and worked perfectly for CRT displays where the electron beam refreshed the frame line by line in the same alternating pattern. It is still used for standard TV, although interlaced video makes less sense for modern LCD and plasma screens. For High-Definition TV (HDTV) both interlaced and progressive (non-interlaced) formats are used. For services on mobile phones only progressive video has so far been standardized.

Other traditional examples of lossy compression are temporal and spatial subsampling. We have already mentioned that low frame rates are commonly used for small screens. Spatial subsampling is useful here as well. Subsampling can also be done selectively on parts of the video data. For instance, as the human eye is more perceptive to changes in the overall intensity of light than changes in color, chrominance information can be spatially subsampled and coded at a lower bandwidth than the corresponding information for luminance.

Substantial gains from lossy compression come from employing more advanced schemes in addition to the above. Most compression standards for both image and video coding rely on Discrete Cosine Transforms (DCT) or similar transforms where spectral representations replace the natural image. By keeping only low-frequency components, the reconstructed image will be distorted although it may be less evident for the human eye. Quantization refers to how frequency components are coarse-grained and truncated. It is used as a parameter for controlling the degree of lossy degradation. The higher the quantization, the higher the compression gain.

5.1.2.1 Digital Video

Digital video signals can contain either raw (uncompressed) video of a certain sampling format or a compressed format suitable for transport and storage. The formats of the video signals can be characterized by several parameters, such as screen size and aspect ratio, frame rate, interlace or progressive scan, color space and depth, and subsampling patterns.

Screen size or screen resolution is expressed in terms of width and height in units of pixels. Screens on mobile phones are typically small and the most common format for video on a mobile device is QCIF (176×144 pixels). It stands for Quarter CIF and is derived from the Common Intermediate Format (CIF) with 352×288 pixels, which is used for video conferencing and as a subsampled TV format. CIF is actually more than a picture format as it also specifies frame rate (29.97 Hz) and aspect ratio (see below). If not explicitly stated otherwise, we will use CIF to refer to the pixel resolution (352×288) without further constraints on frame rate, etc. This is also how the terms CIF and QCIF are used in practice for mobile devices. Today many phones support larger screen sizes than QCIF. Another popular format is QVGA (320×240) which stands for Quarter VGA and is derived from Video Graphics Array (VGA) commonly used in the world of PCs.

Each pixel in a video signal represents the color of a small region. Mobile phones typically use square pixels representing square regions, but other pixel aspect ratios are also common, in particular for TV. The CIF format uses a pixel aspect ratio of 12:11 to represent a screen with 4:3 ratio. Plasma displays often use non-square pixels, where for instance a resolution of 1024 × 1024 pixels may be used for viewing HDTV in a ratio of 16:9.

The Red–Green–Blue (RGB) color domain is often used by displays to represent the color of a pixel, whereas for the purpose of processing and transporting video signals, most applications use the YCbCr domain. YCbCr of ITU-R Recommendation BT.601 [111] (CCIR 601) divides the video signal into three color components: Y is the luminance which represents black-and-white or grey-scale video. It is a weighted sum of red, green and blue that accommodates for the sensitivity of the eye to different colors. Cb and Cr add chrominance and are calculated as differences between Y and blue, and Y and red, respectively. The Y, Cb and Cr components for a pixel may represent slightly different areas and positions of the signals depending on the subsampling scheme. For mobile phones the most common scheme is 4:2:0 which corresponds to a spatial subsampling of Cb and Cr by a factor or 2 in each spatial direction in relation to Y.

The color depth refers to the number of bits each color component is represented by. It may vary between the different color components, but for most practical applications including mobile phones and TV, 8 bits are used per color component. However, for more demanding applications, such as studio and editing, 10 or even 12 bits may be used.

5.1.2.2 Hybrid Video Coding

One of the paradigms of video coding is to take advantage of temporal redundancy between frames to make the coding as efficient as possible. This is particularly true for low bit rates and qualities characteristic for mobile applications where inter-frame prediction plays a crucial role in the coding process.

All video coding standards mentioned in the following sections are so-called hybrid coding standards that utilize temporal motion compensation for inter-frame prediction in combination with spatial transform coding. The main reason for treating time and space differently is the low sampling rate in time versus the high sampling rate in space. This two-step approach has proved to be effective from a coding efficiency point of view and yet practical in terms of reasonable algorithmic complexity. In contrast to coding schemes utilizing transform coding in both temporal and spatial dimensions, hybrid coding can be successfully used for low-delay coding which is crucial for any conversational application.

A typical video coder operates sequentially by coding a video sequence picture by picture. A picture is further divided into blocks that are coded sequentially row by row, starting at the top left corner and ending at the bottom right corner. These blocks are called macroblocks and cover 16 × 16 luminance pixels. All standards reviewed here support the 4:2:0 subsampling scheme where each macroblock also contains corresponding 8 × 8 Cb and Cr chrominance pixels.

The first picture of a video sequence is encoded as a still image and is called a key or intra picture (I picture). For the H.264/AVC codec (see Section 5.1.2.5) it is referred to as an Instantaneous Decoder Refresh (IDR) picture. An intra or IDR picture is self-contained and does not depend on previously coded pictures. They are not exclusively used at the start of a sequence, but they are also advantageous to use at instances where the video changes abruptly, such as at scene cuts, or where it is desirable to have a random access point, from which a decoder can start decoding without having to decode the previous part of the bitstream. The pixel values of intra-coded macroblocks are transformed to a frequency domain, e.g. using discrete cosine transform, and the transform coefficients quantized in order to reduce the size of the resulting bitstream. The degree of quantization is used as an input parameter and can vary during the encoding process.

For the subsequent pictures the hybrid coding scheme is used to utilize temporal redundancy. A prediction or inter picture (P picture) is coded as a motion compensated difference image relative to an earlier picture. By using an already decoded picture (reconstructed picture) as reference, the video coder can signal for each macroblock a set of motion vectors and coefficients. In fact, the motion vectors themselves can often be predicted from surrounding motion vectors and it suffices to signal the prediction error. Note that the encoder also includes a decoder and can thus use the same reconstructed image as the decoder in the prediction process. The motion vectors (one or several depending on how the macroblock is partitioned) inform the decoder how to spatially translate the corresponding region of the reference picture in order to make a prediction for the macroblock under consideration. The difference between the prediction and the original is encoded in terms of transform coefficients. However, not all macroblocks of an inter picture need to be motion compensated. If the change from the reference macroblock to the current macroblock is small, it can be skipped, i.e. not coded but signaled to be copied. On the other hand, if the macroblocks differ substantially, it may be better to code it as an intra macroblock. It should be mentioned that the H.264/AVC standard (see Section 5.1.2.5) also has the possibility to signal a macroblock to be copied with a global (collective) motion.

5.1.2.3 Early Video Standardization

The most popular video compression standards used today come from the H.26x series of Recommendations from the ITU-T and the MPEG-x series of International Standards from ISO/IEC. The basic principles of these codecs are the same. They are block-based hybrid video coding standards combining motion compensation for inter-frame prediction and transform coding for still image coding as well as coding of inter-frame prediction errors. All standards also have in common that they prescribe the decoding process, i.e. how the uncompressed video signal is reconstructed from the compressed format. Some standards have restrictions that apply to the encoding process, but the general principle is that each standard includes a set of tools that the encoder is free to use and combine. There is a lot of room for optimizing the usage of these tools depending on the nature of the video content and the application. As long as the video stream produced by the encoder follows the syntax of the standard, the stream can be decoded with a predicted outcome by any decoder compliant to the standard. Another common denominator is that all the above-mentioned standards use variable-length coding to efficiently represent the compressed bitstream. In addition, some codecs support options to use arithmetic coding for higher efficiency. However, as this option works best in lossless (non-radio) environments and requires higher complexity, it has so far not been adopted for usage in mobile phones.

H.261 [112] was developed as a video compression standard for the H.320 [117] video conferencing standard over ISDN at $p \times 64$ kbps, $p = 1, \ldots, 30$. It was approved in 1990 by the ITU-T and includes as main features motion compensation, DCT transformation, quantization and variable-length coding. For anyone studying video coding, H.261 is perhaps the best starting point as it formed the bases for the MPEG-1 and MPEG-2 standards that were to be developed next. As appropriate for a standard for conferencing H.261 specifies a maximum coding delay of 150 ms. Similar constraints can also be applied to the other ITU-T and MPEG standards, although they typically allow for higher efficiency by utilizing longer coding delays.

MPEG-1 [90] was developed for storage of video in the CIF format at 1.5 Mbps with requirements including random access, fast forward and reverse playback. It was published in 1992 and has been successfully deployed by the Video CD (VCD) standard supported not only by VCD players but also by most modern DVD players. The quality of MPEG-1 corresponds roughly to that of the VHS standard for VCR recorders. In addition to the features of H.261, MPEG-1 uses half-pixel accuracy for motion compensation, intra-macroblock prediction for intra pictures, and bi-directional prediction pictures (B pictures) which can be predicted from both past and future pictures. Prediction from the future can be achieved by altering the picture coding order, whereupon the decoder reorders the pictures back into chronological order before they are rendered. B pictures give high compression gains, but induce delays as they alter the display order and have therefore limited use in conversational video applications.

MPEG-2 [91] is backward compatible with MPEG-1 and is also known as H.262 [113]. It supports additional features to make it suitable for high-quality standard TV and high-definition TV. The commercial deployment of MPEG-2 has been very successful, most notably for the DVD standard and digital TV over cable and air. In MPEG-2 motion compensation is more accurate than for MPEG-1 as macroblocks can be split in half to allow for two independently motion compensated regions. In addition, MPEG-2 supports interlace, additional subsampling formats and higher color depths. It also includes tools for making bitstreams scalable spatially and quality-wise in addition to temporal scaling. As MPEG-2 supports a variety of tools, they have been divided into profiles targeting different applications. One or more levels are defined for each profile with bounds on complexity in terms of screen resolution, frame rate and bit rate. The Main profile Simple level targets standard TV at bit rates up to 15 Mbps.

5.1.2.4   3G Video Standards H.263 and MPEG-4

For video conversation in 3G mobile devices, H.263 [120] and MPEG-4 Visual [93] are the most important video coding standards. They are used in 3G-324M, which is the mobile circuit switched video telephony standard defined by 3GPP in [6] and [7] based on the ITU-T Recommendation H.324 [119] for video conferencing. The baseline version of H.263 was approved in 1995 and has been adopted by 3GPP as the common mandatory video codec that all video-enabled 3G mobile phones must support. There are two extended versions of H.263 published in 1998 (Version 2) and 2000 (Version 3), respectively. All versions are called H.263, although the 1998 and 2000 versions are also referred to as H.263+ and H.263++, respectively. H.263 targets video communication at low bit rates up to 64 kbps but there is no upper limit and it works well up to 1 Mbps or even higher.

H.263 introduces new coding tools that further improve the compression efficiency compared to H.261 and gives in general an additional bit rate reduction of 50%. It includes refinements to reduce syntax overhead in the bitstream as well as other improvements including the usage of half-pixel accuracy in motion vectors. There are many optional tools in H.263 including the usage of up to four motion vectors per macroblock, advanced intra coding, support for multiple reference pictures and unrestricted motion vector lengths. The list is long, but far from all tools are used for video conferencing.

Each tool is specified in an Annex of H.263 and the capability of a tool is signaled by external means, e.g. by using H.245 in 3G-324M. Certain combinations of tools in H.263 can also be signaled using profiles. The capability of a profile of H.263 in the Multimedia Telephony service for IMS is signaled in SIP by using SDP.

There are currently nine profiles defined for H.263. The profiles that are used in 3G systems are the Baseline profile (Profile 0) and the Version 2 Interactive and Streaming Wireless profile (Profile 3). The Baseline profile refers to the syntax of H.263 approved in 1995 with no optional modes of operation, whereas Profile 3 includes support for motion vectors over picture boundaries (D.1), four motion vectors per macroblock (F.2), advanced intra coding (I), de-blocking filter (J) to reduce block artifacts in addition to improved slice handling (K) and modified quantization (T), where the corresponding Annexes are indicated within parentheses.

For each profile it is also possible to give a level indicating the degree of complexity that is supported. There are currently eight levels defined: 10, 20, 30, 40, 45, 50, 60 and 70. Support for any level other than level 45 implies support for all lower levels. Level 45 is a special case and was added more recently after a request from 3GPP. Support for level 45 implies support for level 10. Table 5.4 shows a summary of levels 10 to 45 with maximum bit rate, picture format and picture rate for each level. All levels support the sub-QCIF (128×96) and the QCIF (176×144) formats. The CIF format (352×288) is supported where indicated. The timing between pictures is controlled by a picture clock frequency at $30000/1001$ (approximately 29.97) Hz and is restricted to an integer value times $1001/30000$. Hence, the picture rates in Table 5.4 are not exact albeit very close.

Table 5.4: H.263 levels up to 45.

Level	Max bit rate [kbps]	Max picture format and rate
10	64	QCIF@15 Hz
20	128	QCIF@30 or CIF@15 Hz
30	384	CIF@30 Hz
40	2048	CIF@30 Hz
45	128	QCIF@15 Hz

Levels 10 to 40 of H.263 are the same for all profiles. Level 45 includes for all profiles except 0 (Baseline) and 2 the support for custom picture formats of size QCIF and smaller and custom picture clock frequencies. Levels 50, 60 and 70 support higher bit rates as well as custom picture formats and custom picture clock frequencies.

Based on the baseline version of H.263, MPEG-4 Visual (MPEG-4 Part 2) [93] was developed in parallel with the later versions of H.263 and has a similar performance. MPEG-4 decoders also have a compatibility mode capable of decoding H.263 baseline of 1995. The objective of MPEG-4 was not only to achieve high video compression efficiency but also an object-based standard where a scene may consist of several audiovisual objects. A video picture can be composed of several video object planes corresponding to image regions of arbitrary shape. It should be mentioned, however, that this feature is not much used and in particular not for mobile devices where only single-object rectangular-shaped video is used today. MPEG-4 Visual was first published in 1999 and later versions appeared in 2000 (Version 2) and 2003 (Version 3). MPEG-4 has gained popularity for mobile phones and is also the video compression standard behind the popular DivX and XviD formats supported by many DVD players.

As far as video compression is concerned, the tools of MPEG-4 Visual are similar to those of H.263. This is particularly true for mobile applications which use the Simple profile.

Level 0 (L0) of the Simple profile corresponds to H.263 level 10 and was included in MPEG-4 to satisfy the requirements of the 3G circuit switched Multimedia Telephony service (3G-324M). MPEG-4 Visual Simple profile L0 decoders support one video object per visual stream and up to 99 macroblocks (16×16) per picture and 1485 macroblocks per second, e.g. up to QCIF at 15 pictures per second. The maximum bit rate is 64 kbps. The next level, L0b, corresponds to H.263 level 45. It was added at the request of 3GPP and boosts L0 to 128 kbps. The MPEG-4 Visual Simple profile also includes five higher levels for bit rates up to 8 Mbps. There are also more Visual profiles of MPEG-4. Currently the specification includes N-bit, Main, Core, Simple, Simple Scalable, Error Resilient Simple Scalable, Advanced Coding Efficiency, Core Scalable, Advanced Core, Advanced Real Time Simple, Simple Studio, Core Studio, Advanced Simple and Fine Granularity Scalable profiles.

5.1.2.5 H.264–Advanced Video Coding (AVC)

H.264/AVC is the latest video coding standard from ITU-T and MPEG. Recommendation H.264 [118] is technically identical to MPEG-4 Advanced Video Coding (AVC) [94], also known as MPEG-4 Part 10, and they were developed together by the Joint Video Team (JVT) formed as a partnership between the ITU-T Video Coding Expert Group and the ISO/IEC Moving Picture Experts Group (MPEG). The first version of H264/AVC was published in 2003 and shortly afterwards it was amended with support for higher-fidelity video. Support for more advanced profiles and Scalable Video Coding (SVC) will be finished in 2007 and support for Multiview Video Coding (MVC) is expected in 2008.

The main objective for developing H.264/AVC was to substantially reduce the required bit rate (by a factor of 2) compared to previous standards for achieving good video quality without increasing complexity too much. The initial focus was on low bit rates and screen sizes suitable for mobile devices, but the standard is flexible and spans the whole spectrum of applications from low to high resolutions and bit rates. The standard has been adopted by an impressive range of organizations, including 3GPP for broadcast, streaming, messaging and conversational services. It is used for terrestrial and satellite broadcasts, HDTV and the new DVD formats HD-DVD and Blu-ray Disc, to name a few.

The overall architecture of H.264/AVC is similar to the other ITU-T and MPEG codecs, but a number of new features make it much more effective. It includes motion compensation with quarter-pixel accuracy and variable block sizes (from 4×4 to 16×16), an in-loop deblocking filter for reducing blocking artifacts, a bit-exact DCT-like transform that operates on 4×4 (or 8×8) blocks and advanced prediction of intra blocks. The multi-picture motion compensation can use 16 reference pictures, which increases the chances of finding good matches for motion prediction. However, as delay has a severe impact on conversational services, reference frames should only be used for forward prediction. The option to use context-adaptive binary arithmetic coding increases the compression efficiency further. It works best for coding large data units at high bit rates and is less useful for mobile applications where the bit rate is relatively low and the data is segmented into independently decodable slices for increased error robustness. See Section 5.1.2.6 for more information about slices and segmentation of video.

H.264/AVC distinguishes conceptually between a Video Coding Layer (VCL) and a Network Abstraction Layer (NAL). The VCL specifies the signal processing functionality of the codec, i.e. the decoding process that follows the hybrid video coding design. A VCL encoder outputs slices that contain slice headers and encoded macroblocks. A NAL encoder encapsulates slices from the VCL encoder into NAL units suitable for transmission.

NAL units contain a one-byte header indicating type and relative importance for the decoding process. The rest of the NAL unit is the output from the VCL. Another distinction from previous video coding standards is that H.264/AVC does not refer to time. The actual presentation time of a NAL unit is specified by the transmission layer (or storage format).

The first version of H.264/AVC included three profiles: Baseline, Main and Extended. The Baseline profile does not support arithmetic coding, B pictures or interlace, but plays a central role as it targets applications with relatively low complexity. It is the profile of choice for mobile devices and video conferencing. Later versions of H.264/AVC also include a number of higher-fidelity profiles, in particular the High profile that is used for high-definition TV and the next generation DVD formats.

All profiles have levels for various decoder capabilities in terms of buffer sizes, bit rates, screen sizes, etc. Level 1 is the lowest level. It corresponds to level 10 of H.263 and level L0 of MPEG-4 Visual by supporting bit rates up to 64 kbps and QCIF resolution at 15 frames per second. The main difference lies in the superior quality of H.264/AVC. In the same manner, level 1b corresponds to level 45 and level L0b, respectively, and was added to H.264/AVC after a request from 3GPP. Some examples of picture formats and rates for the lowest levels are given in Table 5.5.

Table 5.5: H.264 levels up to 1.3 for Baseline, Main and Extended profiles.

Level	Max bit rate [kbps]	Example picture formats and rates
1	64	sub-QCIF@30 Hz, QCIF@15 Hz
1b	128	sub-QCIF@30 Hz, QCIF@15 Hz
1.1	192	QCIF@30 Hz, QVGA@10, CIF@7.5 Hz
1.2	384	QVGA@20, CIF@15 Hz
1.3	768	CIF@30 Hz

5.1.2.6 Error Resilience

Error resilience and recovery from data loss are important for achieving robust video transmission. On the one hand, errors can be suppressed by increasing the reliability of the transport channel. The cost for increased reliability is increased overhead (bandwidth) and/or increased latency due to retransmissions. On the other hand, the effect of errors can be minimized in the media plane by using coding schemes that stop error propagation. As video compression relies heavily on prediction that utilizes correlations, error propagation is a serious problem that needs to be addressed. Several tools are available in all video coding standards that constrain error propagation and facilitate recovery from errors.

The simplest technique to stop temporal error propagation is to refresh the decoder by inserting an I (or IDR) picture and cut the prediction chain from earlier pictures. This technique introduces a trade-off between latency and quality degradation due to the relatively large sizes of I pictures compared to P pictures of comparable quality. In order to avoid latency for constant channel bandwidths, the encoder either has to reduce the quality of the I picture or skip several P pictures. Another possibility is to make a partial refresh of a picture and thereby spread the cost of intra coding over several pictures. All codecs discussed so far support segments of one kind or the other. In H.261 each frame of a video sequence is divided into a number of segments called a Group of Blocks (GOBs), where each GOB contains 33 macroblocks arranged in 3 rows by 11 columns. H.263 also uses GOBs, but here they correspond to single rows of macroblocks. Instead of refreshing the entire picture at one time instant, an encoder can distribute the cost over the sequence by refreshing GOBs or just particular macroblocks. The time it takes before the decoder has recovered depends on the refresh algorithm used by the encoder.

In order to limit spatial error propagation, different ways of segmenting a picture can be employed. A slice has no dependences on other slices in a picture, i.e. there is no prediction from one slice to another. The loss of a slice can thus be isolated to one part of the screen. Moreover, by aligning slice boundaries with transport layer boundaries, it is possible to limit the effect of the loss of one transport unit. The MPEG-1 and MPEG-2 video standards support slices that contain a variable number of consecutive macroblocks in scanning order. Slices were also introduced in the second version of H.263 and are supported by Profile 3. An advantage with slices is that they can have variable lengths and end at an arbitrary macroblock position. This makes it possible to align the slice sizes with the data units of the underlying transport protocol. It is for instance possible to optimize the usage of slices for RTP by ending a slice before the maximum payload size of an RTP packet is reached.

Slices in H.264/AVC work similarly and can also be combined with several other tools that increase error resilience. Flexible macroblock ordering, also known as slice groups, and arbitrary slice ordering make it possible to shape and reorder slices. Redundant slices provide alternative representations of slices in case the original slices are lost.

5.1.3   Text

Text communication can be categorized into three different groups depending on the offered level of interactivity:

1. Real-time text communication offers the highest level of interactivity. In this form, the receiver sees how the message is being written character-by-character. The receiver even sees if erroneous keys are pressed and if/when errors are corrected. The receiver may even have a chance to interrupt the writer before the sentence has been completed. This form of text communication can therefore be compared with (full-duplex) telephony.

2. Semi-real-time, or half-duplex, text communication offers a quite high level of interactivity, almost as high as real-time text communication. In this form, the writer constructs either a few sentences or a complete message before sending it to the receiver. The transmission time is very short, almost immediate, and the receiver can respond quite rapidly. Most chat services work this way. This form of text communication can therefore be compared with (half-duplex) walkie-talkie communication.

3. Non-real-time, or simplex, text communication offers a very low level of interactivity. This is the typical email communication where the writer constructs the complete message, which might be very long, and then sends it to the receiver. In the best case, the delivery of the email message is as fast as for chat messages but delivery times of several hours are not uncommon when the mail servers are overloaded.

Multimedia Telephony will enable both the real-time and semi-real-time versions of text communication. For semi-real-time text communication, see Section 5.2.4 for more details.

The full-duplex, real-time version of text communication is needed mainly for hearing and/or speaking disabled users who have used the text telephony service on PSTN but nothing prevents other users from using it. To give a true real-time feeling, the characters are transmitted with RTP.

It has been decided that the protocol for text conversation specified in ITU-T Recommendation T.140 [102] will be used. T.140 is used in several other text communication services such as:

- H.324 [119];

- H.245 [121];

- T.134 [103];

- V.18 [106];

- Packet Switched conversational Multimedia (PSM) [37]; and

- CTM [5].

The widespread use of T.140 should make interworking with other systems and services fairly simple.

T.140 specifies that the characters shall be encoded with ISO/IEC 10646-1 level 3 [92] since this encoding scheme has been developed to support all languages in the world, which makes the service useful on most markets. The character encoding in level 3 is a variable length coding. Most common characters (A, B, C, . . . , a, b, c, . . .) can be encoded with two octets, the so-called simple representation. To support languages that use special symbols, it is possible to combine characters with, for example, accents and diacritical marks. This gives a so-called Composite Character Sequence (CCS), which typically requires four octets even though more complex combinations may require more octets. One example of a CSS is the Swedish letter 'å' when it is encoded as 'a' combined with 'a ring above'. The fact that this character is also available in the simple representation makes the implementation somewhat complicated.

In addition, T.140 also allows for using a subset of ISO/IEC 10646-1. T.140 also includes a few editing and control functions like: new line, erase last character and alert (bell).

The delay requirements for text communication are not as strict as for voice. It is therefore possible to buffer characters for a while before transmitting them instead of transmitting the characters one-by-one. Buffering will reduce the packet rate, which also means that the overhead is reduced. It is however not possible to buffer for too long a time because this would impact the feeling of having a real-time communication. T.140 specifies that the buffering should not be longer than 0.5 s. Since the characters are entered by humans, who type with an irregular speed, this means that a variable number of characters will be encapsulated in each packet. Typically, one can expect between one and ten characters per packet. The payload format for real-time text communication and how the characters are packed in RTP packets is further discussed in Section 5.2.4.

The error handling is somewhat different for real-time text compared to the error handling for speech and video. Error concealment by copying the previous frame, possibly accompanied with a muting of the signal, works quite well for speech and video, even if one loses a few frames in a row. For text, repeated characters are very annoying and the likelihood that this will work is quite low since most languages have few occurrences of duplicated characters. It is therefore important to keep a low character error rate to ensure that most users are satisfied. Studies have however shown that users accept a slightly higher error rate provided that missing characters are marked, which is typically done with a special character. Addendum 1 to T.140 recommends that the graphical character 'white question mark in a black diamond' should be used.

Since it is a human that generates the characters, typically by pressing keys on a keyboard, the characters will be generated in an irregular stream. The consequence is that the detection of missing characters cannot rely on timers and must be assisted by the transport protocol. The error handling is therefore further discussed in Section 5.2.4.

5.2   Protocols

IMS Multimedia Telephony is a packet switched real-time communication service. All media is transported using IP which, for real-time transport, requires some specific handling in order to get real-time behavior of the service. In this section, we describe the media specific issues for real-time transport using IP.

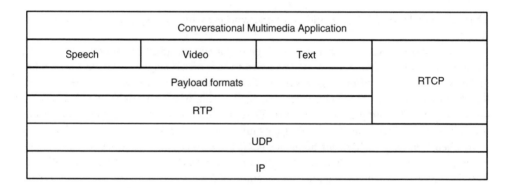

Figure 5.2: The user plane protocol stack for an IMS Multimedia Telephony terminal.

5.2.1   Real-Time Transport Protocol

All conversational media transport in IMS Multimedia Telephony uses the Real-time Transport Protocol (RTP) [177] carried over UDP/IP, see Figure 5.2. RTP provides all necessary means to correctly detect packet ordering and playback/decoder consumption control. It does not guarantee timely delivery but it gives the receiver the information needed to present the data carried, in a real-time fashion. RTP is designed for a wide variety of real-time data, hence its generic header, or fixed header, only contains one part of the necessary parameters needed for media rendering at the end-point. In a two-party communication session, e.g. a normal voice call, the generic header of RTP consists of 12 bytes of data. The respective fields of the header is described in Table 5.6, but overall, the header holds such data as RTP packet number, RTP timestamp, media source identifier, possible contributing source identifier, protocol version, payload type (Section 5.2.5), etc. This information is of generic value for all real-time data sessions, no matter which media format is used, hence the designation generic, or static, part of the RTP header. Although this part of the RTP header is common for all sessions using RTP, the interpretation of the header fields may vary depending on how the application interprets the payload type used. The payload type is a 7-bit field which identifies the format of the RTP payload and determines its interpretation of the application. One example of how RTP payload parsing and interpretation might differ between payload types is the timestamp field. The timestamp field carries information about when the first octet of the current RTP payload was sampled; hence its granularity is dependent on the sample rate. Two different payload types may have different sampling rate which forces the receiving application to correctly map the active payload type towards the correct sampling rate, otherwise the timestamp field will be interpreted erroneously. In order to correctly decode the fixed RTP header, and to enable transport of a large variety of media formats, two other parts are needed; an RTP profile and an RTP payload format.

The RTP profile specifies how to interpret the fixed RTP header in the current application. There are a number of audio and video RTP profile definitions in [176]. This profile is called RTP/AVP and is the default profile for audio and video in IMS Multimedia Telephony. There is another profile under discussion in IETF, e.g. the audio/video profile with feedback (AVPF) which adds other packet formats to the RTCP reports. One specific usage of this profile is mentioned in Section 5.8.6.

Each individual media type has its own RTP payload format and any respective configuration parameter needed for RTP payload packetization and de-packetization during the session is negotiated in the session setup using SIP/SDP. In the following sections, the RTP payload formats for each conversational media type are briefly described. There are a number of usage guidelines, recommendations and restrictions of these media types within the context for IMS Multimedia Telephony. This is treated in Section 5.8. Each respective payload format is not described in every detail; rather the information presented here is somewhat targeted towards IMS Multimedia Telephony use. For more detailed information, the reader is advised to study the respective RFC.

As shown above, RTP is used to packetize real-time media in the forward direction. However, RTP is accompanied by a control version of the protocol, the RTP Control Protocol (RTCP). RTCP is used to provide a back channel to the sender with information about the characteristics of the data transport such as packet loss rates, arrival statistics, etc. Although its basic version is generic for all RTP sessions, there are possibilities to use application specific messages.

There are five different RTCP packet types. The purpose of these messages is to provide means for the transmission of various measurement results between the receiver and the sender:

- Sender Report (SR), for transmission and reception statistics from participants that are active senders.

- Receiver Report (RR), mainly for reception statistics from participants that are not active senders.

- Source description items (SDES).

- Bye (BYE), end of session participation.

- Application specific (APP), application specific data.

Table 5.6: Fields present in the fixed part of the RTP header.

Bits	Number	Interpretation
VER	2	RTP version number. The current version is 2
P	1	Padding bit. If set to 1, the packet contains one or more padding octets at the end which do not belong to the payload. The last octet of the payload contains a count of how many octets at the end should be ignored, including itself
X	1	Extension bit. If set, an extension header follows the fixed RTP header. For more details, see Section 5.3.1 in [177]
CC	4	CRSC count. The number of CRSC identifiers that follows the fixed header. For a two-party session, this is set to zero
M	1	Marker bit. The marker bit is defined by a profile (RTP payload format). E.g. when using the AMR payload format, a set marker bit indicates that the first speech frame in the packet represents the start of a talk spurt
PTYPE	7	Payload type. Identifies the format of the RTP payload hence giving guidance to the receiver on how to interpret the RTP header fields. In IMS Multimedia Telephony, the payload type mapping to specific media parameters is done in the SDP during session negotiation
SEQ NUM	16	Sequence number for the packet. The starting number is chosen at random
TIMESTAMP	32	The timestamp field denotes the sampling instant of the first octet in the RTP packet. The granularity, or clock rate, of the timestamp field is set in the payload type definition. E.g. for audio, the granularity is often the sampling rate, which for the AMR speech codec would give a 160 tick increase for every new RTP packet if one speech frame is encapsulated in each RTP packet. The starting value is chosen at random
SYNC SOURCE ID	32	Identifies the synchronization source of the packet This is a random number unique for all contributing sources in case of a multi-party session
CONTRIB SOURCE ID	32	Used in multi-party session by an RTP stream mixer to identify the contributing sources to the data in the RTP packet. Not used in two-party sessions

The BYE and the APP message are most straightforward to understand. BYE is sent when the receiver leaves the session and APP is application dependent. The SDES message can be used to transmit general information about the user who controls the media stream source. The SR and the RR messages convey the information which holds the results of the data transport characteristics measurements. The SR holds all information that is available in the RR and adds a 20 byte field holding information about sender information.

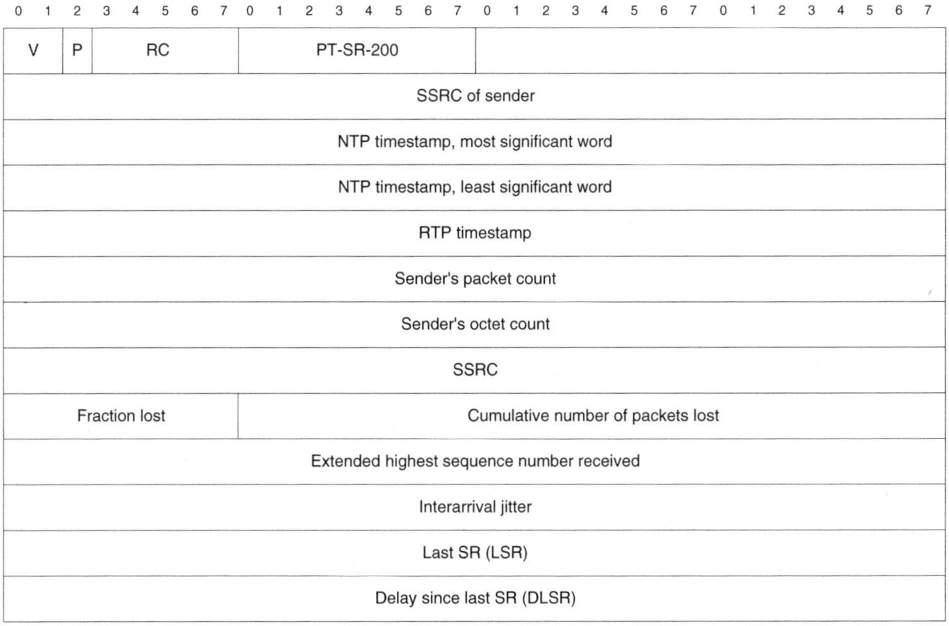

0 1 2 3 4 5 6 7	0 1 2 3 4 5 6 7	0 1 2 3 4 5 6 7	0 1 2 3 4 5 6 7
V P RC	PT-SR-200		
SSRC of sender			
NTP timestamp, most significant word			
NTP timestamp, least significant word			
RTP timestamp			
Sender's packet count			
Sender's octet count			
SSRC			
Fraction lost	Cumulative number of packets lost		
Extended highest sequence number received			
Interarrival jitter			
Last SR (LSR)			
Delay since last SR (DLSR)			

Figure 5.3: An example RTCP Sender Report for a two-party session.

We will not describe every field in the SR and RR but, as shown in Figure 5.3, both the variations in the packet delivery timing (jitter) and the packet loss rate are reported as well as information about the clock of the sender. Hence, RTCP SRs can be used both for media synchronization as well as a trigger for adaptive measures; see Section 5.4.1.

RTCP report is not mandatory to use for all RTP sessions but it is recommended for use in all sessions, especially in an IP multicast environment. The rate of report transmissions is dependent on how much bandwidth the session parties allow the RTCP traffic to consume. A general recommendation is that the RTCP traffic should consume roughly 5% of the total session bandwidth. The reporting interval is calculated according to the report content and the total session bandwidth. Details can be found in [177].

There has been further development done in terms of RTCP report content. In [79], RTCP eXtended Reports (RTCP XR) are described which add another seven report block types to the original RTCP packet formats including a specific VoIP metrics report block. At the time of writing, also so-called codec control message extensions using the AVPF RTP profile are discussed within IETF [185].

5.2.2 Speech

The speech media in IMS Multimedia Telephony is encoded and decoded using either AMR or AMR-WB depending on whether the session uses wideband speech or not. Hence, the second part of the RTP header is defined by the RTP payload format for these codecs, see Figure 5.4. The payload format is the same for both and can be found in [179].

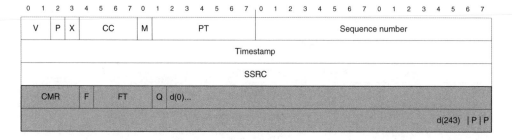

Figure 5.4: An example RTP packet when transporting AMR and AMR-WB. The actual speech data starts at d(0) and ends at d(243). The two last bits denoted P are padding bits. This example shows one AMR 12.2 kbps frame packed into one RTP packet using the bandwidth efficient version of the payload format.

The payload format for AMR holds all necessary information to enable all the features of the codec to be used in an IP environment. Together with the fixed RTP header, interpreted as defined by the payload type definition, the receiver can decode, present and also control its own encoding and transmission according to the control parameters packetized in the RTP packet. It is possible to transmit one or several speech frames in each packet. The number of frames in each packet can vary but a maximum amount of allowed frame aggregation is set during session setup.

There are two different versions of the payload format for AMR: bandwidth efficient and octet aligned. Both versions contain the same information, the same fields, see Table 5.7. The differences between these two versions are that in the case of the octet aligned version:

- the CMR field is padded to one whole octet,

- the TOC field is padded to one whole octet,

- each speech data frame is padded to whole octets before the data frames are concatenated.

No extra padding is needed in the end of the RTP packet. Note that the bits available for transmitting information are the same as for the bandwidth efficient version, only padding bits are added.

The payload specific bits that are added to the RTP header in the payload format mainly transmit encoding mode information for the respective end-points. Since AMR is an adaptive speech codec, the information transmitted in the payload specific part of the RTP header needs to be able to convey information not only about the currently used encoding rate for the transmitted frame(s), but also possible requests to the sender about change of encoding rate. The rate requests are transmitted using the CMR bits, which are a request from the sender to the receiver about what encoding rate to use in the flow going in the opposite direction.

Table 5.7: Description of the AMR specific fields when using the RTP payload format for AMR and AMR-WB.

Bits	Number	Interpretation
CMR	4	Codec mode requests. The request to the receiver of what encoding mode to use for the next encoded frame
F	1	Indicates whether another frame follows the frame to which this TOC belongs
FT	4	Frame type. Identifies what codec mode that was used to encode the current frame
Q	1	Frame quality indicator. Identifies if the frame is error-free or not. If set to zero, the frame contains bit errors. For IMS Multimedia Telephony sessions, a Q value different from 1 can only happen in interworking scenarios. If the speech data is transported end-to-end using IP there will be no bit errors in the data

The other bits, the F, FT and Q, bundled into a so-called table of content TOC field, represent information about the content in the current packet. These are unique for each frame in the RTP packet. Hence, if *n* frames are packetized into the RTP packet there will be one CMR field and *n* TOC fields.

The payload format for AMR and AMR-WB can be used in a large variety of ways and there are a number of features available that are supported by the payload format. This includes the possibility for frame interleaving, multichannel audio, etc. However, not all of these features are used in IMS Multimedia Telephony and are therefore not explained here. For a full description of the features of the payload format, see [179].

In IMS Multimedia Telephony, certain usage restrictions and guidelines apply for RTP transport of AMR encoded media. These restrictions and guidelines are used to enhance the interoperability between IMS Multimedia Telephony clients and legacy systems as well as to provide means for transport channel optimization. This is described in Section 5.8.4. For further details regarding the session negotiation of the parameters needed for a correct RTP transport of AMR encoded data, see Section 5.2.5.3.

There has been an update submitted to IETF of the AMR payload format [178] which is undergoing the last stages of standardization at the time of writing. The changes and updates include a new offer/answer section to clarify the proper procedure at session negotiation as well as various clarifications. Two new media type parameters have also been introduced: mode-change-capability to indicate if the client is able to transmit with a restricted mode-change period and max-red which indicates the maximum amount of milliseconds between the original and the redundant frame, if redundant transmission is used.

5.2.3 Video

Each of the video codecs used in the IMS Multimedia Telephony service has a payload format of its own. H.263 uses RFC 2429 [52], MPEG-4 Visual uses RFC 3016 [130] and H.264/AVC uses the payload format defined in RFC 3984 [186]. They have in common that each RTP packet consists of the fixed RTP header and possibly extra media-type specific headers followed by or interleaved with the video bitstream. With a few exceptions, RTP packets should not contain video data from different frames, i.e. the Timestamp of the fixed RTP header (see Table 5.6) shall apply to the entire packet. It is the RTP Timestamp that defines the presentation time of the data contained in the packet. The Marker bit (M bit) of the fixed RTP header is set to one when a packet includes the end of the current frame and to zero otherwise. This makes it possible to send a frame to the decoder before a packet starting on a new frame arrives.

In the following subsections, we give a short introduction to each format.

5.2.3.1   H.263

The payload format defined for H.263 in RFC 2429 is a replacement of RFC 2190 used for the first version of H.263. There is also an updated version of the payload format under development that eventually will replace RFC 2429 (as a new RFC).

Each RTP packet consists of the fixed RTP header plus a variable-length H.263 payload header followed by an H.263 bitstream. The required fields of the H.263 payload header are described in Table 5.8. As no optional fields are supported by the profiles defined for H.263, these are also the only fields for consideration in the Multimedia Telephony service.

Table 5.8: Description of the required H.263 payload header fields.

Bits	Number	Interpretation
RR	5	Reserved bits. Shall be zero and ignored by receivers
P	1	Indicates the picture start or a picture segment (GOB/slice) start or a video sequence end. Two initial zero bytes in the payload have been omitted if it is set to one
V	1	Indicates the presence of an 8 bit field after the PEBIT field containing information for Video Redundancy Coding. This feature is not supported by any of the H.263 profiles
PLEN	6	Length in bytes of the extra picture header. If no extra picture header is attached, PLEN is zero
PEBIT	3	Indicates the number of bits that shall be ignored in the last byte of the picture header

As slices can be of variable length in non-rectangular slice structured mode, only complete slices should be included in a packet, i.e. slices should not be fragmented over packet boundaries.

The same Timestamp on successive packets indicates that one frame occupies more than one RTP packet. The RTP Timestamp is expressed in units based on a 90 kHz clock. As both the RTP header and the H.263 picture header in the video bitstream carry timing information they should be synchronized. The default is to follow the RTP Timestamp and ignore timing information in the bitstream.

The H.263 bitstream can be used directly in the RTP payload with the exception that the two initial zero octets of a byte-aligned start code are excluded if it occurs at the beginning of a packet. When this happens, the P bit in the H.263 payload header shall be set to one.

In order to increase error resilience it is possible to include a picture header in a packet that does not contain the start of a picture. This makes it possible to decode a slice in a packet, even if the original picture header contained in a previous packet was lost.

5.2.3.2 MPEG-4 Visual

MPEG-4 Visual uses the RTP payload format defined in RFC 3016. RFC 3640 is another payload format for MPEG-4, but it is not used for the Multimedia Telephony service. RFC 3016 can be used when MPEG-4 Audio or Visual streams are used as separate media streams and not as part of MPEG-4 Systems. Note that the usage of MPEG-4 Visual does not imply support for MPEG-4 Systems, as IMS uses SIP and SDP for stream management. In particular, MIME media types and not MPEG System object descriptors are used for negotiation in SDP. If the MPEG-4 Visual bitstream uses short video header mode, which is compatible with H.263 baseline of 1995, then RFC 2429 should be used instead of RFC 3016.

No media-specific RTP header is defined for this payload format and the fixed RTP header is directly followed by an MPEG-4 Visual bitstream. In general it is recommended not to encapsulate more than one picture in a packet, such that the RTP Timestamp applies uniquely to the payload. For more details and packetization guidelines, see RFC 3016 [130].

5.2.3.3 H.264/AVC

H.264/AVC uses the RTP payload format defined in RFC 3984, which specifies RTP encapsulation of H.264/AVC Network Abstraction Layer (NAL) units. The general principle is the same as for H.263 and MPEG-4 Visual, although the specific details of H.264/AVC make it rather complex.

Each RTP packet starts with the fixed RTP header. The rest of the packet is formatted in different ways depending on the packetization mode. For single NAL unit mode, the NAL unit header serves as RTP header and the payload consists of the rest of the NAL unit. For non-interleaved mode several NAL units can be sent in decoding order in one packet. These two modes can be used for video telephony and in particular Multimedia Telephony. The third mode, the interleaved mode, is not suitable for systems with requirements on latency as NAL units can be sent out of decoding order. It is not supported by the Multimedia Telephony service.

H.264/AVC uses so-called parameter sets for information that applies to many NAL units. A sequence parameter set applies to an entire sequence and a picture parameter set to a picture. By sending parameter sets out of band, e.g. in SDP, it is possible to reduce the overhead in the RTP stream while making it more resilient to losses. There is for instance no need to send extra picture headers as described in Section 5.2.3.1 for H.263.

5.2.4 Text

The payload format for real-time text is specified in RFC 4103 [88]. The payload format specifies that it should only be used for communications between humans. The reason is that IP, UDP and RTP are unreliable protocols and humans are intelligent and can therefore quite easily detect errors. It is therefore also assumed that the humans will ask for retransmissions if there are so many errors that it is not possible to understand the message. Other, more reliable, protocols are available for text communication between machines.

The text packed into RTP packets using this payload format consists of one or more so-called *T140blocks*. One T140block may correspond to one character, also called simple representation, which is typically encoded with two octets. One T140block may also consist of a Composite Character Sequence (CCS), which typically requires four or more octets. The payload format defines that:

- the RTP Timestamp is incremented with a clock frequency of 1000 Hz;

- the RTP Sequence Number is incremented by one for each transmitted packet;

- the M bit is set for the first packet in the session and the first packet after an idle period;

- a T140block, and especially not CCS characters, should not be segmented over several RTP packets;

- redundancy is allowed, or even recommended, and RFC 2198 should be used;

- if redundancy is used, then it is required that the redundant T140blocks are identical to the original T140blocks.

As stated in Section 5.1.3, error handling is somewhat special for real-time text. Since packets are not generated at regular intervals, the RTP Timestamp is not sufficient to detect if packets have been lost. The receiver should instead use the RTP Sequence Number.

Some further recommendations for the receiver behavior are also included in the payload format. The receiver should:

- reorder text received out of order;

- mark where text is missing because of packet loss;

- compensate for lost packets by using redundant data.

The payload format also gives guidelines for how to protect the media against packet losses. It defines that RFC 2198 [159] must be used unless another method, for example RFC 2733 [169], is explicitly selected. To handle even severe amounts of packet losses, it further defines that 200% redundancy should be used, i.e. each piece of original text, should be repeated twice in addition to the original transmission.

The use of T.140 is one option for text communication in IMS Multimedia Telephony. However, there are other options available. The basic services for IMS allow a user to start, modify and terminate a session using SIP which implicitly allows the exchange of simple SIP messages. However, this functionality has been further extended to support messaging within IMS Multimedia Telephony.

As stated in [32], two different messaging functionalities are available for IMS: immediate messaging, or page-mode messaging, and session-based messaging. The immediate messaging uses SIP MESSAGE to convey the message. The other messaging type, session-based messaging, uses the Message Session Relay Protocol (MSRP) [64] to transport the message. The main difference between these two message schemes lies in the way that the session is conducted. Messaging schemes which keep track of each individual message as a separate entity are called immediate, or page-mode, messaging while messaging that is a part of a session with a definite start and end is called session-mode, or session-based, messaging. In the following, we will focus the discussion on session-based messaging using MSRP. It is up to the application to decide which messaging scheme to use.

A session which includes session-based messaging negotiates the use of MSRP in the SDP negotiation. Here, the media type of the MSRP session is specified as well as the maximum size of each message. Note that MSRP is not a stand-alone protocol – it requires a rendezvous mechanism meeting certain requirements. In IMS Multimedia Telephony it is aimed to be used with SIP/SDP which is the main candidate for enabling MSRP usage.

Table 5.9: A SIP invite to an MSRP session.

```
INVITE sip:oskar@lulea.example.com SIP/2.0
To: <sip:olle@gammelstad.example.com>
From: <sip:oskar@lulea.example.com>;tag=786
Call-ID: 3413an89KU
Content-Type: application/sdp
c=IN IP4 lulea.example.com
m=message 7654 TCP/MSRP *
a=accept-types:text/plain
a=path:msrp://lulea.example.com:7654/jshA7we;tcp
```

MSRP can exchange arbitrary binary content including text, video and audio and it has two methods; SEND and REPORT. A short text message example is shown in Table 5.9.

Table 5.10: Oskar uses MSRP SEND to send Olle a message. Olle acknowledges the receipt.

```
MSRP a786hjs2 SEND
To-Path: msrp://gammelstad.example.com:12763/kjhd37s2s2;tcp
From-Path: msrp://lulea.example.com:7654/jshA7we;tcp
Message-ID: 87652
Byte-Range: 1-25/25
Content-Type: text/plain

Hi, have you seen Johan's new bike?
-------a786hjs2\$

MSRP a786hjs2 200 OK
To-Path: msrp://lulea.example.com:7654/jshA7we;tcp
From-Path: msrp://gammelstad.example.com:12763/kjhd37s2s2;tcp
Byte-Range: 1-25/25
-------a786hjs2\$
```

In the example in Table 5.10, Oskar wants to send Olle a short text message which can fit into one message. However, if it had been a very large message, the message can be split into several chunks. The message ID corresponds to the complete message which the receiver can use to assemble the chunks into the complete message. The Byte-Range header field denotes the data contained in the current chunk as well as the total size of the message.

The use of MSRP for session-based messaging adds a semi-real-time text communication functionality to IMS Multimedia Telephony. Note that MSRP can also be used for video, audio and image clip sharing; service components within the scope of IMS Multimedia Telephony.

5.2.5   SDP

The end-points need to agree on what media to use in the session and how the media should be encoded. Depending on the media and the encoding, the end-points may also need to agree on a number of session parameters. This section describes how the Session Description Protocol (SDP) [84] is used to configure the media.

5.2.5.1   General SDP Usage

For Multimedia Telephony, the media types that can be used for real-time communication are audio, video or text. The clients also need to define what UDP port to send the media type to. If several media types are used, then one should use one UDP port for each medium.

The participants also need to agree on the transport protocol that will be used for the media. For real-time media in Multimedia Telephony, RTP will be used. The RTP profile to use is the RTP Profile for Audio and Video Conferences with Minimal Control (RTP/AVP) as defined in RFC 3551 [176], see also Section 5.2.1.

In addition to the profile, the participants also need to agree on what payload formats to use for each media type. The payload format is tightly coupled with the codec that will be used for the media. There is typically also a need to include a set of attributes. The attributes can either be specific for one payload format or be common for the media type.

For each media type there is also a need to define at least one RTP Payload Type number. If the media can be encoded or if the media can be packetized in several ways then each configuration needs its own RTP Payload Type number. This number will then be inserted in every RTP packet header and is used to identify what configuration the current packet uses.

An example SDP for audio, video and text is shown in Table 5.11.

Table 5.11: Example of an SDP message.

```
m=audio 49152 RTP/AVP 98 97
a=rtpmap:98 AMR-WB/16000/1
a=fmtp:98 mode-change-capability=2; max-red=160
a=rtpmap:97 AMR/8000/1
a=fmtp:97 mode-change-capability=2; max-red=160
a=ptime:20
a=maxptime:240
m=video 49154 RTP/AVP 99
a=rtpmap:99 H263-2000/90000
a=fmtp:99 profile=0;level=45
m=text 49156 RTP/AVP 100 101
a=rtpmap:100 t140/1000
a=rtpmap:101 red/1000
a=fmtp:101 100/100/100
```

The m= lines indicate what media types are to be transmitted, audio, video or text. The numbers following the media types are the respective UDP port numbers, 49152, 49154 and 49156. The use of the RTP protocol and the AVP profile are defined with RTP/AVP. The list of RTP Payload Types is included in the end of the media line, 97 and 98 for audio, 99 for video and 100 and 101 for text respectively.

Since RTP and RTCP share a UDP port number pair, an even number for media and the next higher odd number for RTCP for respective media, the defined UDP port numbers for the media also implicitly define that the UDP port numbers 49153, 49155 and 49157 will be used for RTCP for audio, video and text media respectively. Even though the clients should typically select UDP port numbers from the Dynamic Port range, 49152 to 65535, they should also handle port numbers selected from the Registered Port range, 1024 to 49151.

The lines beginning with a= define a set of attributes for respective media. The lines starting with a=rtpmap associate the defined RTP Payload Type number with a codec, the codecs sampling frequency and the number of channels that the media will use. For example: a=rtpmap:98 AMR-WB/16000/1 defines that for the RTP Payload Type number 98:

- the codec will be the AMR-WB codec, which also implicitly defines that the RTP payload format for AMR and AMR-WB will be used;

- the sampling frequency will be 16 kHz;

- there will be one audio channel, i.e. mono.

This example also shows that it is possible to define several codecs for the same media type. In this case it is possible to use both AMR and AMR-WB for audio media. It is even possible to define that several configurations that use the same codec can be used. Different RTP Payload Type numbers are selected for different codecs and different configurations to make it possible to distinguish between different media configurations arriving at the same UDP port. Some codecs have been assigned static RTP Payload Type numbers in RFC 3551 [176], but for most recently defined codecs one needs to use dynamic RTP Payload Type numbers. Although it is common to select Payload Type numbers in the 96–127 range, clients need to be capable of using any number in the whole range.

The lines starting with a=fmtp defines additional parameters for the defined payload type. These parameters are defined in the respective payload format specification.

The line a=ptime is common for all audio media and applies to all payload types used for audio. For most codecs and configurations, the lengths of the payloads are self-explanatory; this attribute is used mainly to describe how much data the clients want to receive. It is important that clients do not rely solely on the ptime parameter. Clients should be able to decode the received RTP packets even if they contain more or less data than what the ptime parameter defines.

The line a=maxptime is also common for all audio media but is currently only defined for AMR and AMR-WB encoded media. This parameter is used to define the upper limit of how much data can be encapsulated on one packet. The parameter is typically derived from the receiving client's memory capabilities but could also depend on other capabilities, for example limitations in transport functions.

The configuration parameters may be valid in the transmitting direction, in the receiving direction or be bi-directional as described by the corresponding payload format.

The media attribute lines may be specified in any order, but it is practical to specify them in some logical order, for example the same order as the RTP Payload Types are listed in the m= line. One common modification is to specify the ptime and maxptime attributes directly after the media line.

Many types of media and many configurations for each medium may give a quite large SDP message. In the case where this occurs, it is possible to divide the session negotiation into two or more phases. This approach is a compromise between keeping the

SDP messages small, to reduce the setup time for the first phase, and being forced to use multiple 'ping-pongs' between the end-points, which would increase the setup time. The best compromise is to include the most probable configurations in the first phase and save the less likely configurations for the second, or subsequent, phase(s).

5.2.5.2 SDP Offers and SDP Answers

For 3GPP networks, the SDP offer–answer model defined in RFC 3264 [170] is used to reach a common agreement on how the media should be encoded and how the media packets should be configured.

One fundamental idea with the SDP offer–answer model is that the client that initiates a session should construct an SDP offer that includes all the media types that it wants to use in the session and all the capabilities that the initiating client can support for each media type. One RTP Payload Type is allocated for each configuration that is supported. This SDP offer is then included in a SIP INVITE request that is transmitted to the called party. An example of an SDP offer is shown in Table 5.12.

Table 5.12: Example SDP offer for voice.

```
m=audio 49152 RTP/AVP 98 97
a=rtpmap:98 AMR-WB/16000/1
a=fmtp:98 mode-change-capability=2; max-red=160
a=rtpmap:97 AMR/8000/1
a=fmtp:97 mode-change-capability=2; max-red=160
a=ptime:20
a=maxptime:240
```

In this case, the offerer can use both AMR-WB and AMR with AMR-WB as the preferred codec. It can also support restrictions in mode changes. In the receiving direction, it prefers to receive one frame per packet but can handle up to 12 frames per packet.

Upon receiving the SDP offer, the answerer will then construct an SDP answer. The SDP answer is typically a subset of the SDP offer and is constructed by removing the RTP Payload Types that the answerer does not support and copying the RTP Payload Types from the SDP offer for the configurations that are supported. For some parameters, but not for all, the answerer may change or add configuration parameters to indicate that it can only support a subset of the configurations that the offerer supports. The SDP answer is then included in a SIP response, e.g. a SIP 200 OK message or a SIP 183 Session Progress message, that is transmitted back to the initiating client; see Section 4.4. An example of an SDP answer is shown in Table 5.13.

The SDP offer may also be included in other SIP requests, such as a SIP UPDATE message, for example when modifying an ongoing session.

In this case, the client has no support for the AMR-WB so the RTP Payload Type 98 is removed. It also needs to introduce restrictions both in what codec modes can be used and how mode changes can be performed. The line a=maxptime:80 shows that the answerer is only capable of receiving up 80 ms of speech, i.e. four speech frames, per packet.

The backslash character \ is used at the end of the line in this example only to show that the line folds over several lines. The backslash and the end-of-the-line characters will be ignored in a real implementation.

Table 5.13: Example SDP answer to the SDP offer Table 5.12.

```
m=audio 49152 RTP/AVP 97
a=rtpmap:97 AMR/8000/1
a=fmtp:97 mode-set=0,2,4,7; mode-change-period=2, mode-change-neighbor=1; \
   mode-change-capability=2; max-red=0
a=ptime:20
a=maxptime:80
```

When the initiating client receives the SIP response message with the SDP answer, it knows how it must configure the media stream in order to match the capabilities of the answering client.

A number of SDP offer–answer examples are included in Section 5.8.7. They were derived from different scenarios and highlight the different variants that the clients and media gateways need to handle.

5.2.5.3 SDP Offer–Answer for Voice-Only Sessions

Many factors determine how the SDP offers and answers will be constructed for voice-only sessions. Multimedia telephony clients are required to support the AMR codec for narrowband voice and the AMR-WB codec if the client supports wideband voice.

The client is required to support all codec modes for these codecs but a media gateway may support only a subset of the codec modes. Even if the media gateway itself supports all codec modes, it typically needs to restrict the allowed codec modes to a subset when establishing a session with a GERAN or UTRAN mobile and when Tandem Free Operation (TFO) is used. If all codec modes are supported, then this should be shown by not defining the `mode-set` parameter. For circuit switched GERAN and UTRAN, 16 preferred codec mode subsets are defined in 3GPP TS 28.062 Table 7.11.3.1.3-2 [12] for AMR and six different codec mode subsets are defined in 3GPP TS 26.103 Table 5.7-1 [45] for AMR-WB. When constructing an SDP offer, the media gateway should define one RTP Payload Type number for each subset that can be used in the session. Since none of the subsets included in 3GPP TS 28.062 or 3GPP TS 26.103 includes all codec modes, the consequence is that the SDP will become quite large. A possible solution for this would be to mandate that all media gateways shall support one at least one codec mode set for AMR and AMR-WB each.

In addition to the restrictions in codec mode subsets, a media gateway towards GERAN will need to restrict codec mode changes to every other frame because the in-band signaling in GSM-AMR is restricted to codec mode changes in this manner. Even though the signaling in UTRAN can handle codec mode changes at every frame border, it is likely that some manufacturer has introduced the same restrictions to simplify interworking towards GERAN. One should therefore expect that the media gateway includes `mode-change-period` and `mode-change-neighbor` in SDP offers and answers.

The AMR payload format defined in RFC 3267 [179] allows for using two different versions of the payload format for real-time communication. The two versions are the bandwidth-efficient and the octet-aligned payload formats. If both variants are allowed in the session, then the SDP offer needs to include one RTP Payload Type for each variant.

The access type, HSPA, EDGE or other, may also have an impact on how the SDP offers and answers are constructed; see also Section 5.7.1. The radio bearers in HSPA have been optimized for one frame per packet. Both the clients and the media gateways should therefore define that they want to receive this packetization when operating on this access type. For EDGE, discussions are currently ongoing, but it seems likely that two frames per packet is preferred on this access type to reduce the overhead imposed by lower protocol layers, i.e. the LLC and SNDCP layers. For other access types, for example Wireless LAN (WLAN), it may be beneficial to encapsulate more frames per packet, especially for access types that are packet rate limited. For the case when the access type is not known, the configurations used for HSPA should be used.

5.2.5.4 SDP Offer–Answer for Video

A client may support several video codecs and may therefore include more than one codec and corresponding profiles and levels in its SDP offer. An example is shown in Table 5.14 where two options associated with the RTP Payload Type numbers 99 and 100 are offered.

Table 5.14: Example SDP offer for H.263 and MPEG-4 Part 2 video.

```
m=video 49154 RTP/AVP 99 100
a=rtpmap:99 H263-2000/90000
a=fmtp:99 profile=0;level=45
a=rtpmap:100 MP4V-ES/90000
a=fmtp:100 profile-level-id=9;
          config=000001b009000001b5090000010000000120008845d4c282c2090a28f
```

The first offer includes ITU-T Recommendation H.263 [120] Profile 0 (Baseline) at level 45, which supports bit rates up to 128 kbps and maximum QCIF picture formats at 15 Hz (see Table 5.4 in Section 5.1.2.4). The second offer is MPEG-4 Visual (Part 2) [93] Simple profile at level L0b, which also supports bit rates up to 128 kbps and QCIF at 15 Hz. Here `profile-level-id=9` represents Simple profile at level L0b and may be used for negotiation, whereas the `config` parameter gives the configuration of the MPEG-4 Visual bit stream and is not used for negotiation.

An example SDP answer to the offer in Table 5.14 is given in Table 5.15. It includes only the H.263 codec where the responding client has restricted the level to 10, i.e. bit rates up to 64 kbps. The offerer should not have a problem with a reduced bit rate as support for level 45 implies the support of level 10 as well.

Table 5.15: Example SDP answer to the SDP offer in Table 5.14.

```
m=video 49154 RTP/AVP 99
a=rtpmap:99 H263-2000/90000
a=fmtp:99 profile=0;level=10
```

Another example of an SDP offer for video is given Table 5.16.

The first (preferred) offer is now H.264/AVC [118, 94]. The `packetization mode` parameter indicates single NAL unit mode. This is the default mode and it is therefore not necessary to include this parameter (see RFC 3984). The `profile-level-id` parameter indicates Baseline profile at level 1, which supports bit rates up to 64 kbps (see Table 5.5 in Section 5.1.2.5). It also indicates, by using so-called constraint-set flags, that the bitstream

Table 5.16: Example SDP offer for H.264/AVC and H.263 video.

```
m=video 49154 RTP/AVP 99 100
a=rtpmap:99 H264/90000
a=fmtp:99 packetization-mode=0;profile-level-id=42e00a;
          sprop-parameter-sets=J0LgCpWgsToB/UA=,KM4Gag==
a=rtpmap:100 H263-2000/90000
a=fmtp:100 profile=0;level=45
```

can be decoded by any Baseline, Main or Extended profile decoder. The third parameter, `sprop-parameter-sets`, includes base-64 encoded sequence and picture parameter set NAL units that are referred by the video bitstream. The sequence parameter set used here includes syntax that specifies the number of reordered frames to be zero so that latency can be minimized. The second offer in the SDP in Table 5.16 is H.263 [120] Profile 0 (Baseline) at level 10. It is used here as a fallback in case the other client does not support H.264/AVC.

An example SDP answer to the offer in Table 5.16 is given in Table 5.17. The responding client is capable of using H.264/AVC and has therefore removed the fallback offer H.263. As the offer already indicated the lowest level (level 1) of H.264/AVC as well as the minimum constraint set, there is no room for further negotiation of profiles and levels. It is possible, however, to specify upper limits of bandwidths by using the b=AS and b=TIAS fields of SDP, although we have omitted these fields in the above examples.

Table 5.17: Example SDP answer to the SDP offer in Table 5.16.

```
m=video 49154 RTP/AVP 99
a=rtpmap:99 H264/90000
a=fmtp:99 packetization-mode=0;profile-level-id=42e00a;
          sprop-parameter-sets=J0LgCpWgsToB/UA=,KM4Gag==
```

5.2.5.5 SDP Offer–Answer for Text

Since there is basically one option for real-time text, i.e. to follow ITU-T Recommendation T.140, the construction of SDP offers and answers are fairly straightforward. The only choice is if one wants to use redundancy or not.

5.3 Media Transport Processing

In this section, media transport processing functions will be described.

5.3.1 Definition

Although most functions needed when capturing, processing and delivering real-time multimedia content are the same, or at least similar, in both Circuit Switched (CS) and Packet Switched (PS) transport, there are a few areas where different functionalities are needed. The functions used for handling the media flow consequences of the actual delivery systems are called media transport processing functions. For PS transport, the two major media transport processing algorithms can be summarized as:

- RTP payload packetization and de-packetization,

- jitter management algorithms.

The actual RTP packetization is not regarded as a media transport processing function since it is not a function which is needed to handle the characteristics of the transport but rather a prerequisite to use PS transport. However, the RTP packetization does have built-in properties which can be used under certain circumstances to facilitate the actual transport and give the receiving terminal better possibilities to keep, or enhance, the experienced media quality. Hence, the RTP packetization scheme is treated in several sections (see Sections 5.2 and 5.3.4), described both as a PS transport prerequisite and as a media transport processing function.

Jitter management algorithms are receiver functions which are needed for all PS transport scenarios, not only real-time services. Even non-real-time services such a streaming do require a receiver buffer since both voice, audio and video decoders need to be fed with encoded frames at a steady rate in order to be able to decode and present the media stream in a smooth way. For streaming services however, where no tight real-time demands exist, this can be achieved with a basic solution using a large buffer without any advanced schemes to control the fill level of the buffer during run-time. This is not the case for real-time services with high demands on interactivity and responsiveness. In such services, it is important to have real-time functionality in all parts of the transport chain, see Figure 5.5.

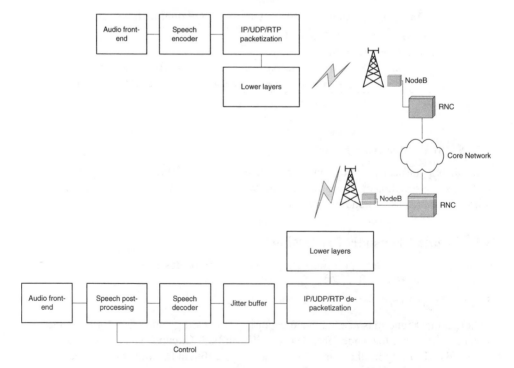

Figure 5.5: Speech media flow in UTRAN PS transport.

Media transport functions are needed for all media formats, although the role of the respective function might differ or have different requirements depending on what media the function is designed to handle. In the following sections, voice and video will be treated in separate sections although the overall functions are the same.

5.3.2 Jitter as a Characteristic of PS transport

In any real-time service, one of the key service aspects is the interactive quality and the experienced responsiveness of the service. The goal for any real-time communication service is to produce a service experience which gives the users a sense of immediacy and nearness; turning a long geographical distance into a short communication path. To what extent a service is able to fulfill that ambition is measured as the conversational quality. Measuring the conversational quality of a service usually requires extensive conversational tests using real users in a controlled environment. There are, however, some objective indications that can be used to get an idea of what conversational quality potential the service will have. These indications are intrinsic media quality and end-to-end delay.

Intrinsic media quality denotes the quality produced by the media codecs. It can be measured by subjective listening/viewing tests and does not take into account any interactivity factor. End-to-end delay, on the other hand, does not take into account any media quality issues, it only comprises the latency built in to the system and hence tries to map to what extent the real-time communication service deviates from a face-to-face conversation.

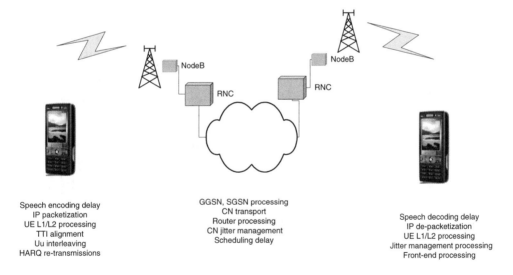

Speech encoding delay	GGSN, SGSN processing	
IP packetization	CN transport	Speech decoding delay
UE L1/L2 processing	Router processing	IP de-packetization
TTI alignment	CN jitter management	UE L1/L2 processing
Uu interleaving	Scheduling delay	Jitter management processing
HARQ re-transmissions		Front-end processing

Figure 5.6: The constituents of end-to-end delay in an HSPA system.

End-to-end delay consists of two parts: a *static* transmission delay and a *variable* delay due to the current situation in the network, see Figure 5.6.

- The static delay is not something that will affect in what way the UEs process the media. It is always present and will seldom change, at least not during an ongoing session.

- The variable delay on the other hand can change often during a session and must be taken into account when designing the UE media transport processing functions. It is this variable delay that is called jitter, or delay jitter:

$$\text{jitter} = \text{short-term variations in the packet receival rate.} \qquad (5.1)$$

Since jitter is an entity which is defined as a variation, actual jitter characteristics can vary significantly, even within one session (Figure 5.7). In HSPA systems, jitter is an inherent part of the system due to the characteristics of a shared channel with fast lower layer retransmissions. Hence, even users with good radio conditions located in cells with low load will experience jitter. The jitter in this situation will be well confined with a rather small amplitude and can be handled in the UE quite straightforwardly. If the same users move into bad radio environments or reside in a congested cell, the jitter amplitude will grow as well as the variation of the jitter amplitude. The situation then becomes more challenging for the UE.

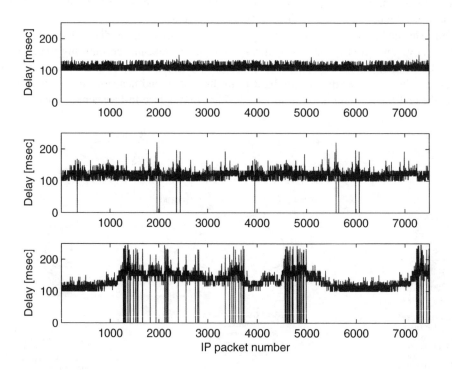

Figure 5.7: Examples of different kinds of jitter. The drops down to zero indicate packet losses. These plots represent actual simulations for an HSPA scenario; E-DCH in the uplink and HS-DSCH in the downlink using a delay sensitive scheduling algorithm.

In some circumstances, e.g. where lower layer retransmissions are slow or where a network congestion situation occurs, another kind of jitter behavior can be observed, jitter spikes. A jitter spike is defined by a sudden dramatic increase of end-to-end delay and the reception of media is totally halted for a brief period of time, typically 100–400 ms in mobile networks. A jitter spike will almost always introduce some sort of loss since the dimensioning

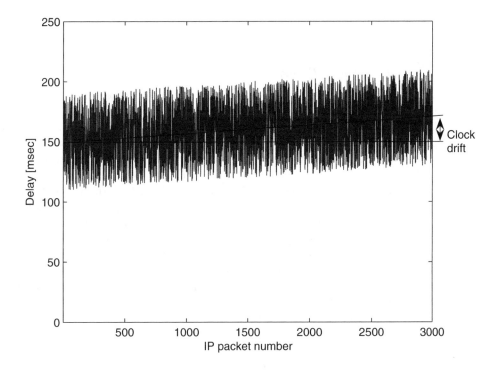

Figure 5.8: Clock-skew manifestation in a jitter profile.

of the jitter management algorithm will in most cases not have a buffer depth large enough to cope with jitter spikes. However, if jitter spikes are a more or less normal characteristic of the transmission channel, there are ways to minimize the effect of the jitter spikes on the conversational quality.

If the session uses a transmission channel where the amount of jitter varies at a rate which is comparable with the packet rate of the sender and if the network allows out-of-order delivery, packets can be delivered out of order. The media decoder cannot consume media packets delivered out of order so any jitter management algorithm used in such circumstances must have the possibility to resort the de-packetized frames so that the decoder can consume them in the correct order.

Although jitter is defined in Equation (5.1) as a short-term variation in packet delivery timing, a long-term end-to-end delay variation can also occur. It cannot be categorized as a static delay since it will slowly change the end-to-end delay during a session, but at the same time, it does not contribute to the short-term variation in packet delivery timing. Instead, it will slowly affect the end-to-end delay during a session and at some stage will affect the way the jitter management function will need to operate. Clock skew between the two media flow terminating end-points is a good example of the source of such behavior (Figure 5.8). Clock skew will appear if the clocks of the respective media flow end-points are not synchronized. The rate of which the media is encoded, processed and transmitted will in that case not be the same as the decoding and rendering rate. Even if the synchronization skew is small, a small difference will eventually build up and in the end either cause buffer under-flow or buffer over-flow. This kind of behavior must be handled in any client where synchronization cannot be guaranteed between the media flow end-points. A well designed jitter management algorithm can solve such a task.

5.3.3   Speech Transport Processing – Jitter

In this section, several different aspects of jitter are discussed in the scope of the speech media component. Since the speech codecs standardized for IMS Multimedia Telephony are AMR and AMR-WB as described in Section 5.1.1.3, they are assumed to be the speech codecs used when applying the jitter management schemes discussed. However, the jitter management schemes are not uniquely tied to these codecs and they may be used in other circumstances equally well.

In order to be able to handle jitter, the UE needs to know what to expect in terms of the characteristics of an undisturbed packet flow as well as a packet flow which has been affected by the characteristics of the current transport channel. For speech, the packets are transmitted at an even rate. In the case of one speech frame per packet, there are 50 packets per second when the speech encoder is in its active mode. AMR and AMR-WB (Section 5.1.1.3) use so-called source controlled rate operation, or discontinuous transmission (DTX), in order to reduce the bit rate when the input to the speech encoder does not provide active speech but silence, background noise, babble, etc. In that case, the packet rate is reduced by a factor of 8 transmitting only one so-called SID update frame and then waiting for another 140 ms before the next SID update frame is transmitted. A SID update frame updates the decoder noise synthesis which is used in DTX mode. It is not uncommon that only about half of the time the speech encoder is operating during a session consists of active speech, the rest of the time the encoder is in its non-active state; it uses DTX. Any jitter management scheme needs to be aware of the use of DTX in order to estimate the jitter and the packet loss rate in a correct way.

5.3.3.1   Measuring Jitter

Jitter is most often measured by using two kinds of timing references: the arrival time of the packet and the presentation time as written in the RTP timestamp field. Using these two entities, two different measures can be computed: the inter-arrival variance (5.2) and the variance in the expected arrival time (5.3):

$$\text{jitter}_{(i,i-1)} = (RT_{(i)} - RT_{(i-1)}) * f_s - f_l \tag{5.2}$$

$$\text{jitter}_{(i,1)} = (RT_{(i)} - RT_{(1)}) * f_s - (TS_{(i)} - TS_{(1)}) \tag{5.3}$$

Here, RT is the arrival time of the packet to the UE, f_s is the sampling frequency, f_l is the frame length and TS is the timestamp decoded from the RTP header. Both jitter measures are expressed in units of samples. In equation (5.3), the first packet is used as a reference. These two definitions will both provide means to estimate the jitter, where the first one operates solely on receiver measurements while the second method requires access to the RTP header data as set by the sender. Note that the timestamp field in the RTP header is written using the granularity specified in the SDP at session setup. If the payload type specified for the conversational media is audio, it is common to set the clock rate of the timestamp field equal to the sampling frequency. For a session using AMR, the RTP clock rate would be 8000 Hz making the granularity of the RTP timestamp field equal to 0.125 ms.

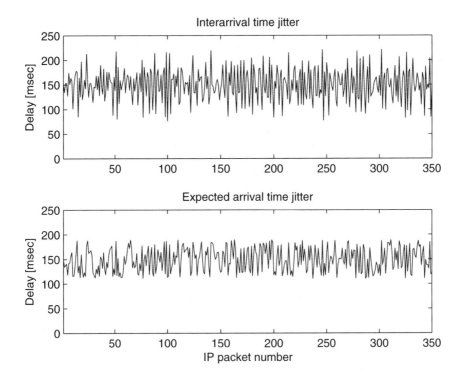

Figure 5.9: The plots are based on the same data with the jitter estimated using equation (5.2) in the upper plot and using equation (5.3) in the lower plot.

Both these estimations will give an estimated packet delivery timing variance which can be used in further media transport processing but they do so in a somewhat different way. As seen in Figure 5.9, the actual measured jitter amplitude differs for a given packet depending on what jitter measure is used. The measure in equation (5.2) only takes into account the current delivery rate variation and compares it with the undisturbed rate. The second measure in equation (5.3) compares both the reception time and the presentation time set by the sender with a reference packet. The second method will therefore mediate local jitter variations with the initial conditions and make the amplitude somewhat smaller.

Due to its long-term comparative nature, the measure in equation (5.3) will not only let you know short-term variations in the delivery timing, but also long-term variations can be measured as caused by e.g. clock skew.

The accuracy of the jitter measurement algorithm is not the only factor that will decide the performance of the jitter management scheme. The mapping between the measurement and the desired target depth of the jitter buffer is the key feature. Hence, it is not sufficient to judge the accuracy of a measurement algorithm by itself, it needs to be viewed in the light of the total control algorithm for the jitter management scheme.

Although jitter is a measure of a variation, the jitter amplitude span is often rather static. In the examples shown in Figure 5.9, the jitter is confined between ± 80 ms with a rather uniform distribution in between. If the jitter suddenly deviates from a so-far static, confined range, a jitter spike occurs. For systems designed for in-order delivery, the amplitude of a negative jitter spike is confined to the presentation time difference between two concurrent speech frames while the amplitude of a positive jitter spike is only limited by any time-to-leave parameters present in any node in the delivery chain for the media. As described in Section 5.3.2, the jitter spike appears suddenly and can be quite challenging for the jitter management algorithm.

For a system where the UEs do not have any knowledge about the jitter the system might induce or explicit information from lower layers, the measurement or detection of a jitter spike is based on statistics. Hence, it is important not only to measure the occurrence of jitter spikes but also to treat that information in an intelligent way, not letting a rarely occurring event color the jitter management scheme too much. Also here, it is clear that the jitter measurement by itself is not enough to control the jitter management scheme – a more elaborate control algorithm is needed.

When presenting the jitter measurements in a traditional histogram, the properties and the distribution of the jitter become visible. Further, the approximated probability distribution which a histogram visualizes gives the UE possibilities to better judge what to expect in terms of jitter in the rest of the session.

5.3.3.2   Correcting Jitter – Controlling the Jitter Management Scheme

There are two dimensions in which any jitter management scheme can work: Packet Loss Rate (PLR) and conversational delay. Since jitter is all about deviations from the expected packet receival rate, a buffer is the remedy to handle this problem (Figure 5.10). The media decoder requires a steady flow of encoded frames and will use concealment operations if that is not possible, even if it would have been possible a few milliseconds later. Jitter by itself does not produce packet losses; it is only when adding the real-time criteria for the service that jitter will induce losses. A jitter induced packet loss, a so-called *late loss*, occurs when the jitter management algorithm decides that an incoming frame is received too late to be used in the decoding process. This is a function of the jitter buffer depth; even more so, the allowed late loss rate is a parameter often used to control the jitter buffer depth together with the current jitter statistics.

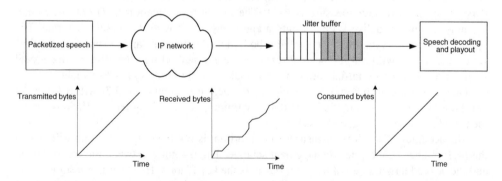

Figure 5.10: A graphical representation of the jitter management scheme.

The control of the jitter buffer is a function of what kind of loss rate the buffer is allowed to induce and what depth the buffer should have. This control process can be either static, i.e. not changing its buffer depth during active speech, or dynamic, i.e. changed whenever needed during the session. The control process separates jitter buffer implementations into two categories, *static* and *adaptive*.

A static jitter buffer has a certain depth which is not changed unless a buffer under-run occurs during the session. The depth of the buffer determines the maximum jitter amplitude the UE can handle without introducing buffer under-runs.

An adaptive jitter buffer varies its target buffer depth during run-time. It can use statistics and the target late loss rate to optimize the target buffer depth for the current session. Adaptation can be done either on a frame basis or on a more fine-grained sample basis. This is discussed in depth in Section 5.3.3.4.

A common misunderstanding is that a jitter management scheme adds to the total end-to-end delay. This is not true. An ideal jitter buffer does not introduce any additional delay in the end-to-end delay budget since the depth of the buffer is equal to the amount of jitter the transport link produces. The buffering depth should only be as large as the jitter on the transport link, in some cases even smaller since a certain amount of so-called late losses might be allowed. Hence, the depth of the buffer is used to absorb variations in the packet delivery rate which stems from the transport mechanism. In real-world implementations, a perfect alignment between jitter buffer depth and link jitter is hard to achieve, unless the jitter of the transport link is very small or does not vary throughout the session. The challenge in the jitter buffer design is therefore to balance unnecessary buffering depth and the risk of introducing late losses.

All-in-all, any jitter management scheme has as its goal to keep the end-to-end delay as low as possible. A so-called static jitter buffer will keep the end-to-end delay constant after the session has started, at least until a re-buffering occurs where the depth of the static jitter buffer might change. An adaptive jitter buffer on the other hand tries to minimize the end-to-end delay at all times.

5.3.3.3 Static Jitter Buffers

The most simple buffer implementation to handle jitter is a static or fixed jitter buffer. These buffers are often seen in streaming applications where no real-time criteria are present for the session. A buffer depth of a couple of seconds is common which makes this jitter management scheme suitable for handling both short-term jitter and slow bit-rate throughput variations. However, a jitter buffer depth of a couple of seconds is not possible to use for real-time services while still keeping the real-time criteria of the session. In 3GPP, the definition of a real-time conversational service using speech states that the mouth-to-ear delay should be less than 400 ms [44]. Seen in that respect, it is clear that the allowed jitter, hence the corresponding jitter buffer depth, cannot be too large since a part of the total delay budget is always consumed by the default transport mechanism.

There are two parameters that constitute the depth of a jitter buffer: the total buffer depth and the target buffer depth (Figure 5.11). The total buffer depth denotes the total allocated memory the buffer may consume. This is set during implementation and sets the absolute maximum depth the buffer can use during any session. The target buffer depth denotes the buffer depth which is the goal to be achieved in the current session. It defines the buffer fill limit which needs to be filled before the decoder is allowed to consume frames in the start of a session or at the start of a talk spurt; the active speech period after DTX. The playout point is a pointer to the frame in the buffer which is next to be consumed by the decoder.

The most simple version of a fixed jitter buffer does not allow the target buffer depth to be changed during a session. Hence, there is no need for any jitter measurement algorithm

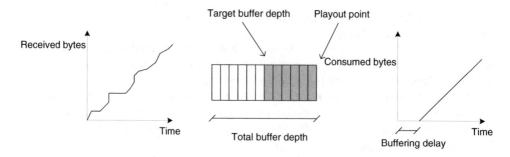

Figure 5.11: Functional view of a fixed jitter buffer. In this case, the target buffer depth will allow the buffer to handle jitter up to 120 ms, or six 20 ms speech frames.

nor a control mechanism. A more advanced fixed jitter buffer scheme is the so-called semi-fixed jitter buffer which is allowed to change the target depth during run-time but only during the start of a talk spurt. This buffer is also known as an adaptive jitter buffer using on-set adaptation.

The major advantage with fixed jitter buffers is ease of implementation. The buffer control is very straightforward, at least in its most simple form, and the speech processing chain synchronicity is kept intact, e.g. when using AMR, each node in the chain is only assumed to operate every 20 ms. This is not the case if any sample-based adaptation schemes are used; see Section 5.3.3.4.

A fixed jitter buffer is characterized by the fact that the playout point is always moved by the frame length in milliseconds relative to a fixed time scale. Hence, the end-to-end delay is kept constant without taking into account the current fill level in the buffer. If a large amount of jitter is present, this will eventually result in a buffer under-run.

5.3.3.4   Adaptive Jitter Buffers

An adaptive jitter buffer is a jitter buffer which in our definition can change its playout point any time during a session. The adaptation can be either *frame-based*, i.e. by inserting or removing full speech frames from the buffer, or *sample-based*, i.e. the decoder will change its frame consumption rate from the jitter buffer by varying the amount of samples that each decoded frame will generate.

5.3.3.5   Adaptive Jitter Buffers, Frame-Based Adaptation

Adaptive jitter buffers which use frame-based adaptation operate by inserting or removing full frames into the buffer (Figure 5.12). Frame insertion is normally only frame repetition while frame removal often uses a frame classification scheme in order to remove frames which degrade the intrinsic media quality the least. Frame classification can be done in a number of ways but all of them require some kind of intelligence to be able to judge the content in one frame and then classify it according to some principle. These classifications algorithms can in principle be very complex or rather simple depending on the sought-after performance of the algorithm. However, the normal case seems to be that implementers choose a quite straightforward classification scheme, e.g. the amount of energy represented by the frame, pitch gain value, etc. Every frame that arrives at the jitter buffer is classified according to the chosen scheme so that, whenever the jitter buffer control algorithm chooses to reduce the

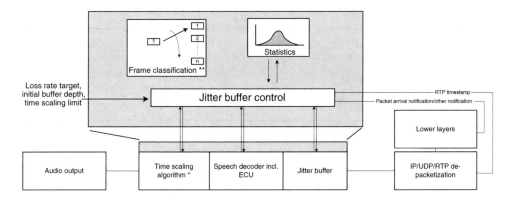

Figure 5.12: Functional nodes in an adaptive jitter buffer implementation. The time scaling unit marked with * is only present when sample-based adaptation is used. The frame classifier marked with ** is normally only present in adaptive jitter buffers using frame-based adaptation.

buffer depth, the frame with the lowest priority according to the classification is dropped. A similar algorithm could of course also be used to control which frame to repeat if the jitter buffer depth needs to be increased.

5.3.3.6 Adaptive Jitter Buffers, Sample-Based Adaptation

Sample-based adaptation is a more elaborate scheme to change the jitter buffer target depth. The basic concept is that depending on whether you want to increase the target buffer depth or decrease it, you change the rate at which the decoder consumes frames from the buffer. Since the whole speech decoding processing chain must produce samples at the given sampling rate, this is achieved by making sure that it is possible to run the decoder faster than real-time and then remove or add samples after the decoding has taken place but before it is sent to the audio rendering device (i.e. the soundcard buffer). The principle is shown in Figure 5.13.

There are a number of different sample add/removal techniques, often called *time scaling* techniques. The naming is quite self-explanatory since scaling time is exactly what is achieved. Each encoded speech frame represents 20 ms of speech and any change of that length will be perceived as time expansion or time compression. One of the most common algorithms to achieve this is called Wave-form Synchronized OverLap and Add (WSOLA). Other such examples are Pitch Synchronous OverLap and Add (PSOLA) and sample synchronization. The basic idea of all these algorithms is to extract or introduce additional pitch periods in the signal in order to change the duration without changing the frequency content, e.g. the pitch. If this is done properly, e.g. by restricting the amount of time scaling applied, the quality degradation is minor, even unnoticeable in some cases. If time scaling is used on signals other than speech, the degradation is more severe. SP-WSOLA, or single packet WSOLA, is a special time scaling algorithm used in VoIP applications since it can perform time scaling based on the information in the current frame and in the previous frames. There is no need to wait for future frames in order to scale the target frame according to some criteria. This will reduce the end-to-end delay compared to other time scaling techniques which require information from both future frames as well as old frames.

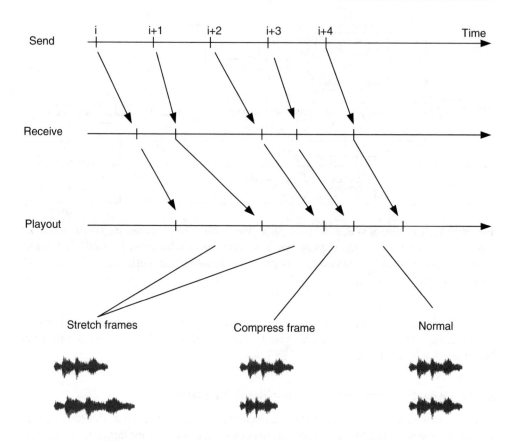

Figure 5.13: Principle of sample-based jitter buffer adaptation.

The use of time scaling enables the adaptation of the target jitter buffer depth to be performed at any time. Although some sound excerpts are more sensitive to scaling operations, the built-in similarity method of the time scaling algorithm will only scale as much as the current frame content will allow (according to the similarity measure). A target scaling factor is set, then it is up to the control of the algorithm to fit the scaling achieved with the target, and the current, jitter buffer fill level. This will most often take a couple of frames, especially if a limit has been set on how much scaling is allowed. Too much scaling, both compression and expansion, will eventually degrade the intrinsic media quality. There are no specific limits in the literature on how much time scaling should be allowed for speech content, but a common practice indicates that the target time scaling operation should be kept within $\pm15\%$ measured over a limited period of time.

A radical use of sample-based adaptation can be seen in so-called *virtual* jitter buffers. Here, time scaling is used to fill the jitter buffer up to its target depth at the beginning of every talk spurt. No threshold or playout point is set where the decoder can start to consume frames. Instead, as soon as there is a frame in the buffer, the decoder starts to decode. The virtual buffering scheme then uses time scaling to adapt the frame consumption rate from the buffer so that no concealment operation is triggered due to buffer underflow. Although this scheme has the potential to increase the conversational quality, there are two major risks involved. The first risk is that the amount of time scaling applied in the beginning of each talk spurt will be considerable hence risking introducing time scaling artifacts. The second risk is that if a late loss occurs at the beginning of a talk spurt due to insufficient application of time scaling, the quality degradation will be significant due to the fact that the loss, hence the error concealment, will be at the beginning of a talk spurt, a speech onset, which is the most challenging part of a talk spurt to conceal. In order to assist the error concealment unit in the decoder, there is a technique known as state recovery, or codec re-synchronization, in which the data in the frame delivered too late is decoded in parallel using a second instance of the decoder. The resulting decoder states are used to update the states in the live decoder, hence reducing the error propagation in the decoder states.

5.3.3.7 Adaptive Jitter Buffers, Adaptation Control

The actual control of the target jitter buffer depth is an important part in an efficient adaptive jitter buffer implementation. Since in most cases, the receiver can only rely on its own measurements, the adaptation control in the UE will determine the performance of the jitter buffer. There are different ways to perform adaptation control but the basic parameters which will govern the overall control of the jitter buffer is the allowed late loss rate and the overall design criteria to keep the target buffer depth as low as possible. These two parameters will define a current, upper limit on the buffer fill level as well as a lower limit. If the current fill level goes beyond the lower limit, the adaptation control will target an upward adaptation using whatever adaptation mechanism is available. If the buffer fill level is above the target depth, the control will trigger a downward adaptation to avoid unnecessary delay build-up in the jitter buffer.

As discussed in Section 5.3.3.1, the basis upon which the jitter buffer control algorithm operates is the jitter measurement. The jitter measurement in itself does not automatically translate into what target jitter buffer fill level should be set since it is impossible to correctly target a suitable fill level during run-time which will eliminate any late losses. It is of course possible to set the depth so large that the probability for late losses is virtually zero, but doing so will in all relevant cases also severely degrade the real-time performance of the service. Hence, the control algorithm needs to make some intelligent decision on what the target jitter buffer depth should be and when adaptation is needed. In short, the fundamental target for the jitter buffer control algorithms is to find an optimal balance between late losses and jitter buffer depth in order to optimize the conversational quality of the service.

In real-life adaptive jitter buffer implementations, the adaptation control is often based on the use of statistics. By building up a probability density distribution during the session, the adaptation control mechanism can estimate the late loss and map that to whatever late loss target the control mechanism has. However, there is no definitive answer to how this probability density distribution should be weighted. Normally, newer packets should have higher priority but if the transport characteristics show some periodic behavior, other weighting schemes might be more suitable. The specific weighting, or the aging process of the probability density distribution, is a parameter which needs to be carefully examined during design and implementation and possibly also tuned according to the target environment in which the adaptive jitter buffer will operate.

Expressing the measured jitter in terms of a histogram makes this situation more visible. In Figure 5.14, the example jitter shown in the upper part of Figure 5.9 is shown.

If the playout point is inserted in the plot of the histogram, the connection between the set playout point in the jitter buffer and the allowed late loss rate is clear. The position of the playout point will in practice define what target late loss rate the current jitter buffer setting

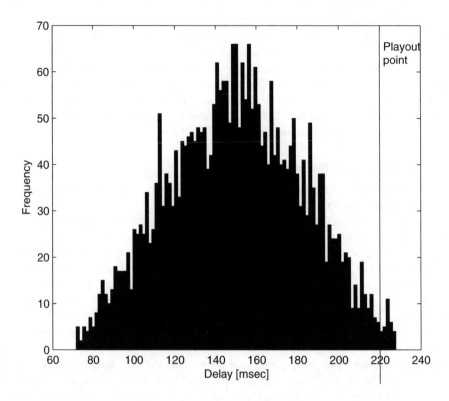

Figure 5.14: End-to-end delay expressed in a histogram. The measurement is based on expected arrival time. Using the playout point highlighted in the picture, the allowed late loss rate would be 1%.

will have. Since the jitter distribution will change, the playout point might also change if the same target late loss rate is to be kept during the session. Hence, allowing a certain late loss rate is equivalent to setting a minimum depth of the jitter buffer below which the buffer fill level never should arrive – adaptation measurements should be triggered before that happens. The minimum depth is in real implementations often set to 1. As mentioned before, aging of the histogram can be fast or slow depending on what the target behavior is of the jitter buffer. Fast aging will mean fast adaptation with low margins for sudden jitter amplitude changes; slow aging will mean higher tolerance for e.g. delay spikes but unnecessary large jitter buffer target depth at more modest jitter amplitudes.

Adaptation control can also be done using more simplified schemes. One such example is an adaptation control scheme similar to the TCP behavior. It monitors the delivery rate of the packets and slowly tries to decrease the jitter buffer target depth until a late loss occurs. An upward adaptation is then performed and the process starts all over again. This will induce a sawtooth-like behavior of the jitter buffer target depth and will only be re-active rather than pro-active. However, the implementation is simpler compared to the statistics-based control algorithm and it will also provide a robust behavior since it is completely independent of any jitter measurement.

There are two ways to measure the depth of the jitter buffer, either when an incoming frame is inserted in the buffer or when a frame is sent to the decoder. In practice, only the latter method is used since the crucial point is that there should always be a frame in the buffer that the decoder can consume. If the minimal level of the buffer, i.e. the level at which upwards adaptation is triggered, is set to less than one frame, the risk of late losses is significant since there is no way to know if the next frame will arrive in the buffer before it is needed by the decoder. However, if a certain late loss rate is allowed, the control algorithm might temporarily allow late losses to occur. Note that by always having at least one frame in the buffer, you have at least the chance of being pro-active and increase the buffer depth before the buffer has been emptied and a re-buffering must take place. The upper limit is set by the target depth.

Re-bufferings occur when the jitter buffer is empty. Then, the decoder starts to use the concealment unit to produce output but no frames are sent from the jitter buffer to the decoder. The re-buffering is done until the target jitter buffer depth has been reached. Hence, it is not uncommon that a re-buffering produces several (3–6) consecutive concealment operations which will significantly degrade the intrinsic media quality and should therefore be avoided as much as possible.

5.3.4 Speech Transport Processing – Packet Loss Concealment

Although error concealment by itself is not tied to a specific transport mechanism (CS or PS), there are some additional methods that are available when using PS transport. These mechanisms are not dependent on some specific functionality below the IP layer; they only influence the IP packet, possibly also including its media payload. There are basically two different versions of providing added error resilience on the application layer when using IP transport: media resilience (such as error resilient media encoding) or packetization resilience. In this section, packetization resilience will be treated. For a more fundamental description of speech decoder error concealment, see Section 5.1.1.5.

RTP transport in general is a non-reliable transport protocol. If an RTP packet is lost, there is no mechanism built into the protocol to enable a retransmission of that particular packet. Although feedback using RTCP does inform the sender that packet losses might have occurred, these reports are not tied to a specific lost packet but rather provide statistics of the current status of the transport link. Hence, error resilient efforts on the RTP layer can improve the performance for RTP transport over lossy links.

In the context of IMS Multimedia Telephony, RTP error resilience for speech is limited to application layer redundancy. The principle of application layer redundancy is very simple; transmit each speech frame in at least two different IP packets. By doing so, the information in each lost RTP packet would still be available in at least the packet preceding the lost packet or, in the case of more advanced application redundancy, in several other packets. The effect is that the tolerance for packet losses on the transport link goes up far beyond the tolerance limit given by the acceptable quality degradation by packet losses without application layer redundancy. Figure 5.16 shows the results from a subjective listening test indicating the benefit of using application layer redundancy to increase tolerance to packet losses. The payload format for AMR allows the transmission of redundant speech frames.

Application layer redundancy can be used in different schemes. In Figure 5.15 an example using 200% application level redundancy is shown. The example of 200% redundancy, where the redundant frames are transmitted in consecutive packets, will make it possible to reconstruct double consecutive packet losses without using any error concealment function

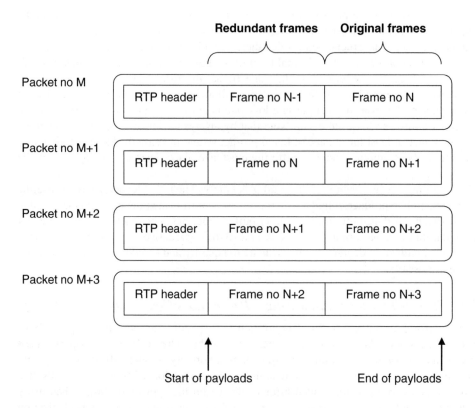

Figure 5.15: The use of 200% application level redundancy when only one new speech frame is sent in each RTP packet. The order of the redundant frames and the new frame must always be the same for the RTP payload parsing to function properly.

from the speech decoder. The drawback is that the latency, or delay, will increase. In order to make it possible to use redundant packets, the buffer on the decoder side must have a sufficiently large fill level. In the example shown in Figure 5.15, the increase in latency would be 40 ms. It is possible to transmit the redundant frames with an offset greater than one frame. This would increase the tolerance to consecutive losses, but the conversational quality would degrade due to an increase in end-to-end delay.

Since the speech codecs used in IMS Multimedia Telephony have rate adaptation abilities, the use of application layer redundancy can be kept more or less bit-rate neutral. This makes application level redundancy used in a responsible way an interesting possibility to use in conjunction with the AMR codecs since the net load on the network will not increase. Table 5.18 shows the payload sizes of different combinations of AMR modes and level of application layer redundancy.

The discussion on application level redundancy for speech has so far only treated so-called full redundancy, i.e. the case when all frames are sent redundantly. However, it is also possible to use so-called selective redundancy where only certain frames are sent redundantly. The selection of what frames are to be sent in that way should be based on how well a concealment algorithm would be able to conceal a possible loss. Frames that are more

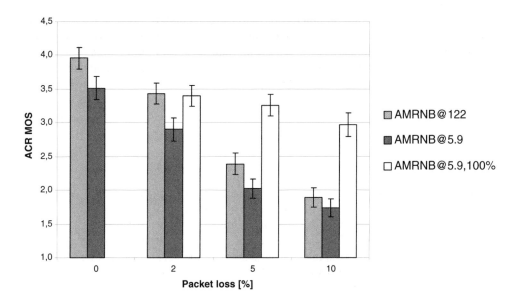

Figure 5.16: The results from a subjective listening test comparing the performance of AMR with and without application layer redundancy using lossy links. As indicated in the plot, there is a clear benefit to use application layer redundancy when the packet loss rate of the current transport link goes above a certain threshold. This is also true even if the source coding bit rate is reduced.

Table 5.18: The net payload size in bytes when using certain mode combinations with and without redundancy. As can be seen, several of these combinations provide a bit-rate neutral, or close to neutral, way of increasing tolerance to packet losses.

Codec	1 frame/packet (no redundancy)	2 frames/packet (100% redundancy)	3 frames/packet (200% redundancy)
AMR 12.2	32		
AMR 6.7		36	
AMR 5.9		32	
AMR 4.75		26	39
AMR-WB 15.85	41		
AMR-WB 14.25	37		
AMR-WB 12.65	33		
AMR-WB 8.85		47	
AMR-WB 6.60		35	53

challenging to conceal should in that case be sent redundantly while more easily concealed frames could be sent without using redundancy.

Finally, one could ask the question if application level redundancy used in 3GPP networks is an effective way of dealing with transport links that show considerable losses. In general, link problems are most often best solved on the link layer, not on the application layer. However, since IMS Multimedia Telephony is a service where telephony grade quality and reliability are required, it is ultimately the application layer which has the last possibility to ensure such performance. Hence, application layer redundancy should be used as a last resort, when no other schemes have been successful to secure link behavior in terms of keeping the packet loss rate low.

5.3.5   Video Transport Processing

Real-time video is rather different in terms of characteristics compared to real-time voice. Since video is made up of a series of still images a variation in delivery timing (e.g. packet delay jitter) does not produce the same dramatic effect as for voice where high jitter can trigger concealment actions. For video, jitter will manifest itself as a variation of display frame rate which only produces minor degradations in the conversational quality. Hence, although some sort of receiver buffer is needed, no mechanisms similar to the jitter buffers used for speech are used for conversational video.

5.4   Media Control

In any multimedia session using IP transport, some sort of media control is used. The most simple control is the control action performed at session setup where a specific set of media-related parameters are negotiated hence also controlling the encoding/decoding process as well as the encapsulation of the media frames into the RTP packet. However, media control can also be applied during the session and in this section two different aspects of such run-time media control are described; end-to-end client media adaptation and user induced session adaptation.

5.4.1   End-to-End Adaptation

The ruling paradigm in IP transport is that the transport is end-to-end, i.e. only the session IP end-points are allowed to packetize encoded data into the IP packets and transmit it to the respective receiver. Although it is technically feasible to have intermediate nodes which re-packetize or transcode the material, the default operation is that the data is sent end-to-end. Hence, media adaptation in end-to-end scenarios is based upon some knowledge about what the receiving end experiences or by some measurement about what local conditions the sending side is experiencing. In this section, we will discuss some aspects of end-to-end adaptation. Our scope is limited to speech and video and no user action is assumed. For user induced session adaptation, see Section 5.4.2.

There are a number of different dimensions in which adaptation can be performed. Below, three dimensions are listed and exemplified.

- **Rate adaptation**. When using rate adaptation, the encoding rate is changed during the session either upwards or downwards depending on the input to the adaptation control algorithm. A typical example here would be if a video telephony session experiences significant packet loss indicating an inability to sustain the bit rate needed for the session to continue at a given quality level. A control message from the receiver to the sender indicating this in some way could trigger a downward rate adaptation possibly enhancing the quality of the received media flow. The ultimate rate adaptation is to set the media encoding rate to zero which is the equivalent to component dropping. Such dramatic measures should of course be avoided but in some circumstances it might be the only way of keeping the overall communication session up and running. It is fair to assume that the users in a video telephony session would prefer to switch to a voice-only session if the bit rate needed for video cannot be met by the mobile network compared to a dropped call.

- **Frame aggregation**. Frame aggregation denotes the number of speech frames aggregated in one RTP packet. For a default speech-only session in IMS Multimedia Telephony, one speech frame is packetized in each RTP packet. However, there are some cases where a frame aggregation putting more than one speech frame in each packet and hence reducing the packet rate could enhance the transport quality of the session. One such example is WLANs without any QoS control. It is widely known that when it comes to VoIP, WLANs are packet rate limited rather than bit rate limited and allowing the frame aggregation to change during the session would lower the packet rate and release resources in the WLAN network. The drawback is that the end-to-end delay would increase proportionally to the change in number of frames per RTP packet.

- **Media robustification**. Media robustification enhances the error resilience in the media bit flow and is a function of either the media encoder or the packetization. Based on some sort of feedback from the receiver, the sender would here apply some resilience measures on the media flow which would provide greater tolerance to packet losses. One example here is the use of application level redundancy as described in Section 5.3.4. Another example is to use error resilience tools available in the video codecs used in 3GPP.

All-in-all there are several different ways of doing end-to-end adaptation. However, in the end, it is important to minimize any adaptation usage since almost all scenarios will, if used in an inappropriate way, lower the media quality of the session.

5.4.1.1 End-to-End Feedback Mechanisms

There are basically two ways to establish a feedback mechanisms on the IP layer when transporting speech and video data using RTP: RTCP (see Section 5.2.1) and in-band mechanisms. The most obvious example is of course the RTP Control Protocol (RTCP) [177] the sole purpose of which is to provide a transport mechanism for transmitting data regarding the transport characteristics of the session back to the sender. RTCP is transported out of band, i.e. it is not embedded in the media RTP payload format. In-band mechanisms on the other hand denote mechanisms that are embedded either in the media stream itself or in the RTP payload format.

5.4.1.2 End-to-End Feedback Mechanisms, RTCP

The purpose of all RTCP packet formats is to provide feedback on the quality of the data transmission. The information carried by e.g. SR and RR includes wall-clock time, information on packet count and octet count, packet loss statistics, inter-arrival jitter information, etc.

It is important to recognize that RTCP provides means to signal measurements, not requests. It is up to the receiver of an RTCP report to turn a received RTCP report into some appropriate action, e.g. lower bit rate, aggregate frames or use resilience measures.

5.4.1.3   End-to-End Feedback Mechanisms, In-Band Signaling

In IMS Multimedia Telephony, the only available in-band signaling for media is built into the payload format for AMR. As discussed in Section 5.2.2, the AMR rate control is supported in the payload format, where the receiver can request a specific encoding rate from the sender on a frame-by-frame basis.[4] The requests and indications of the encoding rates requested and used are signaled in-band in the payload format using the CMR and FT bits. However, this only enables rate control, no other adaptation dimension can be used unless the CMR and FT bits are overloaded with other information. This is not supported by the standard. In order to enable adaptation in dimensions other than encoding rate, a new in-band signaling mechanism has been proposed in IETF [126]. This mechanism makes use of so-called shims which are inserted between the RTP header and the AMR payload (including payload-specific information). In this specific proposal, the added field in the header would enable a request for the use of application level redundancy.

The advantage with in-band signaling is that it is as fast as the actual media transport; no additional delay is introduced. Hence for fast adaptive measures, in-band signaling is especially suitable. In-band signaling can also signal requests but requires standardization to be able to function appropriately.

5.4.1.4   Adaptation Triggers

Any adaptation measure is always preceded by some adaptation trigger. This trigger could either be in the form of a measurement of some sort which the adaptation control mechanism interprets in a specific way (e.g. 4% packet loss on the video flow, reduce video bit rate by 50%), or it could be a request by the receiver to perform a specific adaptive measure (e.g. a request to change AMR encoding mode). No matter if the actual adaptation request, or the measurement report, is signaled over the link, the basic triggering mechanism is the same. Signaling a measurement or a request only differs in which end-point will decide how to adapt to the current transport characteristics. Hence, below, only measurements are discussed since it is the actual measurement that will trigger the adaptation. The signal is only a consequence of the measurement.

One can divide the adaptation trigger measurements into two categories, IP layer measurements and cross layer measurements. IP layer measurements comprise two different sets of measures: packet loss rate and packet inter-arrival timing (jitter). These entities are readily available to any client and can be used to trigger adaptation, both proactive adaptation and reactive adaptation. For example, if the client is using an adaptive jitter buffer, the packet inter-arrival timing can be used to trigger buffer adaptation. Used in a proper way, this will be a proactive adaptive measure since the increase in buffer depth will prohibit otherwise occurring late losses. If the packet loss rate is used however, any resilience measure will be reactive since the losses have already occurred. You might view adaptation due to an observed packet loss rate as a hybrid between reactive and proactive adaptation though since added

---

[4]Note that [23] recommends rate requests only to be sent every other frame due to interoperability issues with GERAN CS telephony services.

resilience measures will reduce the observed quality degradation due to packet losses after the resilience has been applied.

Cross layer measurements can also be used to trigger adaptation. Although this is a much debated area when it comes to packet switched transport due to protocol layer violation, there are opportunities to use information from lower layers to improve the quality of the session. If we look into the speech service in GERAN CS, the AMR codec uses radio link measurements to govern rate adaptation. The C/I ratio is monitored and, when certain thresholds are passed, adaptation requests are triggered. Although there are no equivalent algorithms available in standards today regarding a similar method that could be used for packet switched transport, it is clear that the more information that is available for the adaptation control mechanism in the client, the better adaptation measures are possible to perform. Availability of lower layer information increases the possibility of doing proactive adaptation, i.e. adapting even before the media degrading events have occurred. However, in the end, it is always a question of whether a certain adaptive measure does improve the situation for the session. The cost of implementing adaptation measurements, signaling protocols and standardization of client behavior must be compared to the benefit of the adaptive measure by itself.

5.4.2   User Induced Session Adaptation

User induced session adaptation refers to the actions that a user might take in order to adapt the media consumption or production during a session. The most obvious example is to add or remove the video part in a conversational video session. The session might start as a voice-only session but during the session the need for live video arises. One of the parties, or both, might then start a video flow without interrupting the ongoing speech part. This can be realized in two ways. The first alternative is to modify the current session via a SIP UPDATE which is granted both by the IP end-points as well as by the resource management function in IMS. One simply adds another video media type to the current speech media type in the session SDP and, if granted, the session is updated with video. The other way is to do a SIP re-INVITE, which creates a new session in which both voice and video are present. The difference is mainly semantic, the end result being the same: the session is updated with live video. Another example would be to remove a video stream from an ongoing video telephony session and keep the voice part undisturbed. The procedure is exactly the same as described above.

User induced session adaptation does not by itself introduce media layer consequences apart from the addition/removal of the media flow in question. That is, no additional actions are needed which are not already done at session start-up or close-down. However, there might be occasions where e.g. the addition of a video flow might affect the already ongoing speech media flow. One such example would be if video is added to an IMS Multimedia Telephony session over HSPA using a re-negotiated version of the already existing RAB used for voice where the allowed bit rate is increased. Although the grant of the RAB update should pass through some sort of check if the increased bit rate is supported, if the user moves out of coverage for the higher bit rate needed, the voice flow will also be affected.

These kinds of user actions, dynamical media content session modification, are a key feature in IMS Multimedia Telephony and are listed as one action the service shall support.

5.5   Header Compression

Multimedia telephony is the way to realize telephony in an all-IP cellular network. An all-IP cellular network assumes that voice payload is transported *IP all the way*, i.e. the IP protocols do not terminate before the air interface. *IP all the way* enables IP service flexibility, that is, there are fewer dependences between applications and the wireless access network and a base is created where it is easier to create new services including a voice component. This could be compared with today's circuit switched cellular telephony services, that are vertically integrated and optimized – resulting in very high radio performance but limited flexibility. The requirements in terms of spectrum efficiency and voice quality for the Voice over IP service in tomorrow's *all-IP* cellular systems will be reasonably similar to current requirements on the circuit switched service. This introduces new challenges when realizing multimedia telephony over cellular. One fundamental challenge is to reduce the IP header related overhead over the relatively lossy and scarce radio channel, while maintaining the transparency of all header fields.

A major challenge with any Voice over IP over cellular service is the large headers of the protocols used when sending speech data over the Internet. An IPv4 packet with speech data will have an IP header, a UDP header, and an RTP header making a total of $20 + 8 + 12 = 40$ octets. With IPv6, the IP header is 40 octets, for a total of 60 octets. The size of the speech data depends on the codec but is normally in the order of 15–30 octets. These numbers present a major reason for terminating the IP protocols before the air interface: the IP/UDP/RTP headers require a higher bit rate and would cause inefficient use of the expensive radio spectrum. Terminating the IP protocols before the air interface and thus removal of the information from the IP headers would however seriously hamper the inherent flexibility of the IP protocol suite. It would also cause additional complexity and create the need for additionally vertically optimized solutions. What is needed is a way to reduce air interface impact from headers while maintaining the transparency of the header information. The solution here is a technique called *header compression*.

Header compression is a joint name for algorithms that reduce the size of headers on a per-hop basis. Headers are compressed, transported over e.g. the air interface and decompressed resulting in IP headers that are identical to the header before compression. Header compression has, naturally, to be applied on a per-hop basis since the information in the header fields is needed for e.g. routing. If the compression is efficient enough it may solve the problem of IP protocol overhead while maintaining all the benefits of the IP protocol suite at the receiving end.

While all header information in a voice packet is needed, there is a high degree of redundancy between header fields in the headers of consecutive packets belonging to the same packet stream. This observation is the basis for header compression algorithms. These algorithms maintain a *context* – essentially the uncompressed version of the last header transmitted – at both ends of the channel over which header compression is performed. Compressed headers carry changes to the context; static header fields need not be transmitted at all, and fields where the change is small are updated using few bits. When packets are lost or damaged over the channel, as they can be for cellular links, the context on the downstream side may not be updated properly and decompression of subsequent compressed headers will produce incorrect headers. Thus, header compression schemes must have mechanisms for installing context, for detecting when the context is out of date, and for repairing the downstream context when it is out of date.

Within the IETF (Internet Engineering Task Force) several header compression algorithms are being standardized. The first scheme RFC 1144 was brought forward by

Jakobson [122], and compressed TCP/IP only. The originally envisioned use case for this scheme was improving interactivity when typing in e.g. Telnet services using very low rate modem lines, e.g. few kilobaud. It was later improved by a header compression scheme that allowed compression of both IP/TCP and IP/UDP brought forward by Degermark; this scheme is commonly called IP header compression defined in RFC 2507 [62]. For real-time IP services, Compressed Real-time Protocol (CRTP) [60] was previously the only relevant scheme that could compress all headers of a VoIP packet. CRTP can compress the 40 octet IPv4/UDP/RTP headers to a minimum of 2 octets. For context repair, CRTP relies on there being an upstream link over which the decompressor sends requests for updating headers. While the context is out of date, all packets received by the decompressor will be lost since the headers cannot be decompressed. The round-trip time over the link will thus limit the efficiency of the context repair mechanism. CRTP performance over cellular links has been evaluated in [61], which found that the compression ratio of CRTP is fairly good, but the packet loss rate for CRTP over realistic cellular links is too high. Whenever CRTP experiences context damage at the decompressor, which will occur due to a single lost packet, it must request a context repair. All packets are lost due to malfunctioning header compression during a round-trip time or round-trip times if the request for context repair was lost, until a much larger compressed header containing context repair information is successfully delivered. A viable header compression scheme for Voice over IP over wireless must be less fragile than CRTP but cannot be less efficient. Additions have later tried to solve these short-comings of CRTP in a scheme called enhanced CRTP but this has significantly reduced the compression ratio of enhanced CRTP.

There are a number of basic requirements on a header compression scheme that is to be used over a cellular link:

- Compression efficiency

  - The average header size must be minimized. It is basically never small enough.

  - The header size variation should be minimized since e.g. large updating headers introduce larger requirements for bit rate flexibility on the lower layers and increase the probability for packet loss.

- Robustness

  - The header compression algorithm must be able to efficiently handle packet losses and reordering of packets before the compressor.

  - The header compression algorithm must be able to handle packet loss between compressor and decompressor efficiently, packet losses shall not be increased from the use of header compression and packet loss should not cause the need for verge context updating headers.

  - The header compression algorithm should be robust to moderate reordering between compressor and decompressor.

- Compression reliability

  - The transparency of header compression must be ensured, the decompressed headers must be identical to those before compression.

– The header compression algorithm must be robust to single residual bit errors. That is, the header compression algorithm must be robust against and preferably detect bit errors in compressed headers that have not been detected by link layer checksums.

These requirements and the short-comings of CRTP were the basis when a new header compression was developed and specified in IETF targeting the use case VoIP over cellular access. This scheme and specification is called ROHC – Robust Header Compression defined in RFC 3095 [51]. ROHC has a compression ratio higher than CRTP but even so is fully robust to packet losses between compressor and decompressor. It can compress a 40 bytes IPv4/UDP/RTP header to a minimum size of 1 byte with an average size just above the minimum size. While doing so it is fully robust to packet losses. It does so by taking the properties of the headers of the IP packets as well as the cellular channel into account.

If looking at the individual header fields in a flow of IP/RTP/UDP headers each header field may be classified in one of a few classes: header fields may be static, e.g. never change; it may be possible to infer their values from other layers of the protocol stack, e.g. the size of a packet; header fields may be dynamic but possible to predict, e.g. sequence numbers constantly increasing; or header fields may be random, impossible to predict. Figure 5.17 show a classification of the header fields in an IPv4/UDP/RTP from a flow of VoIP packets.

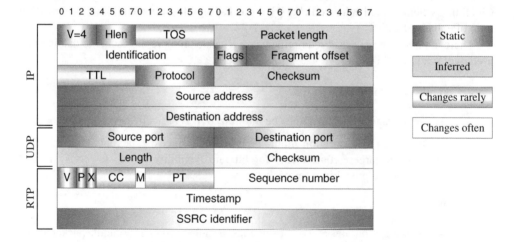

Figure 5.17: Classifications of IPv4/UDP/RTP header fields.

Looking at this picture, it is clear that a large part, approximately three-quarters, of the RTP/UDP/IPv4 header is static and basically never changes. These are thus easily compressed by simply sending these once to install a static part of the header compression context. The same applies for the fields possible to infer, e.g. the packet length field, that it is possible to learn from the link layer that implements header compression. The challenge lies in compressing robustly the header fields which change often. The IPv4 identification field (ID), the RTP timestamp (TS), the RTP sequence number (SN), the RTP marker bit (M) fields change every packet but often in a predictable manner. The UDP checksum also changes in every packet when used but then completely randomly, impossible to predict. Hence, the challenge lies in efficiently compressing the dynamic fields such as TS, SN and ID in a robust and efficient manner. To do so it is important to know the characteristics of these fields, individually and in relation to each other. To learn this you have to look at the syntax of the protocols using these header fields. The characteristics of the RTP header fields SN, TS and M for example are changing differently based on the application payload they are carrying. If the payload carried is voice traffic, the sequence number is increased by one for each new voice packet while the timestamp is increased for each packet by an amount representing the time the carried payload was generated and thus should be played. The M-bit is then set for each packet at the beginning of a silent period. For video payload being carried there is a slightly different behavior. Then an IP packet may carry only a small amount of a video frame and the sequence number is then increased for each packet but the timestamp is only increased when there is a new video frame being carried in the payload.

The IPv4 Identification field is used for possible fragmentation on the IP layer and must be statistically unique for each IP packet. This can be implemented in different ways in different operating systems. The IPv4 ID field may be sequentially increased by one for packet, or it may be increased by one for all UDP packets sent from a host resulting in sequential jumps for an individual flow of packets, or it may be pseudo-random. The first example is naturally easiest to compress and the pseudo-random case the must difficult one. Hence a compression algorithm must be able to handle all these header field properties but may of course support one specific characteristic most efficiently.

Another property important to take into account is the loss and reordering properties of the transport including the cellular access. UDP packets may be lost or reordered before or after the compression point and this must be possible to handle. It can be observed from packet traces that reordering of packets larger than the order of one is rather uncommon on at least typical fixed IP networks. Hence, at least reordering of one packet should be handled efficiently. Packet loss may also occur before and after the compressor. Packet loss for a real service such as VoIP should of course be kept below an acceptable quality driven threshold but an end-to-end loss of 1% speech frames is acceptable for most speech codecs and can be considered as an example of criteria for acceptable quality. Hence, header compression must be able to handle packet loss of this order at least.

To efficiently handle these properties of the header fields and the transport and cellular access, ROHC uses a number of specific compression techniques. These include:

- usage of a header checksum to ensure compression correctness and robustness;

- usage of a master sequence number to enable that only the difference from this master sequence number may be transmitted;

- robust and efficient coding techniques of header fields.

To further improve compression efficiency and flexibility of ROHC it also utilizes different compression profiles to enable tailor-made compression for a specific header field flow and also to enable introduction of new compression profiles within the ROHC framework as new transport protocols are being introduced.

Currently, the ROHC specification framework includes the following header compression profiles:

- uncompressed;

- compression of RTP/UDPIPv4 and RTP/UDP/IPv6;

- compression of UDP/IPv4 and UDP/IPv6;

- Compression of RTP/UDPLite/IPv4 and RTP/UDPLite/IPv6;

- compression of IPv4 and IPv6;

- compression of ESP/IP, i.e. encrypted tunnels.

New profiles such as compression of TCP/IP including TCP options such as SACK and Timestamp are also under development. These and future compression profiles make ROHC both efficient and flexible, enabling ROHC to be the only header compression framework required.

To illustrate ROHC operation, an example from the RTP/UDP/IP profile and the usage of master sequence numbers, robust encoding and header checksums can be chosen.

To encode the RTP sequence number and the RTP timestamp of a VoIP packet, the RTP sequence number is used as a kernel field from which values of other fields, i.e. the RTP timestamp, are determined. The least significant bits of the RTP sequence number are transmitted as a part of the compressed header. The other part of the compressed header is a small cyclic redundancy header checksum. This checksum is the result of a checksum calculation covering the header before compression. This is all the information needed during a talk burst of voice packets assuming that the header compression context has already been installed. The decompressor can determine the value of the RTP sequence number from the least significant bits in the compressed header, even if some packets with compressed headers have been lost. With this sequence number value it is possible to determine the value of the RTP timestamp since it changes predictably from the sequence number. From these values and the installed header compression context a complete decompressed header may be assembled. The known header checksum algorithm is used to calculate the header checksum with the decompressed header as basis. If the resulting checksum value is identical with the value of the checksum received in the compressed header, it is safe to determine that the compression and decompression has succeeded and that the header compression context may be updated and compression/decompresssion may continue. Figure 5.18 illustrates further the use of a header checksum.

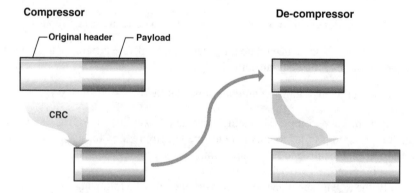

Figure 5.18: Robust header compression using a header checksum.

With mechanisms such as compression profiles, header checksums and robust encoding of header fields, it is possible to compress for example a 40 bytes IPv4/UDP/RTP header to a minimum size of 1 byte with an average size just above the minimum – this, while being robust to packet loss and detecting residual bit errors. In short, mechanisms such as these make it possible to meet the requirements on compression efficiency, robustness and compression reliability stated above. This meets the challenge of transparently reducing the impact of packet header overhead when deploying VoIP over cellular systems.

To be able to use any header compression scheme in any communication system, some basic requirements are put on the underlying link layers. At least possibilities to negotiate that header compression shall be used and flexibility to handle different packet size due to varying compressed header sizes are required. Earlier header compression schemes such as IP header compression also required that the underlying link layer also provided identification of what kind of compressed header each packet represented. In the point-to-point protocol (PPP), which is the typical link layer used with IP header compression, this required an extra 8 bits per packet. However, recent header compression algorithms such as ROHC also require less on the underlying link layers. In ROHC, the compressed packet type identification is carried within the compressed header. All header compression algorithms also require a context identifier (CID) to identify which packet flow a certain packet belongs to. In ROHC the size of this CID may be very small or larger when needed. This has further reduced the number of bits required to carry the compressed headers.

In UTRAN as specified by 3GPP header compression is located in the Radio Network Controller (RNC) and the User Equipment (UE). To support header compression the Packet Data Convergence Protocol (PDCP) has been introduced. 3GPP specifications fully support the use of RObust Header Compression (ROHC) from 3GPP release 4 with the PDCP protocol as supporting link layer. Since ROHC does not require per packet compressed packet type identifier it is possible to run PDCP in transparent mode, that is no PDCP overhead per compressed packet, when compressing headers with ROHC. This further reduces the overhead from IP headers. PDCP also supports IP header compression (RFC 2507) but is then run with a one byte overhead per packet. To initiate header compression, negotiation is done with the radio resource control signaling when a radio bearer is being set up. At this phase the type of compression scheme to use may be determined. CDMA2000 systems as being specified by 3GPP2 can also take advantage of ROHC header compression. ROHC support has been introduced in 3GPP2 specifications and in this case header compression is being performed between the PDSN node and the terminal. In this case header compression is being supported with the PPP link layer which provides full support of header compression but at larger cost in terms of overhead, bits per packet.

The development of robust header compression is continuously ongoing. New compression profiles are being developed meeting new requirements or compressing new protocols. This is important since it can be easily observed that the size-generous header overhead of the IP protocol suite would be a show-stopper for efficient VoIP over cellular if not being addressed. Robust header compression provides however an efficient and future-proof solution to this overhead challenge. Robust header compression is simply essential for all VoIP over cellular services to provide acceptable efficiency.

5.6 Radio Realization

In order to support efficient transport of conversational media over the air interface, it is often necessary or at least beneficial to optimize the network configuration. In this section we discuss the most important radio network parameters affecting the media, and show how they could be optimized for best performance. The actual performance is then evaluated in Chapter 7, for both optimized and unoptimized configurations.

5.6.1   UMTS

UMTS provides several options for realizing the Multimedia Telephony service. The initial packet data access in UMTS was realized using dedicated channels, which provide each user with a dedicated transmission resource with constant data rate. The dedicated channels are also used for the circuit switched voice, and in theory it would appear that they are the most obvious candidate for realizing the conversational multimedia services. However, as will be shown in Chapter 7, the High Speed Packet Access (HSPA) enables more efficient realization.

For UMTS the handling of the media flow depends on the used radio access bearer, as described in Chapter 3. Different radio access bearers correspond to different configurations and can result in widely varying characteristics for the media. In general the radio access bearer is described by at least the following parameters:

- Radio Access Bearer type. In UMTS four different radio access bearers are available, Conversational, Streaming, Interactive and Background bearers. The Conversational bearers are most suitable for the media transmission, while the Interactive bearers should be used to transfer signaling.

- Choice of Radio Link Control protocol configuration, containing parameters such as used PDU sizes, configuration of the retransmission functionality and so on.

- Choice of the used transport channel. Typical choices are either dedicated channel, HSDPA or enhanced dedicated channel.

- Physical layer configuration, containing for example the used coding rate and type (Turbo or convolutional) and TTI length.

In addition to the radio access bearer configuration, the most important systems that may need to be optimized for Multimedia Telephony are as follows:

- HARQ operating point. For HSDPA and E-DCH the choice of the HARQ operating point (i.e. the targeted number of transmission attempts and the desired retransmission rate, see below) determines to a large extent the capacity and the media delay.

- Used scheduling algorithm, if HSDPA is used. The choice of the scheduling algorithm has a significant impact on the system capacity and media quality, as will be shown in Chapter 7.

As the requirements (e.g. with respect to packet loss rate) for the actual media and corresponding signaling are quite different, separate bearers are expected to be used for signaling and media transport. For signaling the used bearer will be of interactive type, and the configuration used for interactive traffic (e.g. web browsing) can be used. The signaling bearer is characterized by a need to perform retransmissions on the HARQ and RLC layers. For scheduling purposes it is expected that the signaling traffic is prioritized over media bearers.

A separate bearer is used for the media transport. In the case of a multimedia communication, it is possible to use separate bearers for each media component. This would allow an efficient separation of different media components, but may lead to unnecessary signaling overhead. It is also possible to use a single media bearer for all media components, in which case adaptation is needed to maintain sufficient media quality for each media component. In general, from the service quality point of view, separate bearers are likely to provide better quality of service.

For the voice component of Multimedia Telephony, the frames generated by the application are small, and the need for optimized parameter settings is great. It is especially important to avoid excessive header overhead or padding. For the video and other components of Multimedia Telephony, the benefit from careful optimization of system parameters is significantly smaller. The video packets are relatively large, and thus the header overhead and padding are much smaller issues. For this reason, we focus the discussion of the optimizations on the voice service.

For the voice service, the first obvious bearer realization choice is to use header compression to reduce the RTP/UDP/IP header overhead. As discussed in more detail in Section 5.5, the best choice for the header compression algorithm for lossy links is ROHC. We will assume that all optimized radio bearers use ROHC, but estimate in Chapter 7 the performance loss caused by not using header compression.

The IP packets are further encapsulated in the Radio Link Control (RLC) Protocol Data Units (PDUs). The header and padding overhead created by the RLC can be significant with improper choice of parameters, and can be reduced significantly by correct choice of the RLC PDU size. For example, the RLC PDU size used for interactive bearers is 336 bits, which allows 320 bits of payload. If the interactive bearer were used for e.g. VoIP service with 12.2 kbps AMR codec (with 256 bit payload and 3 byte compressed header), each RLC PDU would contain the 8 bit RLC header, VoIP packet and 40 bits of padding, resulting in radio link data rate of 16.8 kbps. This may not sound too excessive, but when reducing the codec rate to e.g. 5.9 kbps the resulting radio link data rate would still be 16.8 kbps, which is almost three times the codec rate.

In order to avoid excessive padding, the possible RLC PDU sizes need to be aligned with VoIP packet sizes. For IMS Multimedia Telephony the AMR codec is mandatory with 12.2 kbps narrowband codec and 12.65 kbps wideband codec being the most likely used codec modes. Thus special focus should be paid on AMR 12.2 kbps and AMR-WB 12.65 kbps codecs. However, efficient support of all AMR and AMR-WB modes up to 15.85 kbps and DTX is required.

The RLC payloads should at least cover an IP packet with one speech frame without segmentation and payload sizes should be optimized to the most common sizes to gain padding and RLC overhead (including length indicators).

The following ROHC header sizes would cover most cases efficiently: 3 and 8 bytes when transmitting speech frames, and 4 bytes when sending SID frames. Larger extensions or IR/IR-DYN are expected quite seldom and could then be transmitted using the RLC segmentation.

The preferred configuration is the use bandwidth efficient mode for RTP encapsulation, which should also be chosen as the basis for the RLC PDU size selection. The octet aligned AMR payload format modes would increase the frame size by 8 bits for AMR 5.90, 6.70, 7.40, 10.2 and 12.2 kbps modes as well as for all AMR-WB modes.

In downlink there is a limit on maximum eight different PDU sizes. The RLC PDU sizes defined in 3GPP TS 34.108 [19] are 104, 136, 152, 168, 184, 216, 288, and 336 (or alternatively 328) bits. These fulfill the requirements above reasonably well. In uplink the number of different PDU sizes is not limited to eight and as a result twelve different PDU sizes are available.

Besides encapsulating the IP packets, the RLC is capable of performing retransmissions of packets lost over the air interface by using so-called acknowledged mode. However, the RLC maintains in-order delivery for all packets, and thus a single retransmission of a voice frame leads to all subsequent voice frames being delayed until the retransmission succeeds. For voice application the delayed packets often result in excessive late losses in the de-jittering function, leading to a higher frame loss with retransmissions. Furthermore, the HSDPA includes a second retransmission mechanism (Hybrid ARQ) on the MAC layer, mainly intended to provide robustness against link adaptation errors. This additional MAC layer retransmission mechanism provides the voice application with sufficient protection against air interface packet losses and it is then advisable to use the unacknowledged mode of the RLC, which does not perform RLC layer retransmissions.

The resulting RAB configuration for the voice component of the Multimedia Telephony service are shown in Table 5.19.

Table 5.19: Radio bearer realizations for voice component of the Multimedia Telephony service. For uplink two different radio bearers are defined, one with 2 ms and one with 10 ms TTI. Note that the MAC header fields include only the fixed parts of the header.

	Uplink	Downlink
Logical channel type	DTCH	DTCH
RLC mode	UM	UM
Payload size [bits]	88, 104, 136, 152, 168, 184,	104, 136, 152, 168,
	200, 216, 280, 288, 304, 336	184, 216, 288, 336
	(alt 328)	(alt 328)
Max. data rate [bps]	Depends on E-DCH category	Depends on UE category
RLC header [bits]	8	8
MAC-d header [bits]	0	0
MAC-d multiplexing	N/A	N/A
MAC-d PDU size [bits]	96, 112, 144, 160, 176, 192,	112, 144, 160, 176,
	208, 224, 288, 296, 312, 344	192, 224, 296, 344
	(alt 336)	(alt 336)
MAC-es/hs header [bits]	6	21
MAC-e header [bits]	12	N/A
Transport channel type	E-DCH	HS-DSCH
TTI [ms]	2/10	2
CRC [bits]	24	24
Channel coding type	TC 1/3	TC

As can be seen from the discussion above, the choice of the RLC PDU sizes is relatively complex, and needs to be coupled to the chosen voice codec modes. This restricts the flexibility in deploying new codecs or codec modes. A natural improvement would be to make the RLC PDU size varying so that each VoIP packet can always be packed to a single RLC PDU without padding. Such solutions have not yet been standardized, even though they have been discussed in 3GPP.

The RLC PDUs are transmitted forwarded to the MAC sublayer, which transmits them over transport channels. For packet access, there are basically three types of transport channel.

Dedicated channels are available in the uplink and downlink. In the uplink in addition to the normal dedicated channels, the Enhanced Dedicated CHannel (E-DCH) can be used, while for the downlink the High Speed Downlink Shared CHannel (HS-DSCH, used to carry HSDPA) can be used, as discussed in Chapter 3.

The choice of the transport channel determines to a large extent the system capacity and media quality.

Even though the dedicated channels provide even quality for the media, the resulting capacity in the downlink favors the use of the HS-DSCH in the downlink. As can be seen in Table 5.19, the HS-DSCH is used as conversational voice bearer in the downlink. The drawback of the shared channel is the increased jitter due to scheduling between users, and this jitter needs to be accounted for by the media processing algorithms.

In the uplink there are no shared channels. The fast retransmission mechanism provided by the enhanced dedicated channel improves the uplink capacity, and thus the E-DCH is used as transport channel in the uplink. As for the downlink, the increased capacity creates additional jitter due to retransmissions and needs to be compensated with de-jittering functionality.

For enhanced uplink the possibility of using a short 2 ms TTI is also provided. The short TTI allows a shorter transmission delay for the uplink, which can be turned to shorter end-to-end delay or increased delay threshold for the downlink scheduling, as will be discussed in Chapter 7.

Finally the use of the Turbo coding for channel coding completes the radio bearer description.

In general, each new radio access bearer needs to be tested in order to achieve interoperability between terminals and networks. The bearers that are tested in conformance testing are listed in [19].

In addition to describing the radio bearer optimized for conversational voice transmission, we describe the most important system parameters for optimizing the HSPA system for voice.

The operating point for a retransmission protocol determines the targeted number of transmission attempts and the desired probability for retransmissions (block error rate). In general requiring a very low probability for a packet loss leads to very poor capacity, and thus it is beneficial to operate at non-negligible block error rates.

For HARQ with soft combining of received bits the unsuccessful transmission attempts do not lead to wasted resources, but can rather be useful in ensuring that the initial transmission attempts do not consume too much power (or other resources). For this reason, the HARQ protocols are typically operated at very high block error rates for initial transmissions. It is not uncommon to have close to 100% error rate for the first two or three transmissions, and only for subsequent transmissions is the probability of a successful transmission reasonably low. However, if one of the earlier transmissions succeeds, a significant amount of resources has been saved. The main limitation on the number of transmission attempts comes from the required transmission delay, and with strict delay requirements less retransmission attempts can be used.

For E-DCH the general argumentation about using high block error rates for the HARQ is valid, and typically several transmission attempts are used. As the retransmissions are faster for 2 ms TTI, more transmission attempts are targeted for 2 ms TTI than for 10 ms TTI. For example, if the targeted delay requirement for uplink retransmissions is 80 ms, it would be possible to target a small block error rate only for the fifth transmission with 2 ms TTI, while for 10 ms TTI the block error rate should be small already for the second transmission.

For HS-DSCH the trade-off between delay and number of retransmissions is more complicated. For HS-DSCH there is already a link adaptation mechanism, which tries to optimize the link quality for each user. Using more retransmission attempts makes the link adaptation more complicated, as the scheduler should ensure that the link quality is good not only for a single transmission but also for subsequent retransmissions. For this reason the HARQ operating point for HS-DSCH is typically chosen so that the probability for a successful first transmission is already reasonably high.

The different HARQ operating points for E-DCH and HS-DSCH can also be viewed as a fundamental difference in the way the link adaptation is done in the uplink and downlink. In the downlink (with fast centralized scheduler) it is beneficial to choose the individual user(s) with best link quality, while for the uplink (without fast scheduler) the retransmissions are the most feasible means to provide link adaptation.

For HSDPA the scheduling algorithm has a major impact on both the system capacity and media quality. This effect will be studied in more detail in Chapter 7, but in this chapter we describe the most commonly used scheduling algorithms. In particular we discuss the suitability and possible implementations of the following schedulers for conversational traffic:

- Round-Robin Scheduler;

- Max-CQI Scheduler;

- Proportional-fair Scheduler;

- Delay Scheduler.

Round-robin is the most basic scheduling strategy. For HSDPA, the round-robin algorithm can be realized by keeping a list of active users and maintaining a pointer to the user that was last scheduled. The next scheduled user is then simply the next user on the list.

The only complications for the round-robin scheduling algorithm come from the need to handle users with empty buffers and adding new users to the scheduler. One possible way to handle this is to remove users with no buffered data, and always add new users (and old users with new packets) to the end of the list. As the new packets have no prior scheduling delay in the MAC-hs scheduler, this approach should result in good performance for VoIP.

The scheduling delay of the round-robin scheduling algorithm is determined by the number of users in the queue. The maximum delay is also limited by the number of users, and as long as the total number of users is low enough to satisfy the requirement for the maximum scheduling delay, the round-robin method should provide good media quality. However, the maximum capacity may not be reached with the round-robin scheduler, as the link quality of users is not taken into account.

The other basic scheduling strategy would be to always schedule to the user with best link quality. This approach can be realized by maintaining a list of received Channel Quality Indicators (CQIs) from different users, and selecting always the user with best CQI value. The resulting algorithm could be called Maximum-CQI algorithm.

Maximum-CQI method is typically expected to result (naturally) in the highest total cell throughput (see e.g. [157]). However, the performance for individual users varies a lot, and for users in poor radio environment there is no limitation on the maximum possible scheduling delay. This tends to lead to very large delay for some users in the system, and in practice to a low number of satisfied users (and thus low capacity).

Proportional-fair schedulers schedule the user with highest ratio between some quality and the average transmission rate. We focus on a scheduler that schedules the user according to the maximum of the instantaneous C/I and the average transmission rate. This algorithm takes advantage of the short-term channel variations while at the same time maintaining the same long-term throughput for all users.

While the proportional-fair scheduler maintains the same long-term throughput for all users, it does not provide very good delay characteristics, and, similarly to the maximum-CQI scheduler, can result in some users being starved for prolonged periods of time. This results typically in a poor performance for the conversational services.

The scheduling methods discussed above are general in the sense that they do not take into account the specific requirements of the conversational services. It is also possible to use scheduling methods that have been optimized for conversational services.

The delay scheduler increases the priority of each user based on the delay encountered by the oldest packet. The increase can be realized using any increasing function, but we have focused on a scheduler that increases the priority using a parameterized barrier function which consists of three pieces. Two of the pieces are linear and one is based on a polynomial of the second order. The priority function is shown in Figure 5.19. This type of priority function ensures that the delayed packets have a very high probability of being scheduled.

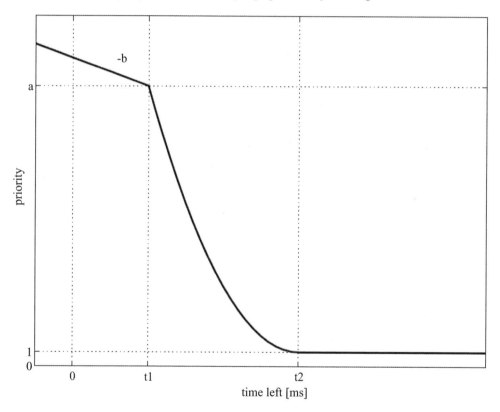

Figure 5.19: The priority-time function used for scheduling.

In addition to prioritizing the delayed packets, the delay scheduler maintains a low buffer level by discarding packets that have been queued for excessive time.

As the delay scheduler takes into account the conversational nature of the Multimedia Telephony service, and it generally provides the best performance in terms of capacity and service quality. However, the delay scheduler is not suitable for interactive or background services (such as web browsing or file download), for which a different scheduler should be used.

A summary of the HSDPA schedulers is listed in Table 5.20.

Table 5.20: Characteristic features of different HSDPA schedulers.

Scheduler	Summary	Suitability for conversational services
Round-robin	Simple to implement and understand	Good
Maximum-CQI	Maximizes system throughput	Poor
Proportional-fair	Fair division of throughput between users	Poor
Delay	Optimized for conversational services	Best

5.6.2   EDGE

The radio realization for EDGE has not been completed at the time of writing. It is however quite clear that one will probably want to encapsulate more than one frame per packet to save overhead in lower layers. Even though this is beneficial for the utilization of the radio bearer, there are also a few drawbacks. Several frames per packet will increase the end-to-end delay because the transmitter will have to buffer frames before it can complete the packet and send it. Another drawback is that one lost packet means that several consecutive speech frames will be lost and consecutive frame losses are typically more audible than single losses. A reasonable compromise is to encapsulate two frames per packet.

5.6.3   Other Networks

IMS Multimedia Telephony must work also for other IP networks, such as:

- land-line IP networks specified by TISPAN [74];

- xDSL, as specified in the ITU-T G.99x series of specifications;

- WiMAX [77];

- WLAN, as specified in the IEEE 802.11 series of specifications;

- packet switched Generic Access networks [11].

These networks should impose no special restrictions on the Multimedia Telephony service, at least not under normal operating conditions. One possible consideration is that these networks may either be bit rate or packet rate limited under different operating conditions. To work well under all possible network conditions, the Multimedia Telephony clients must be prepared to adapt in both bit rate and packet rate.

5.6.4 Example Delay Budget for HSPA

The end-to-end delay is an important compromise in all real-time wireless communication systems. On the one hand, for CS bearers a certain amount of delay is required to give room for sufficient interleaving. Interleaving is needed to handle fading dips and distribute the bit errors over time so that the channel decoder has a fair chance of recovering the transmission block. For best possible channel coding performance, the interleaving should be as long as possible.

On the other hand, long delay is undesirable for real-time services because it is hard to have a fluent conversation if the users have to wait for a long time for the response from the other user. Long delays thus reduce the conversational quality; see for example [109]. The delay is thereby a compromise between channel coding performance and perceived conversational quality.

For circuit switched systems, the end-to-end delay is very close to constant. There is a small amount of sample slip, due to clock drift, which causes a small amount of jitter. The amount of clock drift is however so small that it is usually considered as insignificant for the quality. For HSPA, there is on the other hand a very large amount of delay jitter due to the scheduling in the downlink and HARQ in uplink and downlink. This delay jitter is a necessary component in order to achieve good capacity and coverage.

The delay budget for HSPA thus contains both fixed and varying components, as shown in Table 5.21. The values in Table 5.21 are one possible realization of the delay budget, and are based on [184], but have been modified to better highlight the different sources of the delay. The actual values depend mostly on the implementation, and not on the standardized procedures. The example values listed in Table 5.21 serve as a rough example only, and should not be considered as representative values for typical implementation (as no real implementations are available). The components that contribute to the delay are also explained below.

Table 5.21: Example of a delay budget for HSPA with 10 ms Enhanced Uplink and HSDPA. Note that the actual values are only rough estimates.

Uplink (EUL 10 ms)	Delay	Downlink (HSDPA)	Delay
AMR encoder	35 ms	AMR decoder	5 ms
UE L1/L2 processing	5 ms	UE L1/L2 processing	10 ms
TTI alignment	0–10 ms	—	—
Uu Interleaving	10 ms	Uu Interleaving	2 ms
UL re-Tx	0–80 ms	DL Scheduling	5–100 ms
RNC/Iub/Node B	10 ms	RNC/Iub/Node B	10 ms
Iu + Gi	5 ms	Gi + Iu	5 ms
Sum min. UL	65 ms	Sum min. DL	37 ms
Sum max. UL	155 ms	Sum max. DL	132 ms

The AMR encoder uses a frame length of 20 ms and uses a look-ahead of 5 ms for all codec modes other than the AMR 12.2 kbps mode. It is therefore necessary to buffer 25 ms of speech before the speech encoding can start. The processing time is here estimated to be 10 ms. The buffering and processing times together add up to 35 ms. It should be noted that if the client includes front-end handling functions like noise suppression and acoustic echo cancellation then further look-ahead and processing time may be required. Since the front-end handling is vendor specific, and can be performed in many different ways, the buffering delay and processing times for these functions are not included in this delay budget.

On the receiving side, the processing time for the AMR decoder is estimated to be 5 ms. It should be noted that an ideal jitter buffer does not add to the delay budget. If the jitter buffer allows for a certain amount of late losses then the end-to-end delay can actually be shorter than the maximum packet delay. In this delay budget it is however assumed that the jitter buffer will be ideal.

The layer 1 and layer 2 processing includes the following protocol layers:

- the Packet Data Convergence Protocol (PDCP) [36], which includes ROHC;

- the Radio Link Control (RLC) layer [41];

- the Medium Access Control (MAC) layers [31];

- the Physical (PHY) layer, which includes channel coding and modulation.

It is estimated that the processing time for these functions will be about 5 ms in uplink and about 10 ms in downlink. The longer processing time in the downlink is mostly caused by the decoding process, which is significantly more complex than the encoding in the uplink.

In order to match the transmission to the beginning of the next available transmission time interval, the transmission may have to wait for a while. This is the TTI alignment component. In uplink, when a TTI length of 10 ms is used, this delay can be anything between 0 and 10 ms, and when the TTI length is 2 ms it may be anything between 0 and 2 ms. For the downlink the corresponding number is included in the DL Scheduling time.

The Uu Interleaving consists of the actual data transmission over the air interface. The transmission is interleaved over the whole transmission time interval, resulting in a delay equal to one TTI. This is 10 ms for the uplink and 2 ms for the downlink.

UL re-Tx stands for delay caused by uplink retransmissions. It is assumed that a single retransmission takes 40 ms, and that at most two retransmissions are allowed before the packet is dropped.

Similarly, the DL Scheduling also includes a timer that drops packets that have been waiting too long in the scheduling queue. In this delay budget, the delay for retransmissions is included in the scheduling delay. The total DL Scheduling delay is here set to 100 ms.

The radio access network delay consists of the Node B processing time, the transmission time over the Iu interface and the RNC processing time. A very rough estimate of the network delay is 10 ms. It is furthermore estimated that the Core Network delay (consisting of Iu and the Gi interfaces) will add another 5 ms.

The minimum end-to-end packet delay is thus about 100 ms and the maximum is about 290 ms. These delay components, and thus also the minimum and maximum end-to-end packet delays, are of course implementation dependent. It can also be noted that the major part of the delay jitter comes from uplink retransmissions, downlink retransmissions and scheduling delay. The maximum delay of 290 ms may look discouraging. It will however be shown in Chapter 7 that simultaneous long delays in both uplink and downlink occur quite rarely and the uplink delay and the scheduling delay are for most cases well below the upper limit.

For optimal performance it is important that all nodes from user A to user B are optimized to minimize the different delay components. Minimizing the fixed delay part and the variable delay part in functions within the different nodes allows for using more retransmissions in either uplink or downlink or both, while still maintaining an end-to-end delay within the delay constraint. The performance can be optimized in several ways. A few examples are listed below:

- Faster processors give shorter processing time.

- Streamlining the moving of data in between layers and processing functions so that the packets or the transport blocks are not held somewhere in some queue.

- Prioritizing media packets over other types of packets in the terminals and network nodes so that the media packets are not buffered somewhere while another packet is being processed.

- Aligning encoder and decoder processing with the TTIs so that the TTI alignment time is close to zero. This is already done for CS, at least for the cases where it is possible to ensure that the speech frames are synchronized to the transmission time slots.

3GPP has defined delay requirements for audio in real-time services in 3GPP TS 22.105, [44]. The requirements are that:

- the end-to-end, one-way delay (mouth-to-ear) should preferably be below 150 ms;

- the upper limit for the delay is 400 ms.

Comparing the delay budget with these requirements one can see that the delay budget is well within these requirements. The fixed part of the delay budget, 100 ms, is well below the 150 ms preferred limit. The upper bound of the delay budget, 290 ms, is also well below the upper limit for the delay requirement. One can therefore conclude that there is plenty of room for using retransmissions in order to reduce the packet loss rate and provide satisfactory quality for end-users, in terms of both perceived delay and listening quality.

5.7 Interworking

5.7.1 Speech

Probably the most interesting interworking scenario is interworking between the voice media component in Multimedia Telephony and legacy circuit switched voice calls in GERAN and UTRAN. The reason why this is so important is simply because of the large amount of legacy phones and systems deployed over the whole world. It has been estimated by GSM Association [83], that there are currently over 2 billion GSM subscribers in about 210 countries over all the world. Section 5.7.1.1 therefore elaborates in more detail why it is important to avoid tandem coding.

For proper interworking with GERAN and UTRAN circuit switched services, it is important to understand how the AMR and AMR-WB codecs work, what capabilities and limitations they have and how these codecs are actually used in these systems. Sections 5.7.1.2 and 5.7.1.3 therefore describe the most important aspects for interworking with AMR for GERAN and UTRAN systems respectively. The discussion is here limited to AMR, but the functionality is analogous for AMR-WB. Section 5.7.1.4 discusses the codec configurations that have been defined for AMR and AMR-WB. Section 5.7.1.5 thereafter discusses a number of quite detailed aspects that are important for proper interworking.

5.7.1.1   Importance of Avoiding Tandem Coding

The number of cellular phones and subscribers also has another consequence for how one should use the codecs. When GSM was originally designed at the beginning of the 1980s, the estimated penetration rates were in the order of 5%, even 10% penetration rate was regarded as 'very futuristic' and 'visionary'. It was realized that there would be mobile-to-mobile calls, and to maximize the quality in these calls one needed to avoid tandem coding. Low penetration rates however would mean that mobile-to-mobile calls would be quite rare and tandem-free operation was therefore regarded as less important.

The current penetration rates are however approaching 100% in most Western European countries. In fact, in countries like Sweden and Finland, people are ending their fixed line subscriptions and use only their mobile phones. This makes mobile-to-mobile calls just as frequent as mobile-to-PSTN calls, if not even more frequent. The problem is now that mobile-to-mobile calls require two encoding–decoding processes, so-called tandem coding, in most currently deployed cellular networks. Figure 5.20 shows the principle of tandem coding.

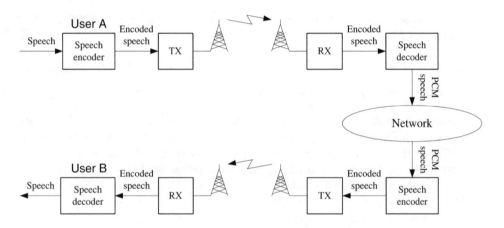

Figure 5.20: Schematic figure of mobile-to-mobile calls using tandem coding.

The original speech from user A is encoded in the first terminal, transmitted over the air interface to the receiver in the base station. The encoded speech is then decoded into PCM data. The PCM data is then transmitted through the network to the cellular network for user B where it is encoded once again, transmitted over the wireless access to the terminal which decodes the speech.

Since the original development of GSM, it has been known that tandem coding reduces the speech quality. Tandem Free Operation (TFO) and Transcoder Free Operation (TrFO) were therefore developed to avoid the degradations introduced with tandem coding. Figure 5.21 shows a schematic figure of how TFO works.

In TFO, the encoded speech is decoded to PCM data in user A's network. However, the least significant bits of the PCM data are 'stolen' and the encoded speech bits are inserted in the stolen bits. Upon receiving the PCM data, including the originally encoded bits, in user B's network the second encoded speech bits can be extracted from the LSBs of the PCM data and those bits transmitted to the terminal for user B. Tandem coding would thereby be avoided for user B, while still allowing the PCM data to be used for legal intercept. The PCM speech will however be somewhat noisy since the stolen bits for the in-band channel add

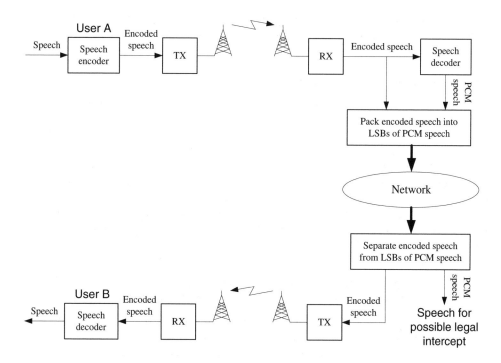

Figure 5.21: Schematic figure of Tandem Free Operation.

noise to the PCM samples. Since it is the LSBs that are stolen, the quality impact is however quite small. The drawback is, obviously, that the transcoders used for speech decoding and encoding are still required in the network.

With TrFO, the encoded bits are never decoded to PCM data. The encoded bits are transmitted through the network to user B as a packet, without the need to decode the information to PCM and without encoding the intermediate PCM data a second time. TFO and TrFO are thus two possible solutions that can be used to increase the quality in mobile-to-mobile calls.

Tandem coding has however not only a quality impact. To compensate for the degraded quality one should in reality reduce some of the other degrading factors, at least when the percentage of mobile-to-mobile calls is significant enough to impact on the aggregated speech quality for the whole system. There are two possibilities:

1. Choose another codec that has a better intrinsic quality. Since the AMR codec modes are state-of-the-art codecs, for their respective bit rate, this would mean that one has to choose a codec with a higher bit rate. A higher bit rate would however reduce the capacity in the system and is therefore not desirable.

2. The other possibility is to reduce the amount of degradations due to channel impairments. The only way to do this is to increase the C/I level for which the service can be used. This would obviously also reduce the capacity in the system.

Tandem coding thus has an impact also on the system capacity. It is important that Multimedia Telephony also offers the possibilities to avoid tandem coding in VoIP-to-CS calls from the start.

5.7.1.2   AMR Operation in GERAN

As described in Section 5.1.1.3, the AMR codec has eight codec modes. When AMR is used in GERAN, there is a limit to how many codec modes one can use in the session. This limit is introduced by the limited in-band channel allocated for signaling Codec Mode Indication (CMI) and Codec Mode Commands/Requests (CMC/CMR).

Codec Mode Indications are used to identify the codec mode that was used for encoding the current speech frame. Codec Mode Commands are transmitted from the system to the client and are used to command the client to switch to a certain AMR codec mode. Codec Mode Requests are transmitted from the client to the network and are used to request that the system switch to the requested mode. The CMC and CMR thus reflect the view that the network is the master and the client is a slave.

Since only 2 bits are allocated to the in-band channel, the consequences are as follows:

- It is only possible to signal four codec modes. An Active Codec Set (ACS) must therefore be selected. The codec must restrict its operation to the selected codec modes.

- CMI and CMC/CMR can only be transmitted in alternate frames. Since CMI defines the codec mode that is currently used for the encoding, and since this information is necessary for proper decoding, the consequence is that codec mode changes must be limited to every other frame border.

Due to fact that the CMI and CMC/CMR are transmitted in alternating speech frames, it is important that the transmitters and receivers are synchronized to avoid CMIs being misinterpreted as CMRs and vice versa. This becomes especially crucial in Tandem Free Operation (TFO) and Transcoder Free Operation (TrFO) since CMI and CMC/CMR can have different phases over the two access links. It was therefore decided to define the 'CMI Phase'. The default transmission phase is 'odd', but the transmission phase can be changed to 'even' in downlink. For the uplink, the transmission phase is 'odd' and is never changed. The capability of changing phase in downlink solves the issue of having different phases in the two different air interfaces in TFO and TrFO calls.

The performance of the in-band channel codec can be improved if the decoder can take advantage of the knowledge that codec mode adaptation is only performed by switching to the neighboring codec modes. It was therefore decided to restrict codec mode changes in GERAN.

These limitations introduce a few requirements for how the speech codec in Multimedia Telephony should operate when TFO or TrFO calls are made to a GERAN mobile:

1. The codec must be capable of operating with a subset of the available codec modes. The ACS may contain four codec modes but may also be smaller.

2. The codec must be capable of aligning codec mode changes to every other frame border in the transmitting direction. It is recommended that codec mode changes are allowed at any frame border in the receiving direction.

3. The codec should change codec modes only to neighboring modes in the transmitting direction. It is however advisable to allow codec mode changes to any mode in the receiving direction.

5.7.1.3 AMR Operation in UTRAN

When the AMR codec was introduced in the UTRAN system, it was decided that the limitations in the GERAN in-band signaling should be removed. There are therefore no limitations on active codec set and codec mode changes in UTRAN.

The development of UTRAN products has however been performed in steps. In the beginning, the only supported codec mode was AMR 12.2 kbps. As time has gone by, different vendors have expanded their supported codec mode set to also include other codec modes. The expansion of the codec mode set has however not been coordinated and one must be prepared to operate with any subset of the AMR and AMR-WB codec modes.

Even though UTRAN allows for more flexible operation than GERAN, it has been decided to harmonize the operation. In the transmitting direction, the operation in the UTRAN system should follow the limitations on codec mode set and codec mode changes of GERAN. In the receiving direction, the speech decoder should however accept all codec modes and any mode changes.

5.7.1.4 Codec Configurations for TFO and TrFO

The limitations in the GERAN and UTRAN give a large amount of possible combinations, which becomes a problem when one wants to use tandem-free operation or transcoder-free operation between two GERAN or UTRAN mobiles or between one GERAN and one UTRAN mobile. To give some guidance to the implementers on what combinations should be supported, it was decided in 3GPP to select and specify some combinations as more important than others.

These configurations are defined in 3GPP TS 28.062 [12] for AMR and in 3GPP TS 26.103 [45] for AMR-WB and include (among other things):

- a codec type, which defines the codec being used and can, for example, be FR AMR, UMTS AMR or UMTS AMR2;

- the AMR and AMR-WB codecs, which includes several codec modes, for example AMR 12.2 kbps or AMR-WB 12.65 kbps;

- a codec configuration, which defines a set of attributes determining how the codec operates, for example an Active Codec Set (ACS) and DTX='on'.

For AMR, one configuration is of special interest. The configuration, Config-NB-Code=1, includes the AMR 12.2, 7.4, 5.9 and 4.75 kbps codec modes and is especially recommended for TFO and TrFO calls. This configuration is often referred to as *the golden compromise mode set* because it gives maximum compatibility for TFO and TrFO connections with optimal speech quality.

For AMR-WB, the most interesting configuration is Config-WB-Code=0, which includes the AMR-WB 12.65, 8.85 and 6.60 kbps codec modes. This mode set is mandatory for UTRAN, which makes it especially interesting for TFO and TrFO calls.

5.7.1.5 Voice Interworking Between Multimedia Telephony and Circuit Switched Systems

To avoid tandem coding in calls between a Multimedia Telephony terminal and a terminal camping on either GERAN or UTRAN, the session setup must be restricted to the operational limitations that apply for each network. These limitations were envisioned when developing the AMR payload format (see Section 5.2.2), and suitable SDP parameters were therefore defined.

- The `mode-set` parameter can be used to restrict the ACS to the codec modes that both end-points can support.

- The `mode-change-period` parameter can be used to restrict mode changes to periods. Note that there is no need to have a parameter to change the CMI period from 'odd' to 'even' in VoIP for the following reasons:

  - This synchronization is made in GERAN. Synchronization will therefore be made on the GERAN downlink.

  - The CMI and CMR signaling in UTRAN and the AMR payload format do not impose the same restrictions in CMI and CMR as GERAN does. UTRAN and VoIP terminals should therefore accept mode changes at any time in the receiving direction.

- The `mode-change-neighbor` parameter is useful when interconnecting with GERAN terminals since changing mode to neighboring modes improves the channel codec performance over the GERAN link.

- A new SDP parameter, `mode-change-capability`, is defined in the new AMR payload format. This parameter is useful when the offerer or the answerer wishes to inform the other party that it has the capability of restricting codec mode changes to every other frame border, if the answerer so wishes.

In addition to these SDP parameters, there are also several other SDP parameters that are useful when setting up the session with a media gateway.

- To save memory in the media gateway, it will probably define quite low values on `ptime` and `maxptime`. These parameters define how many frames will be encapsulated in each packet and has thus an impact on the amount of memory that the receiver must allocate in the receiver buffer. It should not be a problem of the terminal to allocate sufficient memory, even if one desires to send many frames in each packet. For the media gateway, on the other hand, large packets become problematic since the media gateway is responsible for handling a large amount of voice calls, typically in the order of several hundred calls or more.

- The new `max-red` parameter is also useful for the media gateway since it defines the maximum distance, in frames, between an original frame and the last redundant frames. Knowing this, the media gateway can allocate sufficient memory for the redundancy without allocating too much.

- The octet-aligned payload format may be useful for media gateways dimensioned for a huge number of users. In this format, the in-band channel and the speech frames are aligned to whole octets, which simplifies the parsing of the RTP payload and thus reduces the complexity a little. This should not give any major difference for most normal media gateways but it might be useful for media gateways managing thousands of calls.

Another important aspect is the rules imposed by the SDP offer–answer model in that the answerer can only choose from configurations, i.e. RTP payload types, included in the SDP offer when constructing the SDP answer. Therefore, it is beneficial if the SDP offer includes all the possible configurations that the offerer supports.

In some cases, the answerer is allowed to add a parameter to the configuration. For example, if the session is initiated by a terminal, if the terminal supports all codec modes and if the terminal therefore does not define the mode set parameter, then the answerer, which may be a media gateway, is allowed to define a mode set in the SDP answer. If the terminal, on the other hand, includes the mode set parameter in the SDP offer, even if the mode set includes all codec modes, then the answerer is not allowed to change the mode set parameter and thus only has the option either to accept it or to reject it. For a media gateway interfacing GERAN or UTRAN, this means that the media gateway would probably have to reject the SDP since restrictions on the ACS typically apply both for GERAN and UTRAN. To resolve this problem, the offerer would have to try with a second session initiation, which obviously increases the call setup time. There is even a risk that the offerer would not even try with a second SDP offer since it typically does not know the reason why the session was rejected. An improperly defined SDP offer may thus result in unnecessary call blocking, even if the clients and the media gateways both support a common set of codec modes.

To avoid this problem, at least one of the PS end-points, either the terminal or the media gateway, should support all codec modes. Since the media gateway in any case will have to be restricted to the modes supported by the circuit switched session on GERAN or UTRAN, the conclusion is that it is the Multimedia Telephony terminal that must support all codec modes.

Another interworking consideration is what to do when the media gateway receives frames either containing errors or when frames are lost. Frames with errors should only occur in the direction from the CS mobile to the media gateway. In this case, the media gateway has three options:

1. Pack the (erroneous) frame into the RTP packet and set the Q bit to zero, which indicates to the receiver that there are errors in the frame.

2. Drop the received frame and send a NO_DATA frame to the VoIP client.

3. Drop the received frame and send no RTP packet to the VoIP client.

Out of these three options, the last option is discouraged since the receiving VoIP client will think that this was a packet loss in the IP network. If the VoIP client sends RTCP Receiver Reports, which includes a metric for packet loss rate, then the reported loss rate may be inappropriate. In the worst case, dropping the frame without informing the receiver may even trigger adaptation of codec mode and possibly also redundancy.

Options 1 and 2 both have advantages and disadvantages. The option to pass on the erroneous frame may look attractive since the receiving VoIP client will receive the innovation bits that are needed to improve the handling of erroneous frames. This will however require some resources, which is suboptimal if the VoIP network is operating at or close to maximum capacity. The second option may be the best solution since a NO_DATA frame requires very few bits to be transmitted.

In the direction from the VoIP client to the media gateway, one can expect that there are only lost frames. In this case, the media gateway still needs to create a frame to the CS mobile. The typical solution should be to send a NO_DATA frame to the CS mobile.

It should be clear that there are many details that need to be considered when designing media gateways interfacing VoIP in Multimedia Telephony and legacy circuit switched systems. In addition, the Multimedia Telephony terminals also need to be carefully designed to work properly in interworking scenarios.

5.7.2   Video

The video codecs supported by the Multimedia Telephony service are common for video-enabled conversational services and should not cause major concerns. All profiles and levels are standard and it should therefore be straightforward to negotiate support with other clients. No special sub-profiling or restricted capability sets are used, with the exception that H.264/AVC shall not use frames out of order.

As the capability negotiation between clients is end-to-end, it should be possible for clients to agree as long as they have some capabilities in common. The H.263 Profile 0 (Baseline) codec is often a fallback solution since it is commonly used for video conferencing. Support for higher bit rates implies support for lower bit rates, so there is also a possibility to lower the bit rates to reach an agreement. However, if there are no common capabilities it is necessary to have a transcoder in the loop.

The main interworking scenario for a Multimedia Telephony client is with 3G-324M [6, 7], i.e. the circuit-switched counterpart, which then requires a media gateway. It is necessary to make a protocol translation between SIP/SDP and H.245 and also map between RTP payloads and the Adaption Layer Service Data Units (AL-SDUs) of the H.223 multiplexer [114]. In the media plane there is no direct change and the connection to a 3G-324M client via a gateway should be transparent from the IP side. In the overall interworking scenario there are a few aspects that one should remember.

- **Bit rates:** The circuit-switched side is more sensitive to bit rate variations and cannot handle temporary overshoots as well as the IP side. A variable bit rate stream may therefore induce a video jitter, or in worst case an increasing delay and even data loss (depending on gateway behavior), on the CS side. The problem can be reduced through transrating, i.e. transcoding within the same video standard, although this will also affect quality.

- **Losses:** Bit errors on the CS side should trigger error concealment on the IP side and packet losses on the IP side should trigger error concealment on the CS side. By letting bit errors on the CS side induce packet losses, i.e. increasing the sequence number without sending an IP packet, the decoder will be notified of the loss. However, there is also a risk by doing this, as many CS-side bit errors may lead to many packet losses, which can be falsely interpreted as if the IP network is congested. On the other hand, forwarding packets with bit errors from the CS side to the IP side may seriously break unsuspecting IP client decoders. Packet losses on the IP side can also trigger error concealment on the CS side by using optional sequence numbers on the CS side, but this may not be strictly necessary as CS decoders are usually very resilient.

- **Packet sizes:** In order to minimize the effect of data losses, it is advantageous to align video packets (GOBs, slices) with AL-SDUs, which typically have maximum sizes around 1 kilobyte for different client implementations. In practice, however, AL-SDU maximum sizes vary between 256 bytes and 4 kilobytes. It may therefore be a problem to fit them into individual IP packets, but the more likely problem is that IP packet contents are larger than can be fitted into a single AL-SDU and alignment is thus at least partially lost. One consideration is also that the overhead on the CS side is smaller than on the IP side.

5.7.3 Text

T.140 is supported by most commonly used text communication protocols and services as described in Section 5.1.3. Interworking between them should therefore cause no major problems since a minimum of character conversion is expected. The only possible problem is legacy text telephone devices for the deaf (TDD) that use 5-bit Baudot code for the character encoding. Due to the small code space, these TDD devices only support simple letters and numbers and also only upper-case letters [106].

5.8 Media Configurations for Multimedia Telephony

In this section, specific information about the usage of the various media formats including payload formats in IMS Multimedia Telephony is described. Various recommendations and requirements are described which have been standardized in order to allow the service to provide interoperability, client efficiency, transport optimizations and session reliability on the media layer. However, since IMS Multimedia Telephony is still undergoing standardization, not all of these areas are finalized standards-wise at the time of writing. Hence, the reader is advised to look up the current status of the standard in [23].

5.8.1 Speech

The speech media component in Multimedia Telephony can be encoded as either narrowband speech, using AMR, or wideband speech, using AMR-WB. In both cases, the terminals are required to support all codec modes of the respective codecs. The reason for mandating all codec modes for the terminals is because this is needed to ensure interoperability and fast session setup, especially when media gateways are used; see Section 5.7.1.

In an end-to-end IP session, the clients will therefore be allowed to use any codec mode. Since the HSPA radio bearer is optimized for the AMR 12.2, 7.4, 5.9 and 4.75 kbps codec modes for narrowband speech and for the AMR-WB 12.65, 8.85 and 6.60 kbps codec modes for wideband speech, the specification also includes a recommendation that these codec modes should be used, unless the session is limited to some other modes.

In interworking scenarios with circuit switched voice in GERAN or UTRAN, the codec mode set needs to be restricted to whatever mode set the circuit switched service uses, at least if one wants to avoid tandem coding in order to maximize the quality. This also means that the Multimedia Telephony client should be prepared to adapt codec mode in order to optimize the end-to-end performance and not only the performance over the local air interface. Interworking with circuit switched voice may also introduce restrictions in codec mode changes.

5.8.2 Video

The video component in Multimedia Telephony can be encoded by any of the following four video codecs:

- H.263 Profile 0 (Baseline) [120];

- H.263 Profile 3 [120];

- MPEG-4 Visual Simple profile [93];

- H.264/MPEG-4 AVC Baseline profile [118, 94].

H.263 Profile 0 level 45 is required and thus supported by all clients. Support for level 45 (128 kbps) implies support for level 10 (64 kbps). Higher levels may be supported and used for negotiation. H.263 Profile 0 is the legacy video codec for 3GPP systems and is required for the circuit-switched 3G-324M service [6, 7] as well (level 10). As both H.263 Profile 3 and MPEG-4 Visual decoders are capable of decoding H.263 Profile 0, it was chosen as the common denominator for video-enabled services.

The drawback with H.263 Profile 0 is that it cannot deliver high enough quality compared to the other codecs supported by the Multimedia Telephony service. Hence, it is recommended to support at least H.263 Profile 3 or MPEG-4 Visual. H.264/AVC requires somewhat more encoding complexity, but gives even higher quality in return. However, as the encoder complexity can also be controlled by the encoder, it is expected that H.264/AVC will be widely supported.

There are no restrictions on the usage of the above video codecs, with the exception of keeping latency to a minimum and that a decoder shall be prepared to decode immediately it receives data. For H.264/AVC this means that the numbered of reordered frames shall be zero and that a decoder shall not wait for an Instantaneous Decoder Refresh (IDR) before it starts decoding.

5.8.3 Text

The specification for IMS Multimedia Telephony [23] puts no further restrictions or requirements on how text should be encoded in addition to what is specified in ITU-T Recommendation T.140 [102]. When IP/UDP/RTP is used for transmitting text media, the clients should be prepared to both send and receive redundant media since the transport is unreliable with these protocols. Redundancy is not required when text is transmitted with SIP MESSAGE or MSRP since both methods use acknowledgments to ensure that the text is properly received.

5.8.4 Protocols

There are a number of different restrictions and recommendations when it comes to the usage of the payload formats associated with the respective media codecs that are used in IMS Multimedia Telephony. As mentioned in Section 5.2, the payload formats support a number of features, but not all of them are allowed to be used in IMS Multimedia Telephony. The following restrictions and recommendations apply. Note that only payload format related restrictions and recommendations are described, not source codec usage.

5.8.4.1 Speech

When using the AMR codec for speech and packetizing the data into RTP packets according to [179], there a number of recommendations available. Normally, the protocol usage guidelines are divided into one set of recommendations for the sender and another for the receiver. The recommendations for the receiver are quite often less strict than the recommendations, or restrictions, for the sender in order to enable some level of protocol usage resilience. A receiver should be able to decode a media flow even if the sender does not comply with all required restrictions and recommendations for the service.

- The bandwidth efficient version of the payload format should be used unless the session setup determines otherwise.

- Requests for codec mode changes should only be sent every other frame.

- Internal CRC should not be used.

- Only single channel sessions should be used.

- DTX signaling should be used.

The number of frames that are encapsulated in each RTP packet is a function of the current access type the IMS Multimedia Telephony terminal currently is camping on. If the terminal is access type aware, some special recommendations apply, see Table 5.22. If the terminal is not aware, the default mode of operation shall be used. Generic access denotes generic networks without specific QoS often outside operator control.

Table 5.22: Speech frame encapsulation per access type.

Access type	Encapsulation (speech frames per RTP packet)	ptime [ms]	maxptime [ms]
Unknown	1	20	240
HSPA	1	20	240
EDGE	2	40	240
Generic access	1–12	between 20 and 240	240

Overall, the basic, or default, mode of operation is to keep it as simple as possible. One speech frame per packet, source-controlled rate operation (DTX) is used, no interleaving or internal CRC, mono session only, rate requests every other frame. For more details about the payload format usage for AMR in IMS Multimedia Telephony, see Section 5.3.1 in [23].

5.8.4.2 Video

There are very few recommendations or restrictions when it comes to the protocol usage for conversational video in IMS Multimedia Telephony. Currently, the legacy standard for conversational packet switched video [38] only specifies one criterion; keep the packet size below 512 bytes. For H.264/AVC there is also a restriction that the interleaved mode of RFC 3984 shall not be used as it may potentially increase latency by allowing data to be sent out of decoding order. However, it is important to point out that video coding and packetization guidelines for IMS Multimedia Telephony are still under discussion in the standardization community, hence the same recommendation applies as for the jitter buffer specification as discussed in Section 5.8.5. See [23] for the most current details of the standard.

5.8.4.3 Text

For real-time text, no specific restrictions apply for the RTP payload format usage in IMS Multimedia Telephony. The only recommendation is to use redundant transmission, as allowed by the payload format, on error-prone channels. For semi-real-time text communication two messaging schemes apply: immediate or page-based messaging, and session-based messaging. For session-based messaging in IMS Multimedia Telephony, MSRP shall be used.

5.8.5   Jitter Buffer Requirements

Since the jitter buffer constitutes a crucial part of the speech transport and processing chain, the standard for the media handling in IMS Multimedia Telephony [23] applies some specific criteria on its operation. These criteria consist of functional requirements and minimum performance requirements. The functional requirements specify the session critical criteria, i.e. the criteria which, if followed, allow the session to occur, or, if not followed, will force the session to terminate. The minimum performance criteria, on the other hand, specify quality related requirements on the jitter buffer implementation used in the client. The purpose is to set a lower limit which must be met by all implementations used in IMS Multimedia Telephony clients. It is OK to do better than the demands set by the standard, as no bit-exact solution is specified, but any manufacturer must guarantee that the jitter buffer implementation used in its products complies both with the functional requirements as well as with the minimum performance requirements.

At the time of writing, the technical details of both the functional requirements and the minimum performance requirements are undergoing standardization in 3GPP, hence no technical details will be mentioned here. However, it is clear that the functional requirements will specify such things as sorting capabilities, minimum/maximum jitter buffer depth, clock drift handling, etc., while the minimum performance requirements target criteria in three different areas: buffer delay requirements, requirements on induced decoder concealment operations by the jitter buffer, and requirements on jitter buffer adaptation artifacts such as time scaling. Two main design requirements are however specified.

1. The overall design of the jitter buffer shall be to minimize the buffering time at all times while still conforming to the minimum performance requirements of jitter induced concealment operations.

2. If the limit of jitter buffer induced concealment operations cannot be met, it is always preferred to increase the buffering time in order to avoid growing jitter induced concealment operations.

The basic interpretation of these overall requirements is that any jitter buffer used in IMS Multimedia Telephony shall at any given instant in time always try to minimize the buffering depth while still having the concealment operations induced by the buffer beyond a specific limit. Further, the preferred way of dealing with increasing late losses is to increase the buffering depth even if that will make the overall end-to-end delay grow. For the most current information about jitter buffer requirements for IMS Multimedia Telephony, see [23].

5.8.6   Media and Session Adaptation

Media layer adaptation in IMS Multimedia Telephony has not been standardized at all, at least not yet. However, discussions are ongoing and, since rate adaptation is already in place for speech, it is fair to assume that there will be some kind of media layer adaptation in place in the final 3GPP IMS Multimedia Telephony standard for 3GPP release 7. Hence, in this section, we take the liberty to discuss one example of how media layer adaptation could be done for speech in IMS Multimedia Telephony.

IMS Multimedia Telephony is a service which targets telephony-grade quality, reliability, predictability and interoperability. Hence, the service uses standardized codecs operating in order to maximize interoperability and give well-characterized media quality. The codecs for video and speech are both available in the equivalent CS service and their performance is well known.

The speech codecs used in IMS Multimedia Telephony, AMR and AMR-WB, have adaptation as a core component in the design of the codec. The adaptive feature of the codec, enabling rate changes and transmission of rate requests, is supported by the RTP payload format for AMR, hence rate adaptation is already in place in the standard in terms of signaling and requests. The only thing lacking is a solid definition of bit rate adaptation triggers. Further, adaptation of the source coding bit rate can also be extended to include rate adaptation by using frame aggregation. Increasing the number of speech frames in each RTP packet will reduce the IP layer bit rate. Since speech consumes a rather small amount of bit rate, the size of the IP header should not be neglected in terms of bit rate consumption. Although robust header compression will make the IP header shrink to only a few bytes over the radio link, it is a feature only available over the radio link. The transport network will see the full IP header which in comparison with the actual speech data is of considerable size. Hence, aggregating e.g. two frames in one packet instead of one frame will remove 40–60 bytes every second speech frame. This is a considerable rate reduction in the transport network. Finally, bit rate adaptation is not the only adaptation dimension which might be suitable for speech. Application level redundancy, as shown in Section 5.3.4, is also a good media layer adaptation candidate for speech.

The video codecs in IMS Multimedia Telephony, on the other hand, do not have the same adaptive heritage as their speech counterpart. Although some aspects of adaptation are perfectly feasible also for video, no such new adaptive mechanism is specified so far for IMS Multimedia Telephony. In fact, adaptation for video might be a crucial ingredient to achieve the telephony-grade characteristics sought after for video; even more so than speech due to its larger demands on available bit rate and higher sensitivity to packet loss.

Adaptation by itself should always be treated as a way to adapt to the current conditions for the data transport. In the ideal case, the session starts using the negotiated codec and codec rate and stays that way throughout the session; no adaptive measure is used at all. Fortunately this case will be the common case in the large majority of IMS Multimedia Telephony use cases, especially when using HSPA systems. However, due to the telephony-grade demands on the service, scenarios which change the transport characteristics in a way that threatens the targeted media quality need to be managed in order to maintain service performance and reliability. The scenarios in question will mostly manifest themselves as an increase in packet loss, jitter (where applicable) and re-bufferings. In the following discussion, we will assume that an adaptive jitter buffer takes care of changing jitter characteristics.

A basic scenario in IMS Multimedia Telephony where adaptation can come into play consists of an increase of measured packet losses at the receiver. The source of these losses can vary, as exemplified by the following:

- Transport network congestion. Too much traffic is trying to pass a bandwidth limited link. Losses can occur, mainly due to packet drops in routers.

- Radio link failure. Rough radio conditions either due to coverage issues or lack of sufficient cell capacity.

- Hand-over to a radio access technology that cannot provide the needed resources in terms of bit rate to continue the session in its current state.

- Other sources, e.g. losses due to late arrival of packets at the receiver.

It cannot be assumed that the receiver, or the sender after receiving receiver feedback, can know what causes the packet loss, hence any adaptive measure must be valid no matter what the source of the problem is, or, at least, will not degrade the situation further.

5.8.6.1 Media Layer Adaptation – Speech

Media layer adaptation for speech in IMS Multimedia Telephony can be done in three dimensions, bit-rate, frame rate and error resilience. Since there are multiple adaptation possibilities, it is wise to try them one at a time and evaluate the performance of the tested solution before trying another solution. An example adaptation scheme was developed which uses a state machine. State transitions are triggered by performance metrics and each state includes adapting in one of the above defined adaptation dimensions. In this example, we propose a three-step adaptive process to be used for speech in IMS Multimedia Telephony. The solution is based on two design guidelines.

- The session is set up to use the 12.2 kbps mode of AMR (a similar scheme is possible also for AMR-WB).

- Any resilience measure should be bit rate neutral, i.e. the terminal should avoid going beyond the maximum bit rate negotiated during session setup.

The reason for selecting these guidelines is that, for low load levels, one should deliver good quality speech to the end-users. One should therefore start with a high bit rate mode. When the operating conditions get worse, the client can, at best, know the load level on its own air interface. Even if the local system load level were to allow adding redundancy without switching to a lower-rate mode, it is desirable to avoid increasing the bit rate since one can assume that the clients do not have any information on the load level on the remote side.

The adaptive scheme is divided into three steps: source bit rate reduction by adapting to a lower-rate AMR codec mode, frame aggregation to reduce the packet rate, and addition of application level redundancy. These states are used for adapting from high-quality source coding towards error resilience. In addition, the state machine also includes a state to be used when the operational conditions have improved enough to adapt back towards the high-quality source coding state. Figure 5.22 illustrates the four different states of the state machine.

Table 5.23 describes the intention with each respective state. All adaptive measures are indicated as state transitions between the states S1, S2, S3 and S4. For this adaptation scheme to work, there must be some guiding principles on how to measure, signal and adapt. Table 5.24 describes the state transitions including the condition that triggers the state transition and the actions that should be taken in each respective state.

The only trigger we will consider is Packet Loss Rate (PLR) measurements done by the receiver. Even if other measurement schemes and adaptation triggers are feasible, they require knowledge of lower layer information which we will not assume that a generic IMS Multimedia Telephony client will have access to.

In order for the adaptation scheme to work, the aggregation of the statistics on which the PLR threshold is based needs to be done using at least 1–2 seconds of data. In addition, some sort of hysteresis is needed in state transitions in order to avoid oscillating behavior.

This adaptation scheme requires that bit rate, frame aggregation and redundancy can be signaled end-to-end. The signaling for bit rate adaptation is already in place in the AMR payload format, the CMR bits [179], and the response of the sender is defined in the

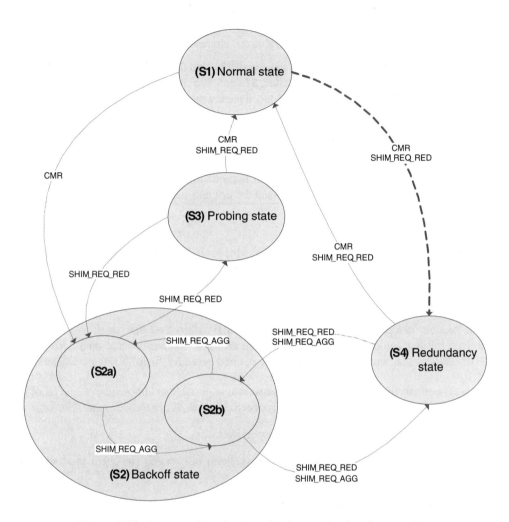

Figure 5.22: State machine for the adaptive mechanism for speech.

interpretation of the bits. The signaling for a change in the frame aggregation or a request for redundancy are on the other hand not in place. Here, different solutions are possible but we would like to introduce the shim concept where a new payload type is introduced for the AMR payload transport which includes up to two extra signaling bytes between the fixed RTP header and the AMR specific payload information.

The addition of the shim field makes it possible to signal the shim messages shown in Figure 5.23. Hence, two different shim messages are needed, one for the redundancy request and one for the frame aggregation request. Note that it is possible to send both requests at the same time.

Hence, whenever an extra shim byte is needed, the sender will switch the payload type to the one which includes the shim, enabling the receiver to parse and decode the

Table 5.23: The four states in the speech adaptation mechanism.

State	Description
S1	Default state. Good channel conditions. Highest coding and packet rate are used
S2	In this state, the encoding rate and possibly also the packet rate are reduced. The state is divided into two substates, S2a and S2b. In S2a, only the encoding rate is reduced. State S2b also adds reduced packet rate
S3	An interim state which is used for probing the use of higher bit rate. The same net bit rate and the same packet rate as S1 are used, but added error resilience makes the probing packets more tolerant to losses
S4	Most robust state. Application layer redundancy is used. The packet rate may either be kept as in state S2 or as in the original state S1

message correctly. Note that the usage of this extra payload type is negotiated during session setup and can be excluded for use within the session if the end-points agree.

5.8.7   SDP Examples

This section includes quite a few examples of SDP offers and SDP answers. They were derived by first defining the scenarios that are the most likely ones and then deriving the SDP offers and answers. The scenarios were defined based on the following parameters:

- The media that will be used in the session. It is possible to choose one or several of the following media types: narrowband speech, wideband speech, video and text.

- What codecs can be used for each respective media.

- Whether the session will be between two clients or between one client and one media gateway.

- The access type(s) that will be used.

5.8.7.1   Narrowband Voice Sessions over HSPA

In the example in Table 5.25, a terminal camping on HSPA is initiating a speech session. The terminal supports only narrowband speech with the AMR codec. Since the terminal is required to support all codec modes and both the bandwidth-efficient and octet-aligned payload formats, it chooses to create the SDP offer shown in Table 5.25.

In this case, the client knows that the access type is HSPA, so it chooses to set `ptime` to 20. Since the client has enough memory to handle receiving up to 12 speech frames per packet, it chooses to set `maxptime` to 240.

It is important that the terminal does not define any mode set because then the answerer is free to respond with any mode set that it can support. If the terminal were to define mode set to any value, then the answerer only has the option either to accept the payload type or to reject it. If the answerer is a media gateway that is interfacing for example GSM-AMR and if mode set is defined, then the media gateway might not be capable to use the defined mode

Table 5.24: All allowed state transitions in the speech adaptation mechanism.

State transition	Conditions and actions
S1→S2a	*Condition:* PLR greater than or equal to 5% or packet loss burst detected *Action:* The codec rate is reduced, for example from AMR 12.2 to AMR 5.9 by using CMR. The state will then be S2a
S2a→S2b	*Condition:* PLR greater than or equal to 5% *Action:* This state transition occurs if the PLR is still high despite the reduction of codec rate. The packet rate is reduced with SHIM_REQ_AGG
S2b→S2a	*Condition:* PLR less than 1% *Action:* This state transition involves an increase of the packet rate. Further, the packet rate is restored to the same value as in S1 using SHIM_REQ_AGG. If the state transition S2b→S2a→S2b occurs, the state will be locked to S2b for some time. This time period should randomized to avoid oscillating behavior
S2a→S3	*Condition:* PLR less than 1% *Action:* Redundancy is turned on (100%) with SHIM_REQ_RED. Further, the packet rate is restored to the same value as in S1 using SHIM_REQ_AGG
S3→S2a	*Condition:* PLR greater than or equal to 2% or packet loss burst detected *Action:* Same actions as in transition from S1→S2a. If the transition S2a→S3→S2a→S3→S2a happens, state S3 is disabled for some time. This time period should be randomized to avoid oscillating behavior
S3→S1	*Condition:* PLR less than 2% and no packet loss burst detected *Action:* Redundancy is turned off by using SHIM_REQ_RED. The codec rate is increased with CMR
S2b→S4	*Condition:* PLR greater than or equal to 2% *Action:* Redundancy is turned on (100%) by means of shim request SHIM_REQ_RED. Further, the packet rate is restored to the same value as in S1 using SHIM_REQ_AGG
S4→S2	*Condition:* PLR greater than or equal to 10%. This is indicative that the total bit rate is too high *Action:* Redundancy is turned off by using SHIM_REQ_RED. State S4 is disabled for some time. This time period should be randomized to avoid oscillating behavior
S4→S1	*Condition:* PLR less than 2% *Action:* Redundancy is turned off with SHIM_REQ_RED. The codec rate is increased with CMR
S1→S4	*Condition:* PLR greater than or equal to 5% or packet loss burst detected AND the previous transition was S4→S1, otherwise the transition S1→S2a will occur *Action:* Redundancy is turned on (100%) with SHIM_REQ_RED. The codec rate is reduced, for example from AMR 12.2 to AMR 5.9, with CMR

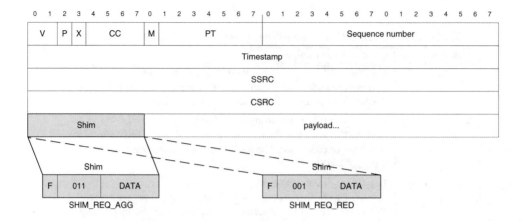

Figure 5.23: Extension of the RTP header to enable adaptation request signaling. The F bit indicates if another shim follows the current. The bits 1 to 3 gives a shim ID giving meaning to the last four bits.

Table 5.25: Example SDP offer for voice created by a terminal camping on HSPA.

```
m=audio 49152 RTP/AVP 97 98
a=rtpmap:97 AMR/8000/1
a=fmtp:97 mode-change-capability=2; max-red=160
a=rtpmap:98 AMR/8000/1
a=fmtp:98 mode-change-capability=2; max-red=160; octet-align=1
a=ptime:20
a=maxptime:240
```

set and would then be forced to reject the whole session. If so, then it might require several initiating trials to agree on a common mode set to use. This would increase the setup time significantly. In the worst case, the session initiator might not try enough mode set variants and the session initiation may fail. This is also one important reason why the terminals must support the complete codec mode sets of the AMR and AMR-WB codecs.

With `mode-change-capability=2`, the terminal shows that it does support aligning mode changes every other frame and the answerer then knows that requesting `mode-change-period=2` in the SDP answer will work properly. The `max-red` parameter indicates the maximum interval between an original frame and a redundant frame.

The SDP answer from another terminal camping on HSPA would probably respond with an identical SDP message showing that it can support all the configurations that the offerer chooses to use.

An SDP answer from a media gateway is much more interesting. The media gateway most probably needs to restrict the codec modes to the codecs that the circuit switched terminal can handle. In the case that the media gateway is interfacing GSM-AMR it must restrict the codec modes to at most four. In this case, it also needs to restrict codec mode changes to be aligned to every other frame border. To improve the likelihood that the codec mode indication is transmitted over GERAN, the media gateway will also set `mode-change-neighbor=1`.

Furthermore, it is likely that the media gateway only supports the bandwidth-efficient payload format. This would give an SDP answer like the one shown in Table 5.26.

Table 5.26: Example SDP answer from a media gateway.

```
m=audio 49152 RTP/AVP 97
a=rtpmap:97 AMR/8000/1
a=fmtp:97 mode-set=0,2,4,7; mode-change-period=2, mode-change-neighbor=1; \
  mode-change-capability=2; max-red=20;
a=ptime:20
a=maxptime:40
```

The media gateway also set `max-red` to 20 to show that it might use redundancy and the redundant frame may be transmitted up to 20 ms after the original speech frame. This is helpful for the receiving client because it defines an upper limit of how long the receiving client may have to wait for redundant frames.

By setting `ptime` to 20 and `maxptime` to 40, the media gateway shows that it wants to receive one frame per packet but can also handle two frames per packet.

When a media gateway initiates the session it is likely that it will create an SDP offer that is identical to the SDP answer in Table 5.26.

Legacy circuit switched UTRAN terminals typically only support the AMR 12.2 mode. If a media gateway interfaces such a terminal, it is likely that the SDP offer is constructed as in Table 5.27.

Table 5.27: Example SDP answer from a media gateway interfacing a legacy UTRAN terminal.

```
m=audio 49152 RTP/AVP 97
a=rtpmap:97 AMR/8000/1
a=fmtp:97 mode-set=7; max-red=0;
a=ptime:20
a=maxptime:40
```

In this case, there is no need to define the `mode-change-period` and the `mode-change-neighbor` parameters for the receiver since there is only one codec mode allowed in the session. For similar reasons there is no need to include the `mode-change-capability` parameter for the transmitting direction.

The `max-red` parameter is also set to 0 to clarify that the media gateway will not use redundancy. It anyway sets `maxptime` to 40 to indicate that it can handle receiving up to two speech frames per packet.

It is possible for a media gateway to support multiple codec mode sets. If the media gateway initiates a session and supports, for example, two mode sets, the SDP offer may be constructed as shown in Table 5.28. In this case, the media gateway declares that it supports the mode sets:

- AMR 4.75, 5.9, 7.4 and 12.2;

- AMR 4.75, 6.7, 7.95 and 10.2.

The terminal receiving this SDP offer will then probably respond with an identical SDP answer since it has to support all codec modes.

5.8.7.2   Wideband Voice Sessions over HSPA

A terminal supporting both wideband and narrowband voice will probably create the SDP offer shown in Table 5.29.

Since wideband codecs give a better quality than narrowband codecs, it is important to list the payload types for AMR-WB before the payload types for AMR. If the payload types were listed with the payload types for AMR first, then the answerer would believe that the offerer prefers to use narrowband and since all terminals and media gateways are required to support this codec they would then create an SDP answer with a preference for the AMR codec and the wideband codec would probably never be used.

It is also important that a terminal offering AMR-WB also offers AMR as a back-up. Then the answerer can immediately remove the payload types for AMR-WB and a narrowband speech session will be used. If the offerer did not offer both wideband and narrowband speech, then the answerer would have to reject all payload types for speech, thereby forcing the offerer to create a second offer. This would increase the call setup time and might in the worst case even mean that a session cannot be established.

In this example, `ptime` and `maxptime` are inserted directly after the media line to show that this is also a valid structure.

There are several possible SDP answers to this SDP offer depending on what configurations the other terminal or the media gateway supports. If the responding terminal supports both AMR and AMR-WB, then the SDP answer will probably be identical to the SDP offer.

Table 5.28: Possible SDP offer from a media gateway.

```
m=audio 49152 RTP/AVP 97 98
a=rtpmap:97 AMR/8000/1
a=fmtp:97 mode-set=0,2,4,7; mode-change-period=2, mode-change-neighbor=1; \
   mode-change-capability=2; max-red=20
a=rtpmap:98 AMR/8000/1
a=fmtp:98 mode-set=0,3,5,6; mode-change-period=2, mode-change-neighbor=1; \
   mode-change-capability=2; max-red=20
a=ptime:20
a=maxptime:80
```

Table 5.29: Example SDP offer from a terminal supporting both wideband and narrowband speech.

```
m=audio 49152 RTP/AVP 97 98 99 100
a=ptime:20
a=maxptime:240
a=rtpmap:97 AMR-WB/16000/1
a=fmtp:97 mode-change-capability=2
a=rtpmap:98 AMR-WB/16000/1
a=fmtp:98 mode-change-capability=2; octet-align=1
a=rtpmap:99 AMR/8000/1
a=fmtp:99 mode-change-capability=2
a=rtpmap:100 AMR/8000/1
a=fmtp:100 mode-change-capability=2; octet-align=1
```

In the case that the responding terminal does not support AMR-WB, then the terminal would have to remove the payload types for AMR-WB. In the case that the answerer is a media gateway interfacing a circuit switched GERAN terminal capable of handling both wideband and narrowband speech, the media gateway needs to restrict the codec modes to a maximum of four. Since 3GPP has declared that the most important mode sets for TFO are

- 12.65, 8.85 and 6.60 kbps for AMR-WB,

- 12.2, 7.4, 5.9 and 4.75 kbps for AMR,

it is likely that the media gateway will answer with the SDP answer outlined in Table 5.30.

Table 5.30: Likely SDP answer from a media gateway for wideband and narrowband speech.

```
m=audio 49152 RTP/AVP 97 99
a=rtpmap:97 AMR-WB/16000/1
a=fmtp:97 mode-set=0,1,2; mode-change-period=2, mode-change-neighbor=1; \
  mode-change-capability=2; max-red=20;
a=rtpmap:99 AMR/8000/1
a=fmtp:99 mode-set=0,2,4,7; mode-change-period=2, mode-change-neighbor=1; \
  mode-change-capability=2
a=ptime:20
a=maxptime:80
```

Another quite likely SDP answer is given in Table 5.31. In this case, it is probably a UTRAN terminal on the circuit switched side and it is likely that it only supports AMR 12.2 for narrowband voice.

Table 5.31: Likely SDP answer from a media gateway for wideband and narrowband speech.

```
m=audio 49152 RTP/AVP 97 99
a=rtpmap:97 AMR-WB/16000/1
a=fmtp:97 mode-set=0,1,2; mode-change-period=2, mode-change-neighbor=1; \
  mode-change-capability=2; max-red=0;
a=rtpmap:99 AMR/8000/1
a=fmtp:99 mode-set=7; max-red=0;
a=ptime:20
a=maxptime:80
```

5.8.7.3   Narrowband Voice Sessions over EDGE

As described in Section 5.6.2, it is likely that for the EDGE access type one wants to encapsulate two speech frames in every packet. Also, redundancy might not work so nicely for EDGE. An SDP offer created based on these assumptions would probably look like in Table 5.32.

In this case, `max-red=0` shows that the terminal will not send redundancy. The attribute `maxptime:240` shows that the terminal can handle up to 12 speech frames per packet, which, given the requirement that no more than four original speech frames can be encapsulated in each packet, implicitly means that it can handle redundancy.

Table 5.32: Example SDP offer created by a terminal camping on EDGE.

```
m=audio 49152 RTP/AVP 97 98
a=rtpmap:97 AMR/8000/1
a=fmtp:97 mode-change-capability=2; max-red=0
a=rtpmap:98 AMR/8000/1
a=fmtp:98 mode-change-capability=2; max-red=160; octet-align=1
a=ptime:40
a=maxptime:240
```

5.8.7.4   Other Narrowband Voice Sessions

For some access types it is more beneficial to reduce the packet rate than to reduce the bit rate. One example of such an access type is WLAN [81]. A suitable SDP offer for such a case is shown in Table 5.33.

Table 5.33: Example SDP offer created by a terminal using WLAN.

```
m=audio 49152 RTP/AVP 97 98
a=rtpmap:97 AMR/8000/1
a=fmtp:97 mode-change-capability=2; max-red=160
a=rtpmap:98 AMR/8000/1
a=fmtp:98 mode-change-capability=2; max-red=160; octet-align=1
a=ptime:80
a=maxptime:240
```

If the media is transported over access types that have no QoS mechanisms, it is beneficial if redundancy is supported to handle occasional large error rates. The SDP example in Table 5.33, in combination with the redundancy levels supported by the Multimedia Telephony specification, makes it possible to use up to 200% redundancy, even if one originally encapsulates four speech frames per packet.

It should be noted that the default packetization time, ptime, does not necessarily have to be 80 ms. It could very well vary depending on the load of the network. If the load is low, and if the terminal could detect this, then it could very well define that it wants to use 20, 40 or 60 ms packetization time.

5.8.7.5   Narrowband Voice Sessions When the Access Type is Not Known

For the cases where the access type is not known and cannot be determined by implicit means, one should use the SDP offers and answers outlined for the HSPA access type. The reason behind this is that HSPA is seen as the most interesting access type for Multimedia Telephony and other access types are less likely to accommodate Multimedia Telephony with the same performance as circuit switched calls.

5.8.7.6   Video-Only Sessions

An example of a video-only session is outlined in Table 5.34. In this case, the offerer wants to communicate using the video encoded with H263 Profile 0 (Baseline) at level 45, i.e. bit rates up to 128 kbps.

Table 5.34: Example of an SDP message with only video.

```
m=video 49154 RTP/AVP 99
a=rtpmap:99 H263-2000/90000
a=fmtp:99 profile=0;level=45
```

5.8.7.7 Text-Only Sessions

When initiating a text-only session, the Multimedia Telephony client will probably construct the SDP offer shown in Table 5.35.

Table 5.35: Example of an SDP offer for text-only sessions.

```
m=text 49156 RTP/AVP 100 101
a=rtpmap:100 t140/1000
a=rtpmap:101 red/1000
a=fmtp:101 100/100/100
```

This example shows that RTP Payload Type 100 is to be used for sending text without redundancy while RTP Payload Type 101 is used for sending text with redundancy according to RFC 2198 [159]. It should be noted that two payload types need to be defined even if one only plans to use the payload type for redundant media. This is because, in the payload type for redundant media (101), one needs to reference a payload type which is used without redundant media, in this case payload type 100. To indicate that the sender will use 200% redundancy, the number '100' occurs three times in the slash-separated list on the a=fmtp:101 line. The meaning of this definition is that payload type 100 will be used in the primary encoding and also in the two redundant encodings.

5.8.7.8 Sessions Including Voice and Video

When multiple media types are included in a session, the SDP offer needs to include one media line, m=, for each media type. An example of this is included in Table 5.36.

Table 5.36: Example of an SDP message for voice and video.

```
m=audio 49152 RTP/AVP 98 97
a=rtpmap:98 AMR-WB/16000/1
a=fmtp:98 mode-change-capability=2; max-red=160
a=rtpmap:97 AMR/8000/1
a=fmtp:97 mode-change-capability=2; max-red=160
a=ptime:20
a=maxptime:240
m=video 49154 RTP/AVP 99 100
a=rtpmap:99 H264/90000
a=fmtp:99 packetization-mode=0;profile-level-id=42e00a;
          sprop-parameter-sets=J0LgCpWgsToB/UA=,KM4Gag==
a=rtpmap:100 H263-2000/90000
a=fmtp:100 profile=0;level=45
```

In this example, both narrowband voice and wideband voice are supported for speech. For video, both H.263 and H.264 are supported.

5.8.7.9   Sessions Including Voice, Video and Text

An example of a session setup with voice, video and text is included in Table 5.37.

Table 5.37: Example of an SDP message for voice, video and text.

```
m=audio 49152 RTP/AVP 98 97
a=rtpmap:98 AMR-WB/16000/1
a=fmtp:98 mode-change-capability=2; max-red=160
a=rtpmap:97 AMR/8000/1
a=fmtp:97 mode-change-capability=2; max-red=160
a=ptime:20
a=maxptime:240
m=video 49154 RTP/AVP 99 100
a=rtpmap:99 H264/90000
a=fmtp:99 packetization-mode=0;profile-level-id=42e00a;
          sprop-parameter-sets=J0LgCpWgsToB/UA=,KM4Gag==
a=rtpmap:100 H263-2000/90000
a=fmtp:100 profile=0;level=45
m=text 49156 RTP/AVP 101 102
a=rtpmap:101 t140/1000
a=rtpmap:102 red/1000
a=fmtp:102 100/100/100
```

Chapter 6

Security

Rolf Blom, Yi Cheng, Vesa Lehtovirta, Karl Norrman, Göran Schultz

Security comes from a number of different functionalities, which cover different aspects and complement each other. A common misconception is that security is solved by the use of encryption. This is not entirely incorrect, as encryption is very important, but so are many other security mechanisms such as integrity protection, user authentication, key management, source origin authentication, security protocols and not least the security architecture applied. We will focus here on security aspects specific to IMS as defined by 3GPP. The main IMS security specification is TS 33.203 [15].

The security architecture of a system should always be based on a trust model. The trust model describes assumptions about which parts/actors in the system are trusted and play according to the rules and which parts/actors that do not, that is, those that can be manipulated or made to reveal confidential data. The 3GPP trust model assumes certain parts of the network and actors are to be trusted, but that does not mean that there are no protection measures applied there, it merely means that such measures are not mandatory and/or standardized for the trusted part of the network. Still, operators are expected to protect their networks and the user traffic in the best way possible. As always, standardization is driven by the need for interoperability between network nodes and operator domains, and security standardization is no different.

Security concerns in cellular environments include passive eavesdropping, active tampering with control messages and user traffic data, replay of valid intercepted messages and traffic, man-in-the-middle attacks, etc. It is also essential to securely anchor the end-point of a connection – in particular the IP address used – to a particular user identity to be able to perform traffic policing, carry out user related policy enforcement and enable user accountability. On the other hand, privacy requirements require user identities to be protected and not easily inferred from open traffic and signaling data. These concerns are typically met by application of traditional security techniques such as: user and network authentication to verify that the communicating parties are who they claim to be; verification of users' right to access services in the network; confidentiality and integrity protection of traffic when it transits untrusted domains to avoid modification and eavesdropping; and filtering of spoofed identities and addresses.

The trust model used in 3GPP is basically that core and access network nodes are trusted to behave according to the defined protocols when communicating with each other, and that threats are mainly anticipated to come from outside the network itself, i.e. attackers are expected to attack the network via the radio interfaces (or the base stations) and on connections between operators and to external, non-operator-controlled networks. Typically all other network nodes except base stations are physically well protected, and it is not considered likely that an attacker would be able to tamper with such nodes and it is up to the operator to decide the need for protection. Traffic that traverses non-trusted links must be appropriately protected. The exact nodes and links that are considered trusted will be mentioned throughout the text as necessary.

In the following we will use the term *access domain* to refer to the radio access and core network parts of a cellular system and let *IMS* or the *IMS domain* denote the whole overlay implementing the IMS services. When trying to understand what protection IMS signaling and user traffic is given, it is therefore important to note that the IMS solution standardized in 3GPP is defined as an overlay on top of the access domain. In particular the current specifications for IMS do not define any confidentiality and integrity protection for the media traffic as it is assumed that the underlying access domain will provide adequate media protection.

In the following we first give an overview of the security components and mechanisms defined for IMS in Section 6.1. Then a description of the security solutions in a 3GPP access domain follows in Section 6.2 and the IMS domain security solutions are described in Section 6.3. Finally, there is an outlook on extensions currently under discussion.

6.1 IMS Security Overview

Access to IMS services requires authorization. The authorization is based on user authentication performed in conjunction with user registration in the systems.

The user authentication scheme called IMS Authentication and Key Agreement (AKA) [15] uses the same principles as user access authentication in UMTS, and hence IMS AKA provides mutual authentication. It also provides keys shared between the user and the IMS domain. These keys will be used for the protection of the SIP signaling between the mobile terminal and the P-CSCF. Note that user authentication is always performed by the user's home IMS network, even in roaming situations. Thus user authentication and authorization are always controlled by the user's IMS home network and is never delegated to the visited network.

IMS users may have many identities by which they can be reached. From a security point of view, it is essential that the IMS ensures that users do not use identities that belong to other IMS subscribers. This is checked at registration.

User credentials and algorithms used for user authentication and registration are held in an IP multimedia Service Identity Module (ISIM) similar to a SIM or USIM used in the access domain. The ISIM holds both the private user identity and at least one public user identity that can be used for the registration. The ISIM is an application on a Universal Integrated Circuit Card (UICC). In the cellular environment the ISIM is normally collocated with the USIM on the same UICC, which means that the IMS user credentials enjoy the same type of protection as the access domain credentials. One of the main differences between USIM and ISIM is the identities stored (where USIM holds identities for access network, while ISIM holds the identities for the IMS network). In IMS the ISIM is preferred, but for migration purposes also USIMs are allowed. This requires some functionality in the mobile terminal, which wraps an ISIM facade around the USIM application.

An overview of the IMS system in a cellular environment is shown in Figure 6.1. The SIP signaling traffic between different operators' IMS domains must be protected. In the figure, this is indicated with links between the SEcurity Gateways (SEGs). The standard for this is described in the Network Domain Security (NDS) Specification TS 33.210 [1]. Operators may also apply protection according to NDS of signaling between network nodes in their own IMS domain.

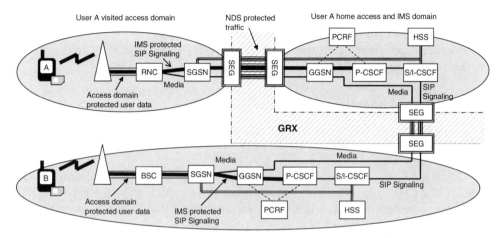

Figure 6.1: Overview of IMS and access domains in home and roaming use cases.

It is important to note that IMS user media traffic as well as user SIP signaling are carried as user data in the access domain. Usually, user data is confidentiality protected by the access domain but that protection is not mandatory; operators may for different reasons turn it off. There is no integrity protection for user data either. Thus, the SIP signaling is never integrity protected and might not even be confidentiality protected in the access domain. To remedy this, and to thwart the types of serious attacks that could otherwise be launched by false signaling, IMS provides mandatory integrity protection and optional confidentiality protection of all SIP messages sent between a mobile terminal and a P-CSCF. The IMS AKA generated keys are used in the establishment of the necessary security associations. In the figure, unprotected links are shown as thin lines.

Media protection, however, does rely on the protection given by the access domain. In UMTS the media is thus normally encrypted between mobile terminal and RNC while in GERAN the protection is between mobile terminal and SGSN. Media sent between different operators can also be protected with IPsec [129] in accordance with the NDS specification.

Before products are available that fully support the 3GPP IMS security features as defined in TS 33.203, early IMS implementations may rely on bearer/network level security. The so-called 'Early IMS' solution [42] works by creating a secure binding in the Home Subscriber Server (HSS) between IMS Public/Private User Identity and the IP address currently allocated to the user at the GPRS level. Signaling and media protection relies on the protection offered by the access network.

6.2   Access Domain Security

The security features of interest to IMS that are provided by GSM and UMTS can be roughly summarized as authentication of subscribers and networks, and protection of the traffic in the access network and between core networks. Since IMS is run in the PS domain, the CS domain is out of scope.

Access security can be examined starting from the way 2G and 3G cellular systems verify a user's identity. The user identity is embedded in a Subscriber Identity Module (SIM) for GSM and a Universal Subscriber Identity Module USIM for UMTS and also anchored in the operator's network in the Home Location Register (HLR) or HSS.

More details about the security protocols and methods chosen in the 3GPP access domain are described in the book *UMTS Security* [183].

6.2.1   UMTS Authentication and Key Agreement

As mentioned, the authentication and key agreement scheme used by IMS is the same as in UMTS. Therefore, a short presentation of the UMTS authentication and key agreement mechanism is in order.

Mobile terminals used in UMTS typically contain a UMTS Subscriber Identity Module (USIM). The USIM contains, among other things, a key, K, which is also present in the HSS in the home operator's network. When a mobile terminal attaches to a UMTS network the user is authenticated by providing proof of possession of the key K. This is done by performing a run of the UMTS AKA protocol [16].

The AKA protocol used in UMTS is an enhancement of the AKA protocol used in GSM. There are two main differences between the two protocols. First, UMTS AKA provides mutual authentication, i.e. the user is assured that the network is legitimate and in addition the network is assured that the user is legitimate. Secondly, UMTS AKA provides replay protection for the authentication.

A conceptual message sequence diagram for UMTS AKA is shown in Figure 6.2. UMTS AKA is a typical challenge–response protocol. It starts with a message from the mobile terminal indicating its identity, the International Mobile Subscriber Identity (IMSI), to the SGSN. After the mobile terminal has presented its identity to the SGSN, the SGSN requests an Authentication Vector (AV) from the HSS in the user's home network. The AV contains a random challenge RAND, keys CK and IK, an authentication token AUTN, and XRES – the expected output from the mobile terminal in response to the challenge. The keys CK and IK, the response RES and the authentication token AUTN can be derived/verified, respectively, only with knowledge of the key K. The SGSN presents RAND and AUTN to the mobile terminal and the mobile terminal gives them to the USIM. The USIM verifies the AUTN and computes the keys and the challenge result RES. The keys and RES are then returned to the mobile terminal. The mobile terminal sends RES to the SGSN, which in turn verifies that RES is equal to XRES. If that is the case, the mobile terminal is regarded as authenticated to the SGSN. Note that the key K is never present in any communication between the mobile terminal and the SGSN during this process.

Network authentication is achieved when the USIM successfully verifies AUTN (which also provides replay protection). The keys CK and IK are used for encryption and integrity protection, respectively, of the traffic in the access network.

Since modification of signaling traffic usually causes greater harm to the network than modification of user traffic, it is important that the integrity of the signaling traffic is preserved. Signaling traffic is always integrity protected in UMTS. Note that the user traffic is never integrity protected, and that the encryption is only optional. This means that an operator may disable encryption in the access network. As mentioned earlier, this fact was very important when the protection of the IMS signaling was designed, as will be discussed below.

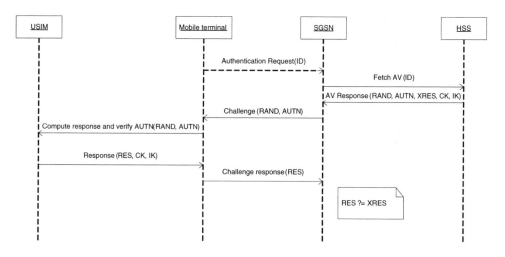

Figure 6.2: Conceptual view of the UMTS AKA protocol. The CK and IK, returned from the USIM in response to the RAND and AUTN, are the encryption and integrity keys respectively.

Note that, on this abstract level, there is nothing said about how the UMTS AKA protocol is transported. As will be discussed below, the IMS registration procedure reuses the UMTS AKA protocol for authentication and key agreement.

6.2.2 Traffic Protection Offered by GSM and UMTS in the Access NW

As a result of an AKA run, the mobile terminal and the SGSN share keying material to provide protection of the traffic in the access network. It is the responsibility of the SGSN to forward the keying material to the network node that performs the actual protection.

In the PS domain, the traffic is protected between the mobile terminal and the RNC for UMTS/UTRAN and between the mobile terminal and the SGSN for GSM/GERAN. Both GSM/GERAN and UMTS offer confidentiality protection of user data while in UMTS signaling also has mandatory integrity protection. The RNCs and the SGSNs are typically located in well guarded sites. Figure 6.3 shows the protection coverage for the user traffic between the mobile terminal and the network.

6.2.3 Internode Security

As described above, the traffic protection provided by the access networks is terminated in the RNC or SGSN. Imagine that the user wishes to connect to another mobile terminal (perhaps attached below another SGSN, maybe even in another operator's network). Even though the traffic will be protected in the target mobile terminal's access network, it is usually necessary

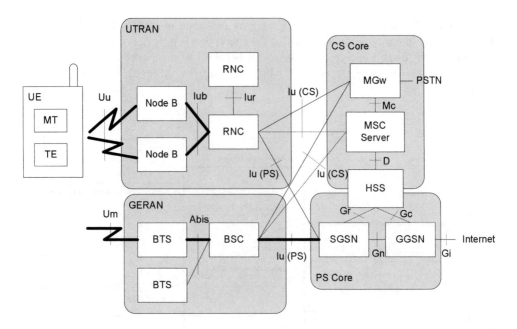

Figure 6.3: As symbolized by the thick black lines, user traffic in GSM and UMTS is encrypted between the mobile terminal and the SGSN/RNC.

to protect the traffic until it reaches the SGSN in the target network. Note that the traffic is typically not routed through the public Internet, but rather over a private GPRS Roaming eXchange (GRX) network, which interconnects the operators' core networks (see Figure 6.1).

How traffic protection works, when roaming, depends on the roaming model applied in the access domains. Today, packet switched roaming is almost always performed in such a way that the user has his point of presence in the home network, i.e. the SGSN in the visited network connects to a GGSN in the home network. This connection should be protected using NDS techniques.

The trust model for roaming assumes that operators that allow roaming between their networks trust each other to a fairly large degree, e.g. SGSNs of one operator may be connected to the GGSN of another operator as described above. To enable authentication of the visited user, the SGSN in the visited network also has to be connected to the HSS in the home network. The reasons for trusting other operators in this respect are mainly based on legal agreements and the fact that roaming agreements between operators are crucial for the business model. Furthermore, it is assumed that e.g. a misbehaving operator would soon find himself without roaming partners, which would give him serious problems.

Even though different operators' networks are connected via a GRX network, the connections are, as mentioned previously, protected. The protection is accomplished by the use of Network Domain Security for IP (NDS/IP) [1] and the Network Domain Security Authentication Framework (NDS/AF) [33]. The NDS architecture for IP-based protocols is shown in Figure 6.4.

SEGs are entities located on the borders of IP security domains. All IP traffic from a Network Element (NE) in one security domain towards a NE in a different security domain is routed via a secure IPsec tunnel established between the SEGs. This interface is referred to

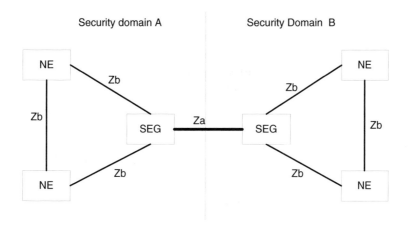

Figure 6.4: Network Domain Security architecture.

as the Za interface and applies Encapsulating Security Payload (ESP) [128] in tunnel mode to provide integrity and optionally confidentiality protection. Within a security domain, a NE may optionally communicate with another NE or with an SEG over a Zb interface. If Zb is implemented, ESP shall be applied to provide integrity protection and optionally confidentiality protection. For both Za and Zb, Internet Key Exchange (IKE) protocol [86] is used to negotiate, establish and maintain ESP Security Associations (SAs).

6.3 IMS Security Mechanisms

6.3.1 Identities

Every IMS user is assigned a Private User Identity. The Private User Identity uniquely identifies the user's subscription and is authenticated at user registration. Private User Identity takes the form of a Network Access Identifier (NAI) as defined in RFC 2486 [48] and TS 23.003 [34]. Associated with the Private User Identity, one or more Public User Identities are allocated to the IMS user. Each Public User Identity takes the form of a SIP URI or a TEL URI (e.g. 'tel:+1-123-456-7890') and is used for routing SIP messages. The relations between the subscription and identities used are depicted in Figure 6.5.

In the IMS operator's network, the HSS stores, for each subscription, the Private User Identity(ies), all allocated Public User Identities, the user's long term key(s), service profiles, and other user information. The long term key is uniquely identified by the Private User Identity.

The Private User Identity, Public User Identities and long term key are made available to the user embedded in the ISIM. If a USIM is used, the identities for IMS are derived based on the access domain identity, i.e. the IMSI, in order to get consistent private and public identities [34]. The IMSI is in this case contained within the NAI for private identity and the contained in the SIP URI for the public identity during initial registration. The ISIM not only provides a secure storage of the subscription data but also performs on-chip cryptographic calculations. Hence the long term key never needs to be exposed to outside of the UICC. Compare this to the discussion regarding USIM in Section 6.2.1.

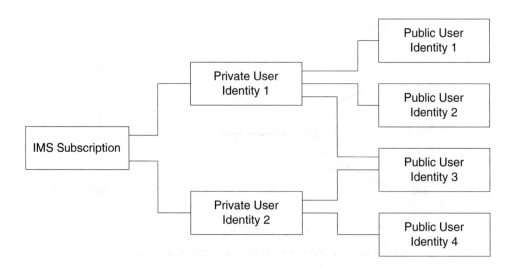

Figure 6.5: Relation of IMS Subscription, Private and Public User Identities (3GPP Release 6).

IMS allows that more than one Private User Identity is allocated to an IMS subscription. This means that a user who has multiple mobile terminals (each with an ISIM) does not need to have a separate subscription for each terminal. A common IMS subscription can cover some or all of the terminals. Furthermore, within one subscription it is allowed to share Public User Identities among the Private User Identities, as shown in Figure 6.5.

The concept of a Public Service Identity (PSI) is also used. Like a Public User Identity, a PSI takes the form of a SIP URI or a TEL URI and is used to route SIP signaling, but in contrast to Public User Identities, which are allocated to users, PSIs are allocated to services hosted by Application Servers. PSIs are not associated to any Private User Identity.

6.3.2   Source Authentication of SIP Signaling

Source authentication of SIP messages, which allows a receiver to verify the identity of the sender of a SIP message, is implemented by letting P-CSCF at the transmitting end insert a so-called P-Asserted-Identity in all SIP messages. This P-Asserted-Identity is removed from the SIP header if and when the message is sent to or received from a node which is not trusted. The idea is that if communication between trusted nodes is protected any node receiving a P-Asserted-Identity can trust that the message was sent from that user. The trust model in IMS is thus that you have transitive trust between network elements within the trusted domain. Hence, if the network has asserted the identity then the mobile terminal will also be able to trust that identity.

6.3.3   Authentication and Authorization

In IMS, a user must register before being allowed to access services. At the initial registration, the mobile terminal and the S-CSCF perform mutual authentication by running the IMS AKA protocol [15]. The IMS AKA uses the long term key that is associated with the Private User Identity and that is shared between the ISIM and the HSS.

Figure 6.6 shows the IMS authentication procedure. The message sequence is the same as in the IMS registration shown in Figure 4.1, but here the relevant message and the authentication data are explicitly depicted.

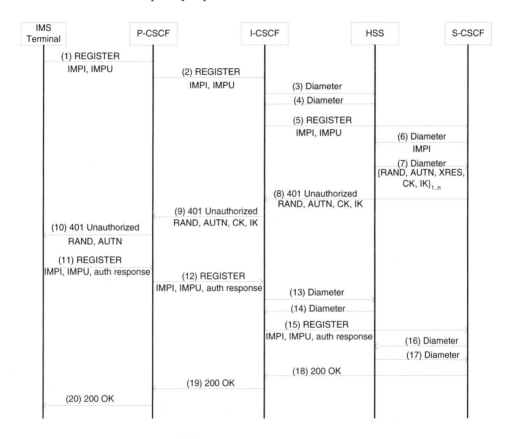

Figure 6.6: IMS authentication.

In the initial REGISTER request (1) the mobile terminal indicates the Public User Identity (IMPU) it wants to register and the Private User Identity (IMPI). Normally the underlying access domain provides confidentiality protection for the link between the mobile terminal and a secured network node. This confidentiality protection helps to protect the privacy of the IMPI when transmitted over the radio interface. Upon receiving the registration request, the S-CSCF fetches AVs from the HSS, messages (6) and (7), in case it does not already have an AV available for that user. The IMS AKA AVs have the same format and content as those used in UMTS AKA (see Section 6.2.1). The S-CSCF sends RAND and AUTN as an authentication challenge in the response (8). CK and IK are also sent in the response but the P-CSCF removes them before forwarding the message to the terminal (10).

Upon receiving the authentication challenge, the IMS terminal instructs the ISIM (or USIM) to verify the AUTN (see Section 6.2.1). If the verification succeeds the network is considered authenticated and the ISIM outputs RES to the terminal. The terminal uses RES and some other parameters to calculate an authentication response as described in RFC 3310 HTTP Digest AKA [137]. The authentication response is sent to the network in a new

REGISTER request (11). Note that this is different from the UMTS AKA in which the RES is sent back to the network. The reason to use HTTP Digest AKA is that the AKA parameters need to be carried in SIP/HTTP messages, and the HTTP Digest framework fits that need nicely as it provides e.g. a challenge mechanism which can be used to carry AKA challenges.

In case the terminal fails to authenticate the network, an error message is instead sent to the network to indicate the failure.

Upon receipt of the new REGISTER request containing the authentication response, the S-CSCF retrieves the active XRES for that user, calculates the corresponding HTTP Digest AKA authentication response, and compares the result with the one received. If the values match the user is successfully authenticated and the Public User Identity is registered in the S-CSCF.

Re-authentication of the same user may or may not be required at re-registration, depending on the security policy of the operator.

6.3.4   IMS Signaling Security

SIP signaling between the IMS terminal and the P-CSCF (Gm reference point) is protected by IPsec, as discussed in the following subsections. In the common roaming scenario described in Section 6.2.3, where the mobile terminal has its point of presence in the home GGSN, the SIP signaling is protected all the way from the mobile terminal in the visited network to the P-CSCF in the home IMS network.

SIP signaling traffic between other IMS nodes is protected by NDS mechanisms on a hop-by-hop basis, see Section 6.2.3.

6.3.4.1   Integrity Protection

It is mandatory to provide integrity protection of SIP signaling between the IMS terminal and the P-CSCF. IPsec ESP in transport mode as specified in RFC 2406 [128] is defined by the standard. Two pairs of unidirectional SAs are set up during the authenticated SIP registration. One SA pair is for traffic between a client port at the mobile terminal and a server port at the P-CSCF and the other pair is for traffic between a client port at the P-CSCF and a server port at the terminal. The integrity key IKESP is obtained from CK and IK (using a key expansion function) and is the same for the two SA pairs. The integrity algorithm is either HMAC-MD5-96 [134] with a 128-bit key or HMAC-SHA-1-96 [135] with a 160-bit key.

The anti-replay service is enabled in the mobile terminal and the P-CSCF on all established SAs.

6.3.4.2   Confidentiality Protection

If the local policy in P-CSCF requires confidentiality protection, each of the four SAs as described above also includes encryption keys and algorithms. The encryption key CKESP is also obtained from CK and IK (again, using a key expansion function) and is the same for the two SA pairs. The encryption algorithm is either DES EDE3 CBC as specified in RFC 2451 [158] or AES CBC as specified in RFC 3602 [78] with 128-bit key.

6.3.4.3   Security Association Establishment

The mobile terminal and the P-CSCF establish IPsec SAs during an authenticated IMS registration. In the first REGISTER request (1) (Figure 6.6), the terminal includes a Security-Client header field. This header field contains a list of integrity and encryption algorithms the terminal supports, the Security Parameter Index (SPI) values, and the port numbers selected by the terminal. The P-CSCF temporarily stores those parameters and removes the Security-Client header field before forwarding the request to the I-CSCF.

After receiving the 401 (Unauthorized) response from the I-CSCF, the P-CSCF adds a Security-Server header field to the message (10). The Security-Server header field contains a list of the integrity and encryption algorithms the P-CSCF supports as well as the SPIs and port numbers selected by the P-CSCF.

From the algorithm list given in the Security-Server header field by the P-CSCF, the terminal selects the first integrity and encryption algorithm combination which the terminal also supports. Putting the selected algorithms, SPIs, IP addresses, ports, and keys (CKESP and IKESP) together, the terminal has now established two pairs of SAs with the P-CSCF. Starting from the second REGISTER request (11), all SIP signaling over the Gm reference point is protected by the established SAs.

In the second REGISTER request (11) the terminal also includes a Security-Verify header field and in that field repeats the P-CSCF algorithm list, P-CSCF SPIs and ports, and terminal SPIs and ports. The P-CSCF will check whether the contents of the Security-Verify header field are identical with the corresponding parameters previously given in the Security-Server and Security-Client header fields. If not the P-CSCF aborts the registration procedure.

6.3.5 Security Aspects of Policy Enforcement in IMS

The policy and charging control function discussed in Section 3.5.11 does not only assure that the user gets the appropriate QoS, gets charged appropriately, etc., but it also serves to protect the mobile terminals from attacks from other mobile terminals and from external networks. It also prevents mobile terminals from sending traffic uncontrollably (e.g. if the mobile terminal is infected by a virus).

As noted in Section 6.2.3 regarding security domains, the underlying network uses strong perimeter protection, and IMS traffic is not generally allowed to flow freely between the mobile terminal and any other node on the public Internet, or even to another mobile terminal connected to the same network.

During the setup of a SIP session, the P-CSCF acts as an Application Function (AF). The P-CSCF requests from the Policy and Charging Rules Function (PCRF), over the Rx interface, that the flows described in the SIP INVITE message be allowed to pass through the Gateway (GW) node, which is a GGSN in the case where the underlying network is a 3GPP network. If the decision by the PCRF is that the flows are allowed to be established, it transfers filter rules (coupled with other data, such as QoS and NAT rules) to the GW, over the Gx interface, which has the effect that these flows will be allowed passage. The GW next instantiates the rules received. If a session involves two mobile terminals that are attached to networks of different operators, it is the task of each P-CSCF to request that the flows be allowed to pass their respective GW.

This mechanism gives the operator strong control over what traffic is allowed to be initiated by the mobile terminals in the network and, as noted previously, also protects the mobile terminals from attacks.

6.4 Outlook

The development of IMS security continues. Some forthcoming security features in standardization are convergence of fixed and mobile IMS networks, media security and countermeasures against SPam over Internet Telephony (SPIT).

6.4.1   Fixed–Mobile Convergence

There are several recent initiatives under discussion in different standardization bodies to adopt IMS in fixed networks. This is the case for example in ETSI TISPAN and PacketCable. It seems that the fixed network initiatives may not be able to directly adopt the IMS security features. One reason for this is that not all the same mechanisms may be directly applicable in both cellular and fixed networks due to different environmental requirements. One such example is NAT traversal capabilities.

Having both fixed and cellular IMS networks aligned as much as possible has several benefits, for example lower development and operational cost, and the possibility to provide similar services regardless of the type of used access network. Therefore, the goal of some initiatives is to eventually combine the cellular IMS networks with the fixed IMS networks to one merged IMS core network that could be accessed from both kinds of networks. This work is called Fixed–Mobile Convergence (FMC).

6.4.2   Media Security

Today, IMS security provides protection only for signaling messages. When IMS is used over cellular networks it is assumed that the security features of the underlying cellular network provide sufficient protection of the media.

As a result of the development of FMC, it is very probable that IMS will not be accessed solely via cellular networks in the future, and then it cannot be assumed that the security features of the underlying cellular networks protect the media. Therefore, work to protect media on the IMS level has been started in standardization bodies, such as 3GPP and TISPAN. The high level goal of media security is to provide protection of the media independent of the underlying network. A possible candidate protocol for real-time media protection is Secure Real-time Transport Protocol (SRTP) [50], which has already been adopted as the method to protect RTP media by several multimedia broadcast systems.

6.4.3   Spam over IP Telephony

Spam is regarded as a major problem in the Internet today. Typically spam means large amounts of unwanted emails that advertise some product. Spam may also carry worms or viruses.

A new form, SPam over Internet Telephony (SPIT), is emerging and can be practiced in different forms. It could for example be automated advertising voice messages or words injected into an ongoing conversation. SPIT may be experienced as more intrusive than email spam, as SPIT disturbs the receiver in real time.

Like spam, SPIT may also cause congestion in networks and in the worst case it might even be able to bring a network down. One fundamental reason behind the general problem of spam and SPIT is the lack of proper user authentication and access control.

Many of the threats opening up for spam and SPIT in the Internet world are not directly applicable to the IMS world. The trust model and business model of IMS mitigates a number of the general spam and SPIT threats. One example is the strong user authentication in combination with the strong trust relationships between different IMS operators. This ensures that the IMS operators can detect and correctly identify any misbehaving users and then act based on this. Policy control is another important mechanism for countering SPIT. While some of the fundamental problems of SPIT are being addressed in IMS, other mechanisms need to be studied to provide additional means to counter the threats.

Chapter 7

Performance

Tomas Frankkila, Janne Peisa, Per Synnergren

The performance evaluation of the conversational services in cellular systems has typically focused on the system capacity, defined as the maximum number of satisfied voice users that the system can support. Based on a long history of voice communication in the fixed telephony networks, it has been possible to model and define user satisfaction based on relatively simple quantities such as frame loss rate and mouth-to-ear delay. The resulting system capacity has been analyzed for almost all current and planned cellular systems and many aspects of the current cellular system design have been optimized to support as many circuit switched voice users as possible.

The performance evaluation of the multimedia service is a much less thoroughly studied topic. The definition of the service quality for the voice component of the multimedia service can naturally be obtained from the corresponding definition for the voice-only service, but the quality definitions for the other components, such as video and text, are not as well established. Furthermore, in order to study the quality of the whole multimedia service, it is necessary to understand how the interactions between different service components influence the overall service quality. A typical example of the interaction is the synchronization of different media components, such as voice and video.

In this chapter the performance of the Multimedia Telephony service over cellular networks is evaluated from both service quality and system capacity points of view.

The system capacity and coverage for the voice service is presented and compared to the corresponding capacity and coverage for circuit switched telephony. It is shown that, with evolved access technologies such as HSPA, it is technically possible to match or exceed the circuit switched capacity.

The delay and packet loss characteristics have a significant impact on the media processing algorithms. Examples of the delay and packet loss distributions arising from typical access and core network configurations are shown. Also typical call setup delays are presented and compared to the CS system.

Substantial parts of this section would not have been possible to write without extensive help from our colleagues. We would like to thank Rickard Sjöberg, Mårten Ericson, Stefan Wänstedt and Stefan Wager for kindly providing data for selected plots.

IMS Multimedia Telephony over Cellular Systems   S. Chakraborty, T. Frankkila, J. Peisa and P. Synnergren
© 2007 John Wiley & Sons, Ltd

7.1   Application Models

In this section we describe how the multimedia telephony service was modeled in performance evaluation. The focus is on data generated by the voice component of the Multimedia Telephony application, not on the functions of the application as such (for more detailed description of the applications, see Chapters 2 and 5).

Calls are generated with exponential inter-arrival times according to a Poisson distribution in a random location chosen uniformly over the whole system. The choice of the inter-arrival time determines the system load.

Each voice user connects to the network and makes a single voice call before exiting the system. The call length is also distributed according to an exponential distribution, with mean value of 80 seconds. Selected results were also generated with a shorter mean value of 30 seconds as well as with fixed call length, but no substantial impact on any results was found.

During the voice call the user alternates between active and idle states. The durations of both active and idle states are generated by an exponential distribution. The average length of both active and idle states are chosen to be 1 second. As the length of the active and idle states is equal, the resulting activity factor is 50%.

During talk spurts voice frames corresponding to chosen AMR mode (typically 12.2 kbps) are transmitted every 20 ms, while during the idle periods no data is transmitted at all (i.e. the transmission of the SID frames is not modeled).

The voice frames for AMR 12.2 kbps mode consist of 256 data bits (32 bytes). For CS bearers, the AMR frames were directly transmitted over the air interface, while for VoIP they were further packed to RTP, UDP and IP packets, resulting in an IP packet of 72 bytes.

The model used for ROHC compressed RTP, UDP and IP headers down to 3 bytes. The compressed IP packets are then further segmented (either in the UE or in the RNC) to RLC PDUs according to the VoIP bearer definition described in Section 5.6. The RLC PDUs are concatenated to form MAC-e or MAC-hs PDUs in the uplink and downlink respectively. The MAC PDUs are finally transmitted over the air.

7.2   Service Performance Requirements

There are numerous requirements on the expected service performance. For example, a typical cellular operator has requirements on service availability, retainability, quality and so on. In this section we will focus on a few selected requirements on areas of service quality experienced by the users and minimum system performance.

The performance requirements for the circuit switched voice are relatively high. For example, the service availability and retainability should often exceed 99.999%, corresponding to only less than 5 minutes of downtime per year. Similar requirements are available for the speech quality and mouth-to-ear delay. These high requirements are not necessarily shared with all VoIP applications. For example, typical users of the currently popular free VoIP applications would not expect to receive the same delay to all destinations, and would most likely not be bothered by a service outage of a few minutes per year.

For conversational services the quality perceived by the end-user consists of (at least) two parts. First, the intrinsic quality of the media stream determines the maximum obtainable quality. For example, for voice telephony the speech quality is determined by the sampling rate, used codec, possible media operations (such as transcoding), frame loss rate and so on. Similarly for video telephony the used codec has a large impact on the quality of the service.

Second, the interactiveness of the service is determined by the end-to-end delay of the media, for example by the mouth-to-ear delay for the voice telephony. When the media quality and interactiveness are combined, it is possible to evaluate the quality of the conversational service.

7.2.1 Voice Performance Requirements

Since voice is expected to be a very important part of all telephony services, we shall look into the quality of voice telephony in more detail.

The quality of voice service is typically expressed as Mean Opinion Score (MOS), which is obtained from listening tests as explained in [100]. The MOS scale varies from 5 to 1, with 5 being best and 1 being worst, as shown in Table 7.1.

Table 7.1: Mean opinion score.

MOS	Quality
5	Excellent
4	Good
3	Fair
2	Poor
1	Bad

Each codec has an intrinsic speech quality, determined by the actual speech coding operation, and typically strongly depending on the resulting bit rate. The intrinsic quality of the codec sets the upper limit on the achievable MOS rate. The speech quality for various codecs and other media processing is discussed in more detail in Chapter 5.

The intrinsic speech quality is lowered by various impairments. In the following we look at the two most common sources of impairments, frame loss and delay. Other impairments include interruptions due to e.g. hand-over, echo, non-perfect echo cancellation and erroneous speech level.

The main impairment on the speech quality comes from the frame losses (erasures). The frame losses typically occur randomly due to packet loss on the radio link. As frame losses are relatively common for both circuit switched mobile telephony and Internet, most of the modern codecs designed for either CS mobile telephony or VoIP over the Internet provide specific Packet Loss Concealment (PLC) methods to cope with packet losses, as discussed in Chapter 5. Typically the speech codecs can tolerate packet losses up to a few percent with relatively modest impact on the speech quality. For AMR the dependence of the speech quality is shown in Figure 7.1. It can be seen that the speech quality remains good for frame loss rates up to about 2%.

The impact of the delay on the voice service is slightly more controversial than the impact of the frame erasures. In general, very low delay is not perceptible, but there is a limit after which the delay starts to be annoying. In Figure 7.2 we show the delay impairment of the ITU-T E-model [110]. In the ITU-T model the conversational quality is degraded already from about 100 ms, even if this is not visible in the figure. The delay becomes noticeable when the one-way delay exceeds 150–170 ms. When the delay exceeds roughly 280 ms, the quality is no longer good, i.e. the MOS score is no longer above 4. This delay value can then be used as the requirement for the maximum allowed one-way delay. Another way to define a requirement is to use the fact that the MOS scores in most subjective tests have a 95%

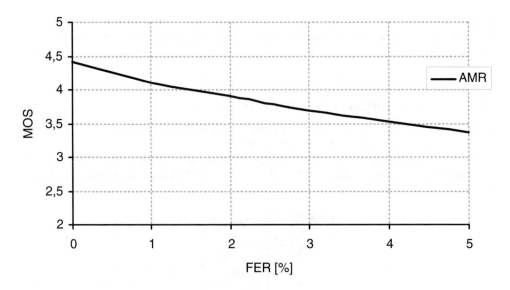

Figure 7.1: The effect of voice frame losses on AMR speech quality.

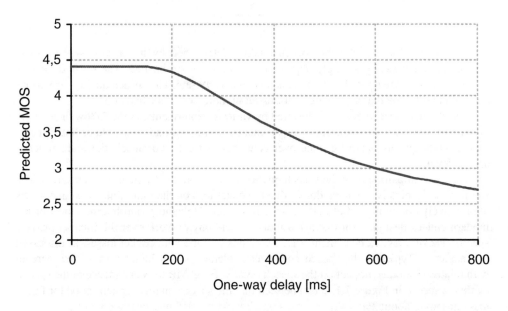

Figure 7.2: The effect of one-way delay on the conversational quality based on ITU-T E-model [110].

confidence interval of about 0.15–0.20 MOS. Applying this margin to the curve in Figure 7.2 gives the result that the quality is equivalent for all delays from 0 ms up to about 220–230 ms.

It should be noted that the exact delay limit and the consequent rate of degradation of the quality are still debated. It has especially been argued that users are willing to accept one-way delay significantly higher than 280 ms. However, the values provided by the E-model are widely used as delay requirements by the CS operators, and in order to facilitate comparison with CS speech quality we have mostly taken the delay limits to match the values provided by the E-model for performance evaluations in this chapter. Still for selected performance measures such as capacity we also show results obtainable with higher allowed delay.

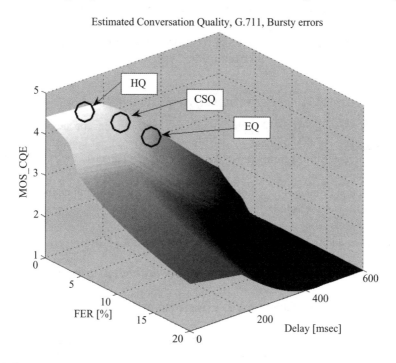

Figure 7.3: The quality of conversational voice service for different conversational quality classes as a function of Frame Erasure Rate (FER) and mouth-to-ear delay. HQ = high quality. CSQ = circuit switched equivalent quality. EQ = economy quality.

When designing a service, one first has to decide what service quality level one wants to aim for. The requirements for the quality related metrics are then derived. In Figure 7.3 the combined effect of the frame erasures and delay is shown for G.711 codec. The best performance is (naturally) reached with no frame erasures and smallest delay, and increasing either frame erasure rate or delay reduces the quality.

It is possible to design a service and a system that delivers a quality at almost any point on this surface. One could, for example, aim for a High Quality (HQ) service, which would require very low frame erasure rate and very short delay. Aiming for this quality level would probably require more resources than providing the CS equivalent Quality (CSQ) service, but might also allow for service differentiation.

Another option would be to aim for a service which has lower quality than the CS equivalent service, both due to higher frame erasure rate and due to longer delay. This operating point is referred to as Economy Quality (EQ), since one can expect that the users would not accept paying the same amount of money for EQ as for CSQ or HQ service. As will be shown in Section 7.3, this service can be provided with less air interface resources than the current CS service, and thus might the allow the operator to provide a lower quality service with lower price.

The target for Multimedia Telephony service is however to replace the existing CS services in GERAN and UTRAN. With an operating point which results in a MOS of roughly 4.0, it is expected that the service would provide similar quality to the circuit switched PLMN services.

Other operating points are also possible. For example, the current free VoIP applications typically emphasize the speech quality over the interactiveness by performing a high level of buffering to combat the jitter in the Internet. This kind of a service does not really exist in the CS side, but is possible to realize for Multimedia Telephony.

7.2.2   Summary of Voice Performance Requirements

In practice it is necessary to simplify the performance requirements of Section 7.2.1 in order to objectively evaluate the application performance in system simulations. Typically it is easiest to set predetermined limits on selected user plane characteristics such as delay and packet loss.

The satisfaction criteria is based on the design target for a CS equivalent service and the following metrics were selected:

- The end-to-end frame erasure rate should be below 2%.

- The end-to-end delay should preferably be 220 ms, with an upper bound of 280 ms.

A service fulfilling these requirements is expected to match the quality of circuit switched voice calls.

A frame erasure rate requirement of less than 2% can be used when end-to-end simulations are performed. For the cases where we only simulated uplink or downlink, this requirement was halved, i.e. only 1% FER per link was tolerated.

If the resulting packet loss from either dropped packets or packets delivered too late was more than 2% for end-to-end simulations, the users were considered to be unsatisfied. Similarly if more than 1% of the packets in uplink or downlink simulation was lost, the user was unsatisfied.

7.2.3   Video Performance Requirements

While requirements for the voice service have been studied extensively in the literature and can be rather clearly defined, the corresponding requirements for the video telephony performance are much more open for implementation.

For voice it is sufficient to provide data rates of slightly above 10 kbps for good speech quality. For video the required data rates are much higher. The exact data rates depend strongly on the video size, but for current mobile screens a good quality video requires data rates of 64–128 kbps. Providing such data rates consistently also at the cell border is more problematic, especially in networks which have been deployed for voice coverage. Due to limited availability of sufficient data rate, most used codecs can adjust their data rate to the current conditions.

As the frame rate of typical mobile video codecs is smaller than 50 Hz, the video is not as sensitive to jitter as voice. The lower the frame rate, the less impact the jitter has on delay.

7.2.4   Multimedia Performance Requirements

In previous sections we have discussed the requirements for voice and video components of the Multimedia Telephony separately. The requirements for other possible service components, such as text messaging, are not as strict as the requirements for voice and video, and thus the requirements for Multimedia Telephony can be derived from voice and video requirements.

When combining various services to create a multimedia service, two additional types of performance requirements can be observed. First, the synchronization of the service components should be maintained; and second, in the case of poor link quality it is beneficial to keep the more important service component.

The synchronization of voice and video has been studied in e.g. [180], where a series of video sequences were shown to 107 test subjects. Different skews between audio and video were tested and the conclusion was that lip synchronization can be tolerated with a skew of maximum 80 ms. In [63], the European Broadcasting Union (EBU) recommends that sound shall be at most 40 ms ahead of video and at most 60 ms after. Even though these tests were done for standard definition video, and thus are not directly applicable for Multimedia Telephony, it can be expected that the requirements for Multimedia Telephony are similar if lip synchronization shall be maintained.

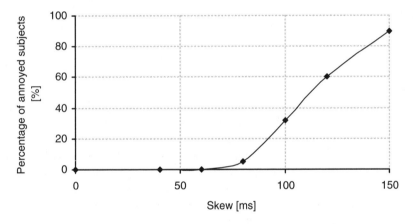

Figure 7.4: Percentage of annoyed subjects as a function of the offset in synchronization between voice and video components. The voice component is played out before the video component always. The authors would like to thank Rickard Sjöberg for providing data for this plot.

In Figure 7.4 we show a schematic outcome of an informal test in which a video clip with audio ahead of the video was shown to test subjects. It can be seen that the limit at which the synchronization starts to be annoying is around 80 ms (as expected) and that just 150 ms offset in the synchronization is perceived annoying by almost all subjects.

7.3   Capacity

The actual capacity of a deployed cellular system is typically determined by a single bottleneck component. The bottleneck can be virtually anything, from the processing capacity of a single node to the available transport capacity in the core network. However, most capacity limitations can be lifted relatively simply by adding additional resources to the bottleneck component. For example, an operator might purchase additional hardware to ease the processing at the bottleneck node, or install additional transmission capacity between two network sites.

In this section, the focus is on the air interface capacity, determined for a given radio network deployment. The reason to focus on the air-interface capacity is two-fold. First, the spectrum available for an operator is typically a limited resource, which is not easily increased. Second, and perhaps more important, the load on the air interface influences the transmission characteristics for IP transmission, which has a direct impact on the perceived service quality.

We evaluate the air interface capacity by dynamical network simulations using simulators (such as Rasmus [156]) developed by Ericsson.

The simulator models the network with a fixed number of base stations while the mobile terminals (corresponding to single users) are created according to a Poisson process, as described in Section 7.1.

The physical link level data transmission is modeled by block error probability curves obtained from separate physical layer simulations. Different link layer block error probability curves are available for different channels with different propagation delay spreads (for example for 3GPP Typical Urban, etc.).

For each slot the received energy per chip, E_c/N_0, for each base station and each terminal is calculated. The received energy is combined for the whole TTI (2 or 10 ms) to obtain the total received energy for the transport block. The total received energy is then converted to signal-to-noise ratio, and the link layer curves are used to determine the probability for a successful transmission for each transport block.

If HARQ is used, the energy for the subsequent retransmissions is combined with the total energy and the error probability is calculated based on the total received energy. Thus the probability for erroneous transmission decreases with the number of transmissions. If HARQ is not used (e.g. for CS calls), the erroneous packets will result in packet losses.

The users make one call with exponential call length. After finishing the call, the user satisfaction is evaluated according to the criteria described in Section 7.2.2. In practice, for VoIP user, the end-to-end voice frame loss, consisting of frame losses from both packets lost in the system and packets delayed more than a fixed delay threshold, must be below 2%.

The network algorithms, such as the HSDPA scheduler and power control, are modeled in detail. The radio link layer and higher layer protocols are also modeled.

The actual air interface capacity depends strongly on the system design and configuration. It is not possible to obtain reliable numbers on the absolute capacity (especially since neither the Multimedia Telephony service nor the networks with advanced radio access techniques have actually been deployed). All the results in this section have been normalized to the circuit switched capacity, which has been obtained by tuning the system parameters to obtain as high capacity as possible.

7.3.1 Simulation Settings

The actual network deployment in all presented results consists of seven or twelve base stations, each containing three sectors (or cells). The base stations are located on a hexagonal grid with site-to-site distance of 1500 meters, corresponding to cell radius of 500 meters.

Figure 7.5: The two network deployments used in capacity simulations. Each base station has three sectors, resulting in 21 and 36 cells for seven and twelve base stations respectively. The site-to-site distance is 1500 meters. The direction of each antenna is shown with a line.

In order to avoid wrap-around effects, the base stations are reflected to the other side of the network. The used networks are shown in Figure 7.5.

Users were considered to be satisfied if less than 2% of the voice frames were either lost or excessively delayed. The actual delay limit has a significant impact on the system capacity (as will be discussed later), but when comparing the VoIP capacity with CS capacity, the total mouth-to-ear delay should be similar for the two cases.

The system capacity was defined as the highest load for which less than 5% of the users are dissatisfied. The capacity was initially evaluated for uplink and downlink separately, after which the system capacity was defined as the smaller of uplink and downlink capacities. The separate evaluation of the uplink and downlink capacities also allows one to identify how suitable the HSDPA and enhanced uplink are for voice transmission.

The physical channel is based on link simulations done with 3GPP Typical Urban channel. While the impact of the used channel model on the actual absolute capacity can be significant, varying the channel model will not affect significantly the conclusions drawn based on relative capacity.

Perhaps the single most important system algorithm for capacity is the used HSDPA scheduling algorithm. Of the scheduling algorithms discussed in Section 5.6, the algorithms chosen for further analysis show clearly the different aspects of different schedulers:

- Round-robin. This scheduler selects the next user based on a predefined order, and does not include any information from the channel. It leads to a very stable behavior, but does not result in optimal capacity.

- Maximum-CQI. This scheduler selects the next user based on the highest current channel quality. It leads to a very high system throughput and instantaneous data rates, but does not serve users evenly.

- Delay scheduler. This scheduler has been optimized for voice service. It tries to ensure that no packet has been delayed too long, and drops packets that have been delayed extensively.

The most important simulation parameters are combined in Table 7.2.

Table 7.2: Most important simulation parameters used in capacity simulations.

Parameter	Value
Channel model	3GPP Typical Urban
Cell radius	500 m
Call length	30 s, exponentially distributed
Activity factor	50%
Mobile speed	3 km/h

7.3.2   Overview of Voice Capacity

It is interesting to note that it is possible to run a Voice over IP service over the interactive bearer available in all existing deployments of UMTS. As the link level retransmissions are performed from the RNC with the RLC protocol (see Section 5.6) it is not possible to maintain the same delay (or quality) as for CS voice service, and the lack of header compression (and other system optimizations) makes the service much less efficient than the CS voice. For those reasons we will not study in detail the performance of the VoIP service over interactive bearers, but will rather focus on what is technically possible with the conversational PS bearer described in Section 5.6.

In Figure 7.6 we show an overview of the voice capacity potential for current and evolved UMTS access over both unoptimized and optimized conversational PS bearers.

Starting with original dedicated channels without header compression or any voice specific system enhancement it is possible to implement a VoIP service with quality equal to or better than the CS voice. However, service realization will consume substantially more resources than providing voice service over the CS bearers. This is shown as the leftmost bar in Figure 7.6.

It is possible to increase the capacity simply by using the HSPA, which will increase the capacity for all services. However, the resulting system capacity is still far below the CS voice capacity. The main reason for this is the header overhead introduced by the RTP, UDP and IP headers, and implementing header compression will increase the capacity close to the CS capacity.

Both HSPA and header compression are already standardized in 3GPP, and will also benefit services other than VoIP. Therefore they can be implemented in a system without a specific VoIP optimization. Together they will bring the VoIP capacity closer to the CS capacity, but will not match the CS capacity.

Other enhancements allow VoIP capacity to exceed the CS capacity, but are much more closely tied to the properties of the voice service. For example, it is possible to boost the capacity significantly by using a scheduler optimized for voice traffic ('delay scheduler') or by optimizing the radio access bearer for the used voice service and codec as described in Section 5.6. These two improvements allow VoIP over HSPA to exceed CS capacity,

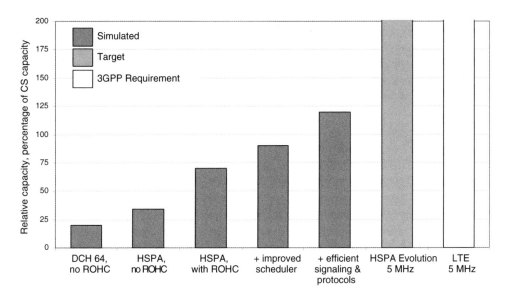

Figure 7.6: The technology potential for IMS voice capacity. The five leftmost values are obtained with detailed system simulations. The sixth is a possible target for the HSPA Evolution, while the rightmost values are 3GPP requirements for Long Term Evolution.

but require tight coupling of the service and the radio access. We will study the impact of the scheduler in more detail.

Also shown in Figure 7.6 are the targets for the evolution of the radio access (see Section 3.3.2 for a description of different evolution paths). For Long Term Evolution, the agreed value is to support 200 voice calls in 5 MHz bandwidth, which corresponds to more than double the current CS capacity. There is no agreed target or goal value for the HSPA Evolution, but in general the performance should be similar as for LTE.

For HSPA evolution, the main improvements in the VoIP capacity is expected to come from the reduced overhead from control channels. For example, for HSDPA, it might be possible to reduce or eliminate the need to transmit the scheduling information on the HS-SCCH. This would not only save the power needed for HS-SCCH, but also allow easy scheduling of more than four users per TTI. For E-DCH, the current system needs to send power control feedback to control the downlink power. However, with HSDPA only downlink, the need for this power control is greatly reduced, and it might be possible to significantly reduce the amount of energy spent on the uplink control channels.

7.3.3 Downlink Voice Capacity

The percentage of satisfied users as the function of the system load is shown in Figure 7.7 for several different schedulers in the downlink. The result has been normalized to the capacity of the delay scheduler and the system is optimized for voice service. The different scheduling strategies show clear differences in resulting capacities. Unlike for interactive services, for which an unfair treatment of users tends to result in maximum system capacity (see e.g. [157]), the lowest capacity is obtained for maximum-CQI scheduler, which places a

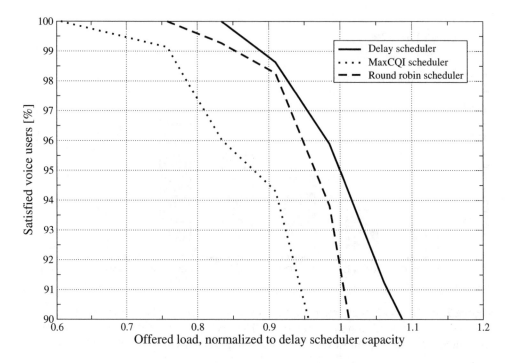

Figure 7.7: The speech capacity over HSDPA for different schedulers. The capacity has been normalized to the capacity with delay scheduler.

high weight on the channel quality. The highest capacity is reached with the delay scheduler, which prioritizes packets based on queuing delay and discards packets which have been excessively delayed. The simple round-robin scheduler also performs well, with capacity between max-CQI and delay schedulers.

The capacity of the HSDPA is higher than the capacity of the bearers. This is caused by a combination of two different factors.

First, the HARQ used for HSDPA provide the possibility to transmit with less redundancy than the fixed convolutional coding used for the CS bearer. Any remaining errors can be corrected with retransmissions, and thus on average less power can be used to transmit a speech frame. However, the average number of retransmissions used in the simulations was quite low (of the order of 10%) and thus the gain obtained with HARQ is not sufficient to overcome the extra overhead introduced by the (compressed) RTP/UDP/IP header.

Second, the channel dependent scheduling allows one to improve the quality of the physical link by scheduling more often to the users that have a good channel. However, it is not sufficient to only look at the channel quality when scheduling (as can be seen from the capacity of the maximum-CQI scheduler); one needs to ensure that no user is queued for too long a time. The delay scheduler, which both benefits from the HARQ gain and combines the limit on the scheduling delay with the improved link quality obtained by scheduling to users with good links, shows how the total voice capacity can exceed the capacity of CS bearers.

The benefit of increased scheduling delay is further illustrated by Figure 7.8, which shows how the increased scheduling delay can be transformed for a higher capacity. The scheduling delay of 150 ms results in a mouth-to-ear delay comparable with the delay of the CS bearers.

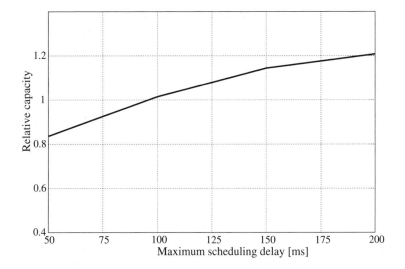

Figure 7.8: The impact of the maximum allowed scheduling delay on the VoIP speech capacity. The capacity is normalized to the CS capacity. The authors would like to thank Mårten Ericson and Stefan Wänstedt for providing data for this plot.

By increasing the allowed scheduling delay, it is possible to increase the capacity. This would obviously reduce the quality since the delay would be longer.

By adjusting the scheduling delay, the network operator can also adjust the trade-off between quality and capacity. As shown in Section 7.2, the perceived conversational quality depends on the mouth-to-ear delay, and as seen from Figure 7.8 the capacity also depends on the allowed scheduling delay. Thus by gradually increasing the scheduling delay, it is possible to increase the capacity while gradually reducing the delay. Unlike for the CS bearers, the balance between capacity and quality can be optimized for the particular needs of the operator.

7.3.4 Uplink Voice Capacity

For the enhanced uplink, the system capacity is shown in Figure 7.9 for both 2 ms and 10 ms transmission time intervals. Both 2 and 10 ms TTI provide capacity that exceeds the circuit switched capacity.

For uplink the potential to obtain an increased capacity by scheduling to users with better link quality is limited due to delay associated with the scheduling of the mobile terminals from the Node B. In the simulations shown in Figure 7.9 it is assumed that the users transmit autonomously without being scheduled.

The main increase in the capacity for the enhanced uplink comes from the HARQ gain. Compared to the HSDPA configuration, more retransmissions were targeted, which increases the HARQ gain. The gain obtained from targeting a high number of retransmissions is evident from the performance difference between 2 ms and 10 ms TTI. The 2 ms TTI configuration targeted four transmission attempts while the 10 ms TTI targeted two transmission attempts. This results in total transmission time of 16 ms for 2 ms TTI and 20 ms for 10 ms TTI, resulting in the capacity of the 10 ms TTI being slightly higher than for 2 ms TTI.

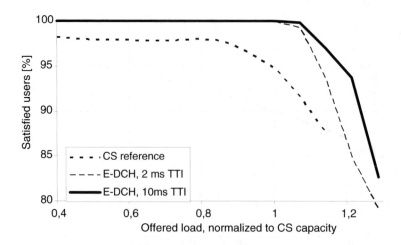

Figure 7.9: The speech capacity for circuit switched users and VoIP users over Enhanced uplink. The results have been normalized to the CS capacity. For both 10 ms and 2 ms TTI, the VoIP optimized system configuration is used. The authors would like to thank Mårten Ericson and Stefan Wänstedt for providing data for this plot.

In general the air interface capacity of the system is defined as the minimum of the uplink and downlink capacities. For both CS bearers and HSPA the limiting factor is the downlink capacity, which is significantly (perhaps of the order of 20%) smaller than the uplink capacity. Thus the overall air interface capacity will not improve even if the 2 ms TTI is used for the enhanced uplink. It is worth noting though that the VoIP capacity of the dedicated channels in the uplink is so small that, if E-DCH is not used in the uplink, it will not be possible to reach the CS capacity.

As a summary of the voice capacity results, it is possible to state that the VoIP service over HSPA can provide equal or better capacity than the existing CS speech service with equal quality. Further the VoIP service has the possibility of trading the conversational quality for increased capacity gradually, which is not possible for the CS service.

7.3.5 Video and Multimedia Capacity

The capacity for services other than voice, expressed in absolute number of satisfied users the system can support, depends strongly on the required data rate.

Based on results of [157], the total maximum throughput of an HSPA cell is limited to roughly 1.5–2 Mbps in the downlink for web surfing traffic. Even though the absolute value of the maximum cell throughput varies depending on the used applications (for example, the throughput for VoIP is significantly smaller than the throughput for web surfing), it is possible to obtain a very rough estimate of the effect of the required application data rate on system capacity by simply allocating the total cell throughput evenly between all users. Each user will receive a throughput of

$$R = \frac{R_{\max}}{N}, \tag{7.1}$$

in which R is the throughput received by the user, $R_{\max}$ is the maximum cell throughput and N is the total number of users. If the rate R is deterministic, this can be simply solved for capacity:

$$N = \frac{R_{\max}}{R}. \tag{7.2}$$

Even though the absolute values obtained by this method are not particularly accurate, especially for low data rate services, the general behavior for applications with similar interactivity requirements is relatively accurate.

As an example, it is possible to evaluate the fraction of the multimedia users the system can support to the number of voice users the system can support. First, note that for voice roughly half of the time is spend in the DTX mode, with very low bit rate transmission. Thus the number of active voice users the system can support is 50% of the number of supported voice users. Each active voice user transmits with data rate of 12.2 kbps. If we assume that each video call consists of a 64 kbps data stream, it is possible to estimate using equation (7.2) that the number of video calls the system can support is only 19% of the active voice calls, and 8% of the total voice calls. Even if the actual number of supportable video calls were expected to be somewhat higher, it is clear that the system capacity for video and multimedia calls is significantly smaller than for voice calls. This argumentation is equally valid for uplink and downlink.

As the capacity for video calls is much lower than the capacity for speech calls, it can be expected that the user will often encounter situations in which the system can support the voice component of the multimedia call, but not video. Such situations can occur simply by increased load in the cell the user is currently occupying, or by doing a hand-over from a cell with low load to a highly loaded cell. In those situations, it would be better from a user performance point of view to eliminate the video component of the multimedia call, and continue with just the voice component. This adaptivity of the service is further described in Section 5.4.1.

7.4 Coverage

In this section we focus on the coverage of the Multimedia Telephony service. Coverage is, in addition to the actual air interface capacity, one of the most important properties of the system. It determines to a large extent the network deployment.

7.4.1 Voice Coverage

The current networks have typically been deployed so that the system has (close to) full coverage for the CS voice. For other services, such as CS video telephony or IP connectivity, the coverage is smaller and the service may not be available in all locations.

As the most likely introduction of the Multimedia Telephony service will be in already deployed networks, the coverage for at least the voice component of the Multimedia Telephony service should be as close as possible to the coverage of the CS speech service.

The actual coverage depends very strongly on the exact network deployment and environment. The results shown in this chapter address only a homogeneous network, and the absolute values for e.g. cell radius should be considered only as an indication of the coverage potential.

We focus on the uplink coverage, as the system coverage is typically determined by the uplink coverage due to low maximum transmission power of the mobile terminal. We follow the procedure outlined in [75] to estimate the uplink coverage:

- By assuming that the UE transmits at the maximum power, and by estimating how much the signal strength fades with distance, it is possible to calculate the received energy per TTI at the base station.

- The received energy is combined for all retransmissions to obtain the total received energy per voice frame. The number of retransmissions is determined by the delay requirement for the service.

- Finally the total received energy per voice frame is converted to voice frame loss probability. The coverage is determined by requiring the frame loss probability to be below 1%.

The received energy per TTI (e.g. E_c/N_0) for mobile i can be obtained (in decibels) by

$$\frac{E_c}{N_0} = G_i + P_i - I_{tot}, \tag{7.3}$$

in which G_i is the path loss of mobile i, P_i is the (maximum) transmission power and I_{tot} is the total interference, consisting of the intra-cell interference (caused by other mobiles in the same cell), inter-cell interference (caused by other mobiles in neighboring cells) and thermal noise. Of these, only the path loss depends on the distance from the base station. We will only be interested in the relative coverage of the enhanced uplink over the existing CS bearers, and by noting that the interference is the same for both CS bearers and enhanced uplink, we can write the difference in path loss as

$$\Delta G_i = \Delta \frac{E_c}{N_0} - \Delta P_i. \tag{7.4}$$

Note that even though the maximum UE transmission power is the same for E-DCH and CS bearers, the relevant quantity in equation (7.4) is the maximum available energy for user plane data transmission. Due to different physical layer control channel implementation, the power used for control information is different for DCH and E-DCH, and thus the maximum available power also differs.

The path loss can be estimated by the Okumura–Hata path loss formula [87]

$$G_i = \beta + \alpha \log_{10}(d), \tag{7.5}$$

in which β and α are constants (with respect to distance) depending on frequency and antenna height (above ground). For UMTS, we can assume 2.1 GHz frequency and 30 m antenna height, which result in $\alpha = -29$ (the value of β will be irrelevant for the relative coverage, as will be shown below). In Figure 7.10 we show the corresponding path loss.

Combining the Okumura–Hata path loss formula (7.5) with equation (7.4), we obtain

$$\alpha \Delta \log_{10} d = \Delta \frac{E_c}{N_0} - \Delta P_i. \tag{7.6}$$

Solving for maximum distance from the base station, we obtain

$$\frac{d}{d_{CS}} = 10^{[\Delta(E_c/N_0) - P_i]/\alpha}. \tag{7.7}$$

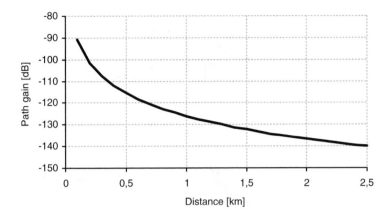

Figure 7.10: Path loss as a function of the distance from the base station for WCDMA at 2.1 GHz frequency.

Thus the coverage can be optimized by increasing the total received energy. The difference in maximum available transmission power depends on the power allocation to data transmission and physical layer signaling, and cannot be easily modified to increase coverage.

With E-DCH the simplest way to increase the received energy is to increase the number of retransmissions. This will increase the received energy at the expense of the total delay. By increasing the number of retransmissions from one to two doubles the received energy, and correspondingly increases the coverage by 3 dB. As discussed in Section 7.2, the maximum allowed delay is determined by the service requirement on the interactivity. In the examples below, we have used maximum three transmission attempts for 10 ms TTI and eight for 2 ms TTI. These values provide a maximum total delay of roughly 120 ms for both TTIs, which is low enough to allow the VoIP delay to match the CS delay, and still provide significant gain for the coverage.

The E_c/N_0 values for 1% residual block error rate (corresponding directly to voice frame loss rate) can be obtained only based on link simulations. The values from [75] are listed in Table 7.3 for two different compressed IP header sizes. These values also include the effect of different maximum available transmission power.

Table 7.3: The required E_c/N_0 values for 1% voice frame erasure rate. The required energy depends on the size of the voice frame, which depends on header compression.

	E_c/N_0 (317–318 bits)	E_c/N_0 (611–634 bits)
CS AMR 12.2 kbps	−18.6	N/A
E-DCH 10 ms TTI	−18.0	−17.5
E-DCH 2 ms TTI	−16.0	−15.1

Based on the values in Table 7.3, and normalizing the cell radius of CS bearer, one obtains the coverage of VoIP over E-DCH shown in Figure 7.11. It can be seen that the coverage for the E-DCH is slightly worse than the coverage for the CS voice. However, as discussed

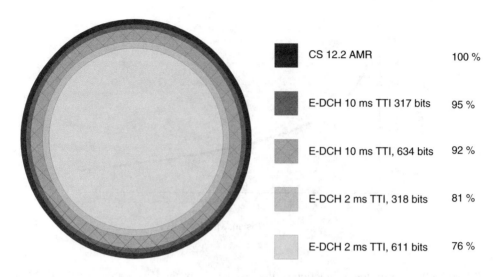

Figure 7.11: The maximum cell size, normalized to the CS cell size.

in [75], this conclusion depends strongly on the chosen parameter values (for example, in [75] it is shown that, by changing the power allocation for the physical control channel, the coverage for E-DCH exceeds the coverage for CS voice). What is more interesting to realize is that with E-DCH the delay can be traded for coverage. Thus it is possible to maintain a good voice quality even in large cells with just reduced interactiveness.

7.5   Transport Characteristics

In the previous sections, the focus has been on the maximum number of users that can be served in the system. We now shift the focus towards the performance for individual users, and evaluate the conditions encountered by users in typical Multimedia Telephony deployments.

7.5.1   End-to-End Characteristics

We first examine typical mouth-to-ear delay for typical VoIP over HSPA deployment. The delay budget for Multimedia Telephony was discussed in Section 5.6.4. We summarize the example delay budget in Figure 7.12. In this example, the total end-to-end delay varies between roughly 100 and 290 ms.

The main contributor to the maximum end-to-end delay is the HSDPA scheduling delay. As discussed in Section 7.3, the maximum allowed downlink scheduling delay determines to a large extent the maximum system capacity, and the large allowance for scheduling delay corresponds to a system optimized for capacity. It is also possible to reduce the scheduling delay if high capacity is not required.

The only other significant delay component to the maximum delay is the HARQ retransmission on the enhanced uplink. For small cell sizes, as shown in Section 7.4, it is sufficient to use at most one retransmission, while for maximum coverage more retransmissions are needed. The expected delay for small cells can be of the order of 40 ms (corresponding to one retransmission), and for the large cells the uplink retransmission delay could even exceed the downlink scheduling delay.

The two major delay contributors (HSDPA scheduling and E-DCH retransmissions) will need to be optimized together in order to reach a good balance between capacity and coverage, while preserving the interactiveness of the conversational service. The example in Figure 7.12 is one possibility, suitable for small cells and high capacity requirement.

Figure 7.12: An example of the components contributing to the mouth-to-ear delay for voice over HSPA. Note that the values are only rough estimates highlighting the different components, and do not correspond to actual implementation values.

From the user's perspective, the transport characteristics should be hidden as much as possible by the media processing methods described in Chapter 5. However, in order to design suitable media processing algorithms, it is necessary to have a clear idea about how the delay and packet loss distributions look over HSPA access.

The main reasons for packet loss in the link layer are aborted retransmissions on both uplink and downlink as well as exceeded delay in the HSDPA scheduling. Neither of these mechanisms occurs very frequently, and the resulting packet loss should be well below 1% in correctly configured systems.

For both uplink and downlink, the retransmissions delay not only the retransmitted packet but also all subsequent packets due to link layer reordering. As the retransmissions are frequent on the radio link, the link layer protocols have been designed to provide in-order delivery of the packets so as to avoid extensive problems with applications that are not well prepared to handle packet reordering. For VoIP the reordering may be problematic, as the voice application would prefer to receive packets continuously and decide itself if an error concealment is needed to cover for a missing packet. The simplest solution for this is to configure the link layer retransmission protocols to abort the retransmissions after a fixed number of attempts, and to release the subsequent packets. This mechanism is one of the main sources of packet losses on the link layer.

For HSDPA the other source of packet losses is the dropping of packets in the Node B. The main reason for dropping packets in the Node B is the hard hand-over mechanism used to change the serving HSDPA cell. For circuit switched calls and for E-DCH soft hand-over (in which the mobile is connected to two or several base stations at the same time during the hand-over) is used. For HSDPA the switching is decided in the RNC and is based on the measurement reports received from the UE. The mobile measures the channel quality, and reports preconfigured events (such as change in the strongest cell) to the network. The network can act on the measurement report directly or may combine several reports to determine the best possible switching moment. However, as this decision is done in the RNC, which does not know the exact status of the HARQ retransmissions and the buffer level in the Node B, it is possible that, when the switching is done, there are still uncompleted retransmissions or packets buffered in the old Node B. Those packets will be dropped, and lead to voice frame losses.

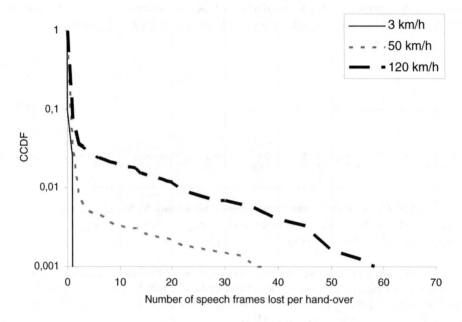

Figure 7.13: The complementary cumulative distribution function of the number of voice frames lost per hand-over when using HSDPA. The results are shown for three mobile speeds, 3 km/h, 50 km/h and 120 km/h. The authors would like to thank Stefan Wager for providing data for this plot.

In Figure 7.13 we show the Complementary Cumulative Distribution Function (CCDF) of the number of voice frames lost per hand-over for various mobile speeds. It can be seen that for all speeds the number of lost frames per hand-over is small for the vast majority of the hand-overs. However, for higher speeds there are hand-overs when several consecutive frames are lost resulting in an audible impairment even with the AMR error concealment. Those hand-overs occur mostly in situations in which the link is fading so fast that the cell reported to be strongest in the measurement report is no longer the best cell when the actual hand-over happens. However, such hand-overs are rare, and will not occur for the majority of calls.

The other reason for packet losses in the Node B is deliberate dropping of packets. In general, it is a good idea to avoid transmitting data that is too old to be usable for the application. For HSDPA the benefit of abandoning old packets is twofold. First, the air interface capacity is increased as resources are not wasted on transmitting useless packets. Second, as old packets are dropped, the queue size in the Node B is reduced, which decreases the overall delay for all connections.

In addition to those link layer packet losses, additional packet losses will come from the congestion in the IP transmission network. It is estimated that in a well dimensioned IP network the typical packet loss rate is even lower than the packet loss rate from the link layer mechanisms. However, if the IP network is congested, the packet loss from the IP network can easily exceed 1% and thus become the main contributor for the total packet loss.

The distribution of the media frame delay ('jitter') for an individual user is determined largely by the same contributors as the absolute delay: HSDPA scheduling and enhanced uplink retransmissions.

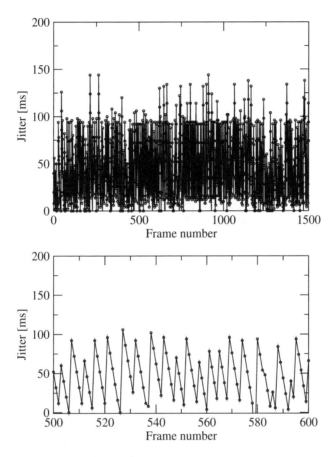

Figure 7.14: The delay variation (measured as the deviation from the smallest delay) between voice frames for the user with median maximum delay. The lower plot is an enlargement of the upper plot, showing that, even though there is no significant time correlation in longer time scales, the HSDPA scheduling and HARQ retransmissions cause the delay for subsequent frames to be clearly correlated. Negative delays correspond to lost packets.

Typically the time correlation of the delay caused by retransmissions and scheduling is limited to relatively small time scales. Even though the delay for subsequent frames will depend strongly on the individual scheduling decisions or retransmissions, the resulting time correlation is of the order of tens of milliseconds. There is no significant correlation over longer periods of time (e.g. up to one second). This behavior is apparent from Figure 7.14, which shows the delay variation, measured as the deviation from the smallest delay, between voice frames for the user with median maximum delay. The scheduler used was the delay scheduler, and the system was loaded to roughly 90% of maximum capacity. There is no clear correlation in the upper plot, but the enlargement in the lower plot shows clearly that the HSDPA scheduling creates a strong dependence in the delay of the subsequent frames. This is mostly caused by the delay scheduler, which favors frames that have already been delayed, and, once the scheduling decision has been made, bundles all frames available for that user to one transmission.

The characteristics of the jitter generated by HSDPA scheduling vary a lot depending on:

- system load;

- user location and radio conditions;

- network configuration and algorithms.

In the following we will analyze the impact of each of these on the resulting delay.

The system load has a large impact on the jitter. For a system with low load there will be virtually no scheduling delay at all, while for high load the jitter can be of the order of 100 ms. This large variation in the expected jitter sets some requirements for the media processing algorithms. On the one hand, in order to guarantee a good service performance even in the loaded system, the algorithms need to be prepared to handle jitter up to 100 ms. On the other hand, in order to provide the best possible performance in a system with low load, the algorithms must not create unnecessary delay. One solution suitable for this kind of situation is to use adaptive algorithms, as described in Chapter 5.

As an example, we show the difference between medium and high systems on the jitter distribution for the delay scheduler in Figure 7.15. The medium load corresponds to roughly 70% of the system capacity (as defined in Section 7.3), while the high load corresponds to slightly above 90% of the system capacity. For both loads, the complementary cumulative distribution function of the jitter for users with smallest and largest maximum delay as well as for users with median maximum delay are plotted.

While the jitter for users with smallest and median maximum delay is very similar for both loads, the difference in the jitter for a worst users is large. For a worst user in the high load case the packet loss is also high, which can be observed from the value where the CCDF crosses the y-axis. This is typical behavior for voice traffic over HSDPA. Some users (or even the majority of users) will observe very good delay performance, while the performance of the worst users will depend strongly on the system load. This will be examined in more detail in Section 7.6.

The variation between different users even with the same load is also apparent from Figure 7.15. The difference between best and worst users is as large as the difference between different loads. Furthermore, the users' radio conditions can change quickly, e.g. as a result of moving behind a large building. Thus the media processing algorithms should not only be prepared for large variation between users, but also for large variations for an individual user during a call. Again the most obvious solution is to use adaptive methods.

The characteristics of the delay variation introduced by the HSDPA scheduling depend naturally also on the chosen scheduler. In Figure 7.16 we show the complementary cumulative distribution function for round-robin and delay schedulers. The schedulers were chosen as (according to Section 5.6) they are most usable for VoIP.

In general the delay for most users is smaller with the round-robin scheduler than with the delay scheduler. This is caused by the tendency of the delay scheduler to select frames that are already late, which leads to increased average (and median) delay. However, the maximum delay of the delay scheduler is limited, and for the worst users the tail of the round-robin scheduler is longer. This long tail will eventually result in an increased delay for all users, resulting in smaller capacity than for the delay scheduler.

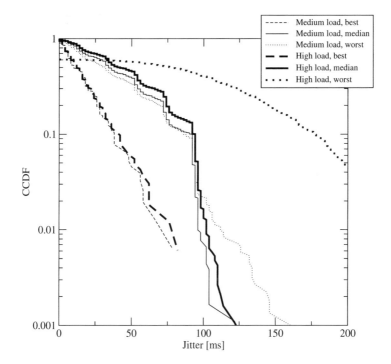

Figure 7.15: The complementary cumulative distribution function of the HSDPA scheduling jitter distribution for the delay scheduler for different system loads and different user locations. The medium load corresponds to roughly 70% of the maximum load, while the high load corresponds to slightly above 90% of the maximum load. The worst, median and best users correspond to users with smallest, median and largest maximum delay respectively. Note that for the worst user in the high load case significant packet loss occurs, which can be seen from the value where the CCDF crosses the y-axis.

7.5.2 Characteristics for Media Gateways

So far we have focused on the end-to-end delay characteristics, which are important both from the overall service quality point of view as well as for the media processing algorithm design for the terminals. Even though the end-to-end packet switched communication emphasizes the terminal processing much more than the media processing in the network, it is still necessary to perform virtually all the same processing functionality in the network media gateways for interworking scenarios. However, the delay characteristics are somewhat different for the media gateways.

For a call that is completed completely within a packet switched network, the media gateway is only needed for special services, such as conferencing. However, the media gateway providing CS interworking functionality must eliminate jitter resulting from the packet access. As most of the mobiles will initially not be using the Multimedia Telephony service, the interworking scenario is important.

The main difference between the end-to-end delay profiles and the delay profiles for the media gateways is the lack of downlink scheduling delay in the jitter that media gateway needs to compensate. This results in significantly smaller delay variation.

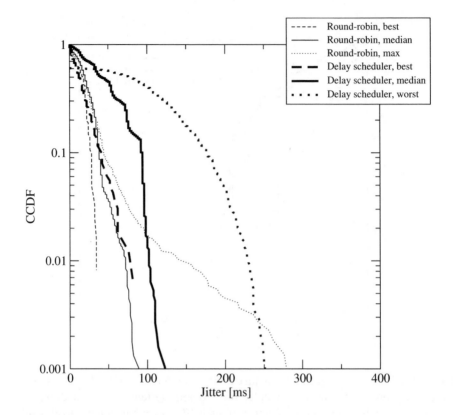

Figure 7.16: The HSDPA scheduling delay for different schedulers.

In Figure 7.17 we show the delay variation arising only from the E-DCH retransmissions and the transmission delay in the radio access and core networks. As can be seen, the variation is much smaller than the end-to-end delay variation.

Due to smaller delay variation it is possible to use less complex algorithms to provide de-jittering in the media gateways than in the terminals. Especially, it is possible to avoid the processionally expensive time scaling, and rely on other adaptive methods to provide adequate packet loss.

7.6   Service Quality

When evaluating the capacity of the HSPA system, the circuit switched UTRAN system is used as a reference, regarding both capacity and quality. Using a reference system makes it fairly simple to judge the validity of the results, i.e. to judge if the results are good enough or if further improvements are needed. The packet switched system also needs to match the performance of the circuit switched system in order to be commercially successful.

It is also important that one compares both capacity and quality. The reason is that it is quite easy to show good capacity numbers if one sacrifices quality. To make a proper judgment of the system performance one must compare the HSPA system with the reference circuit switched system on equal terms. This section describes a few quality evaluation methods and also the method that was finally selected for how to evaluate the quality of

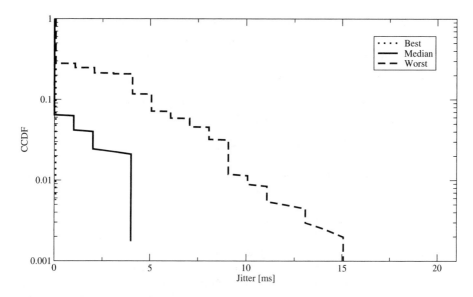

Figure 7.17: The complementary cumulative distribution function for uplink delay of E-DCH. The values are obtained with 10 ms TTI. This delay is important for the media processing in the media gateway providing CS interworking.

the calls in the system simulations. This section includes also the results of the performance evaluation.

7.6.1 Quality Assessment Method

Several speech quality evaluation methods were considered for the evaluation of the speech quality. Each one however has its own advantages and disadvantages. One important aspect of this speech quality evaluation is that the capacity limit is determined when 95% of the users are satisfied. Thereby one must evaluate each and every call to conclude if the user is satisfied or not. Another important aspect is that the quality needs to be verified for a number of different operating conditions to get a feeling for the overall performance and not for just one case.

The most common method to evaluate the quality is with subjective listening-only tests with human persons. ITU-T has defined several methods for such evaluations, for example the ACR test methodology [100]. The process for such an evaluation would be as follows:

1. Select a number of speech segments, typically 8–16 seconds long.

2. Encode them with the chosen speech codec.

3. For each call and for each operating condition:

    (a) Pack the encoded speech into IP packets.

    (b) Send the packets through the simulation environment so that packet losses and delay jitter are applied to the media stream for each call.

   (c) The frames are received in a simulated VoIP receiver application and decoded. The decoder in the VoIP client includes a jitter buffer that equalizes the jitter and also includes an error concealment algorithm that tries to hide the packet losses.

   (d) Store the decoded speech to files.

4. Evaluate the speech quality for each segment in a listening-only test, for example with an ACR test. This gives the Mean Opinion Score (MOS).

5. Calculate the average quality for each operating condition.

6. Compare the score with the score for the corresponding operating condition for the reference circuit switched system simulation.

The main problem with this methodology is that the number of calls in a system simulation is typically several thousand for each operating condition and each call is typically long enough to include two or more segments. The large number of segments that would need to be judged makes it impractical to do a listening test based on all calls. Furthermore, listening-only tests do not evaluate the delay component of a conversation and it would therefore be quite easy to get good scores simply by applying a longer jitter buffering time.

To properly evaluate both the quality of the sound and the impairments introduced by long delay one would have to do a conversation test [100]. In a conversation test, two persons are involved in a conversation and judge the quality in real time. With this methodology, one would have to use real-time VoIP applications to encode the speech, send the speech packets through the system simulator to another VoIP application which then decodes the received packets and presents the synthesized speech to the other person. This kind of test would evaluate both the quality of the sound, the impairments introduced by packet losses and the impairments introduced by long end-to-end delays. The main drawback is that each pair of persons can only evaluate a very limited number of operating conditions, typically no more than 15–20 conditions. Since thousands of calls are simulated in system simulations, this methodology becomes impractical for large scale evaluations. Another problem with this methodology is that the VoIP clients and the system simulator must work in real time, which is very hard to accomplish given that one must simulate many calls in parallel in order to simulate interference from other users correctly.

Since none of the subjective methods are practicable for the desired evaluation, one must rely on objective methods that use software tools to judge the quality. ITU-T has defined a few methods and software tools for objective quality evaluation based on speech. The most frequently used method is PESQ [108]. One nice property with this method is that it judges how transmission impairments impact the speech signal. With PESQ, speech is sent through the simulated system in a non-real-time fashion and the synthesis is recorded to files. The sound in the recorded files is then compared with the original sound file and the difference, which is the degradation, is judged. PESQ was designed to simulate ACR listening-only tests and therefore does not judge the degradations due to long end-to-end delay.

An alternative to objective methods that base their quality assessment on recorded speech is to make an objective evaluation based on transport parameters such as Packet Loss Rate (PLR), late loss rate (LLR), Frame Erasure Rate (FER), end-to-end delay and delay jitter. The packet loss rate is the percentage of IP/UDP/RTP packets that are not received by the receiving VoIP application. The late loss rate is the percentage of frames that are dropped by the jitter buffer in the receiving VoIP application because they are received too late to be useful for decoding. The packet loss rate and the late loss rate can then be converted to the Frame Erasure Rate (FER) that the speech decoder experiences. The frame erasure rate for the VoIP system can then be compared with the frame erasure rate for a circuit switched system for the corresponding operating condition.

In the literature, one can find several proposals for how the frame erasure rate and the delay can be converted to an estimated conversation quality by using the ITU-T E-model [110]. For example, in [163] it is suggested that:

- the quality degradation, as a function of FER, is determined from a listening test and the equipment impairment function is approximated;

- the degradation function due to delay, the delay impairment function, is also approximated.

These two functions are used to derive the R value. The R value is then calculated for short segments, which allows for monitoring how the quality varies over time. A further evolution of this methodology would be to convert the R value to estimated conversational quality, MOS_{CQE} by using equation (B-4) in [110].

Such a method could be used to compare the quality for different operating conditions, even if the end-to-end delays are not the same. A method that calculated the overall conversational quality would be necessary if the delay budget, see Section 5.6.4, indicates that the delay for the VoIP service was significantly longer than for the reference circuit switched service. However, since the delay budget in Section 5.6.4 shows that the end-to-end delay for the VoIP service is on a par with the end-to-end delay of the circuit switched service, or might even be slightly shorter, it is possible to use a simplified method where the transmission parameters are compared directly without first converting them to an estimated conversational quality. This simplified method was selected for this evaluation and the analysis follows the steps defined below:

1. A traffic model is determined. In this model, the lengths of the calls, the active speech parts and the silence parts are randomized according to a selected distribution. This allows us to use the correct voice activity factor (VAF).

2. IP/UDP/RTP packets containing dummy data are used in the system simulation.

3. For each transmitted packet, packet losses and transmission delay are recorded in a log file.

4. The speech quality evaluation is performed as a post-processing function, where for each call:

   (a) The packet loss rate is calculated.

   (b) The transmission delays for the packets are compared with a fixed threshold. A late loss is declared if the delay exceeds the threshold. This simulates using a fixed jitter buffer.

   (c) The packet loss rate and the late loss rate are added. If one speech frame is encapsulated in each IP/UDP/RTP packet, then the PLR and the LLR can be added directly. If not, then the PLR first needs to be converted to FER by taking the frame bundling into account.

(d) The FER and the transmission delay limit are then used to conclude whether the user is satisfied or not; see satisfaction criteria in Section 7.2.2.

(e) The packet loss distribution is verified. The idea here is to double check that no long loss bursts occur, since the error concealment in most modern speech codecs tends to hide single and double losses quite well but longer loss bursts tend to give more severe degradations. If long loss bursts occur, then further evaluation would be required.

(f) The packet transmission delay distribution is also analyzed by calculating minimum, maximum, average, and median delays. The 99, 98, 95, 90 and 75 percentile delays are also calculated. This information is useful when judging how different jitter buffer algorithms would perform. The 99 and 98 percentiles are of particular importance since they show the delay that one could have if the receiving application actively dropped the 1% and 2% latest received packets respectively. The other statistical metrics may on the other hand be very useful in the adaptation of the jitter buffer; see Section 5.3.3.

5. The percentage of satisfied users is calculated.

6. If the percentage of satisfied users is at least as good as for the reference system, i.e. the circuit switched system, then it is concluded that the VoIP system is at least as good or even better.

This analysis is performed for several different load levels to assess how the performance evolves as the load is increased. The analysis is also performed for all the schedulers outlined in Section 7.3.1.

Note that the simulation settings for the service quality evaluation differ from the settings used for capacity simulations. The main difference is caused by the number of HS-SCCH channels used in the system (four were used for the capacity simulations, while only one was used for service quality evaluation). The resulting maximum capacity for the simulations for service quality evaluation is slightly smaller than for capacity simulations, but still above the CS capacity.

7.6.2   Performance with the Delay Scheduler

One of the key performance metrics is the Packet Loss Rate (PLR) because it defines the lowest possible limit for the Frame Erasure Rate (FER). The frame erasure rate may occasionally be higher than what the PLR suggests because received packets can be thrown away in the jitter buffer, if the packets are received too late.

The packet loss rate for different users for different load levels is shown in Figure 7.18. Since the user population is different in the simulations for different load levels, the x-axis is normalized to 0–100% to make it possible to directly compare the performance for different load levels.

The figure shows that the vast majority of users have a packet loss rate well below the PLR limit for most load levels. It is only when the load increases to the CS limit and above that the packet loss rate exceeds the limit for some of the users. The figure also shows only the users that have packet losses since the y-axis is logarithmic.

The capacity limit is normally defined as the load where 95% of the users have a packet loss rate below the limit. From Figure 7.18, it is hard to see at what load level this limit is passed. An enlarged version of the most interesting region is therefore shown in Figure 7.19.

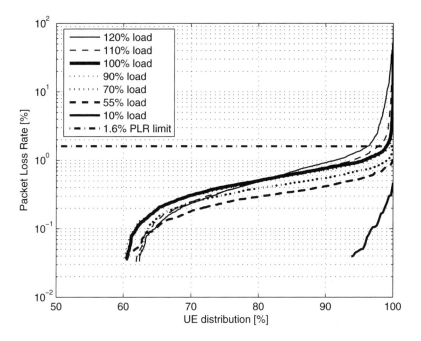

Figure 7.18: Packet loss rates for different users for different load levels with the delay scheduler. The users are sorted in increasing PLR order.

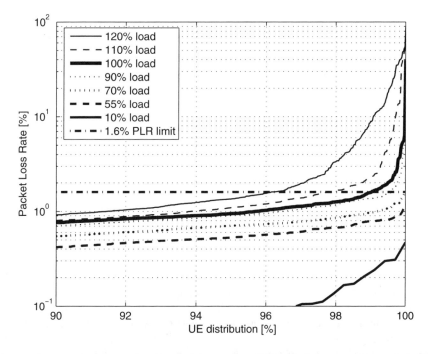

Figure 7.19: An enlarged version of Figure 7.18 zooming in on the most interesting region.

More detailed information about the PLR distribution is given in Tables 7.4 and 7.5. These tables show that about 60% of the users have no packet losses, not even for very high load levels. Figure 7.18 shows that, for the users that have packet losses, the loss rates are small enough not to severely impact the speech quality.

Table 7.4: Percentage of users with no packet losses and the percentage of users with a packet loss rate below the PLR threshold.

Load [%]	PLR = 0%	PLR $\leq$ 1.6%
10	93.51%	100.00%
55	60.45%	100.00%
70	62.35%	100.00%
90	60.14%	99.60%
100	60.46%	99.02%
110	61.88%	97.78%
120	62.38%	96.13%

Table 7.5: Percentage of users with a packet loss rate in the specified ranges. For example, at 90% load, 99.92% of the users have a packet loss rate below 5% and 0.08% of the users have a packet loss rate between 5% and 10%.

Load [%]	PLR$\leq$5%	5%<PLR$\leq$10%	10%<PLR$\leq$15%	15%<PLR$\leq$20%
10	100.00%	0.00%	0.00%	0.00%
55	100.00%	0.00%	0.00%	0.00%
70	100.00%	0.00%	0.00%	0.00%
90	99.92%	0.08%	0.00%	0.00%
100	99.89%	0.11%	0.00%	0.03%
110	99.48%	0.22%	0.15%	0.15%
120	98.48%	0.57%	0.50%	0.45%

Notice that the used PLR limit is set to 1.6% which is a tougher requirement than the 2% FER used for CS. This is to give room for some late losses that occur due to delay jitter. As can be seen in Table 7.4, 99% of the users have a packet loss rate below the defined limit at 100% relative load. This gives the receiving client plenty of room for the application to do a trade-off between:

1. keeping the delay low, which will increase the late loss rate and thereby also the frame erasure rate,

2. reducing the frame erasure rate, which requires allowing a longer delay.

The PLR distribution in Table 7.5 shows that almost all users have packet loss rates below 5%. This is beneficial because then application layer redundancy methods can be used with great success; see Section 5.3.4. Application layer redundancy would mean that a lower codec mode rate has to be used, but the session could at least survive for severely degraded channels with reasonable quality.

The packet loss rates are, of course, very important for the quality because PLR defines the minimum possible frame erasure rate. The frame erasure rate is however the sum of the

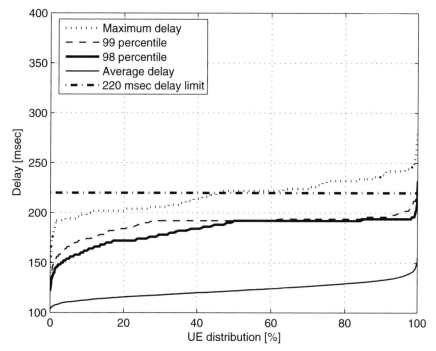

Figure 7.20: Maximum, average packet delays and 98 and 99 percentiles for 55% relative load with the delay scheduler.

frames lost due to lost packets and the frames that are thrown away because they arrive too late. One must therefore also analyze the packet delay statistics. The series of Figures 7.20 to 7.23 show how the maximum and average delays and the 98 and 99 percentiles increase for increasing system load. The 220 ms delay limit is also shown in all the figures. Note that the load is normalized to the CS capacity.

As can be seen in this series of figures, the maximum packet delays rapidly increase above the defined delay limit. The 98 and 99 percentiles however stay below the delay limit, suggesting that the tail of the packet delay is quite small and that the packets that exceed the delay limit can be thrown away by the receiving client.

Both the 98 and 99 percentiles are important because the amount of packets that can be dropped for a user depends on the packet loss rate that the user experiences. The average delays are also shown because this metric is often used by adaptive jitter buffers.

Another observation in these figures is that the delay distribution is quite equal for most users. One can therefore conclude that the delay scheduler manages to distribute the available resources quite well among the users. This is important because the best way to maximize the capacity is to ensure that no single user is allocated too many resources. This also means that, once the capacity limit is exceeded, the amount of distortions that will occur due to packet losses and late losses will be fairly well distributed over the whole set of users and quite a few users will get very bad performance.

The equally distributed degradation is also important because this improves the robustness against non-optimal deployment. For example, if the system has coverage holes, then the equal distribution means that a user moving into a bad coverage area will automatically be

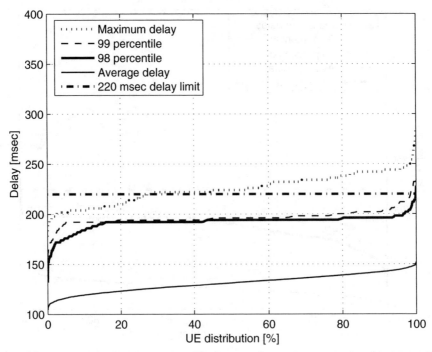

Figure 7.21: Maximum, average packet delays and 98 and 99 percentiles for 70% relative load with the delay scheduler.

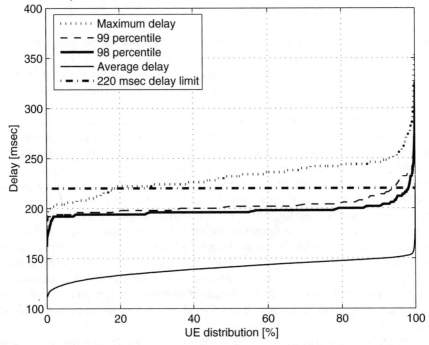

Figure 7.22: Maximum, average packet delays and 98 and 99 percentiles for 100% relative load with the delay scheduler.

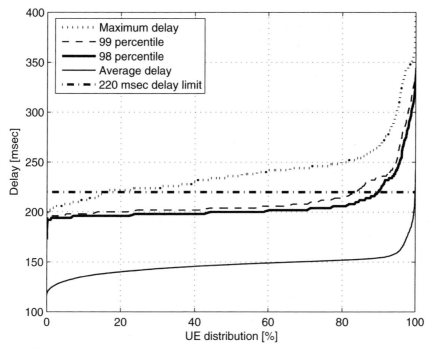

Figure 7.23: Maximum, average packet delays and 98 and 99 percentiles for 120% relative load with the delay scheduler.

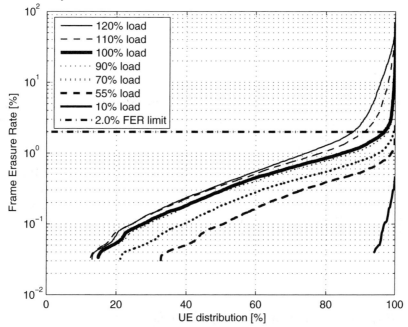

Figure 7.24: Frame erasure rates for different users for different load levels with the delay scheduler. The users are sorted in increasing FER order.

allocated more transmission resources. The system will, in other words, try as long as possible to deliver enough resources for the user to have reasonable quality.

The resulting frame erasure rate for different users is shown in Figure 7.24, when a fixed delay threshold of 220 ms simulates the usage of a fixed jitter buffer. As can be seen in this figure, most users have a frame erasure rate well below the 2% threshold even for high relative loads.

Tables 7.6 and 7.7 provide more detailed information on the FER distribution. Table 7.6 shows that, for a load that corresponds to the maximum capacity of the CS system, almost 97% of the users are satisfied. This makes HSPA marginally better than CS.

Table 7.6 also shows that, for HSPA, most satisfied users will have a frame erasure rate that is well below the limit. This is because there are still resources available that can be used for additional retransmissions. The HARQ function will then, in most cases, be able to properly decode the transmission block, which reduces the loss rate. One can therefore conclude that most satisfied users in HSPA will have a quality that is better than what they would have on CS.

Table 7.6: Percentage of users with no frame erasures and the percentage of users with a frame erasure rate below the FER threshold.

Load [%]	FER $= 0\%$	FER $\leq 2.0\%$
10	93.51%	100.00%
55	32.74%	99.91%
70	21.16%	99.71%
90	14.70%	97.71%
100	14.90%	96.89%
110	12.70%	90.90%
120	13.13%	87.85%

Table 7.7: Percentage of users with a frame erasure rate in the specified ranges.

Load [%]	FER$\leq$5%	5%$<$FER$\leq$10%	10%$<$FER$\leq$15%	15%$<$FER$\leq$20%
10	100.00%	0.00%	0.00%	0.00%
55	100.00%	0.00%	0.00%	0.00%
70	100.00%	0.00%	0.00%	0.00%
90	99.52%	0.28%	0.17%	0.03%
100	99.23%	0.42%	0.27%	0.08%
110	96.15%	1.75%	1.06%	1.04%
120	93.44%	2.32%	2.17%	2.08%

7.6.3   Performance with the Max-CQI Scheduler

The packet loss rates and frame erasure rates for the max-CQI scheduler have also been evaluated. The packet loss rates for different users and for different load levels are shown in Figure 7.25.

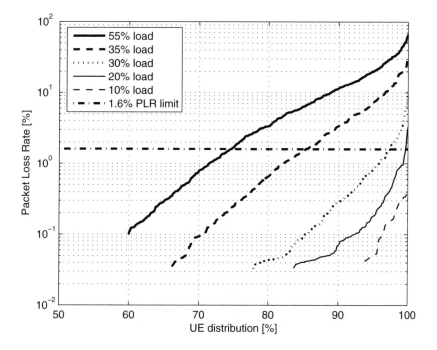

Figure 7.25: Packet loss rates for different users for different load levels with the max-CQI scheduler. The users are sorted in increasing PLR order.

As can be seen, the max-CQI scheduler gives very high packet loss rates for many users, even if the load is significantly lower than for the loads used in the analysis of the delay scheduler; see Section 7.6.2.

One can also see that most users have either very good performance, i.e. very low packet loss rates, or very bad performance, i.e. very high packet loss rate. This is because of the scheduler that only schedules the one or the few users who have the best channel conditions. A scheduler working in this way would probably be quite useful for data services, if or when one wants to deliver high peak rates to users that have good operating conditions. However, for real-time voice services, one cannot take advantage of the high peak rates, due to the small packet sizes. There is simply not enough data to transmit. These things make the max-CQI scheduler unsuitable for VoIP.

Figures 7.26 and 7.27 show how the packet delay statistics change for increasing load levels. Further results are available in the Appendix.

These figures show that many users will have packets with very long delay. In fact, already at 30% relative load, the worst average packet delay is over 500 ms, and at 55% relative load, the worst average packet delay is over 4 s. Combining the late losses due to late arriving packets with the packet losses gives the frame erasure rates for different loads as shown in Figure 7.28.

Compared with the delay scheduler, one can see that the max-CQI scheduler gives much worse performance. Not only does it require much lower load to keep 95% of the users satisfied, however, it also gives large differences between the users that are satisfied and those that are not.

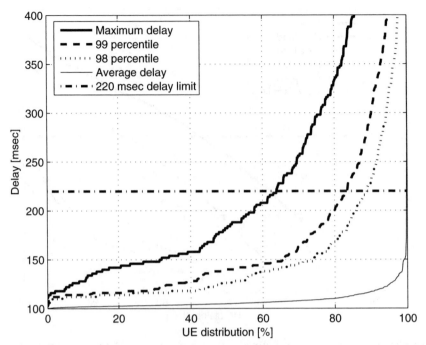

Figure 7.26: Maximum, average packet delays and 98 and 99 percentiles for 30% relative load with the max-CQI scheduler.

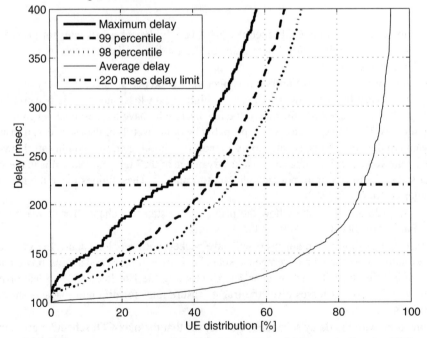

Figure 7.27: Maximum, average packet delays and 98 and 99 percentiles for 55% relative load with the max-CQI scheduler.

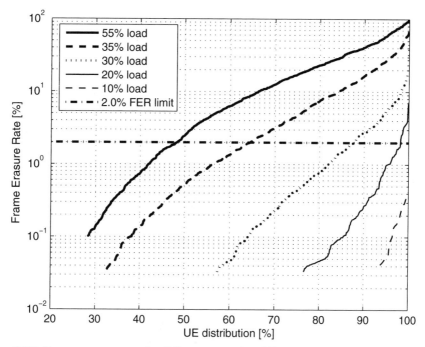

Figure 7.28: Frame erasure rates for different users for different load levels with the max-CQI scheduler. The users are sorted in increasing FER order.

Figure 7.25 shows that the capacity limit for VoIP should be at about 30% relative load since about 95% of the users have a packet loss rate below the threshold. However, the high frequency of long packet delays gives a large amount of late losses, which significantly reduces the number of satisfied users. As Figure 7.28 shows, the capacity limit is closer to 20% than 30% relative load.

As Figure 7.28 also shows, the frame erasure rate exceeds 10% for many users. One can also see that the performance is almost 'binary'. The users have either very good performance or very bad performance. This indicates that mobile users will have either good sound quality when they have good radio conditions or very poor sound quality when the radio conditions are worse. In fact, one can expect that the sound will be completely muted for long periods when the mobile is in a spot of bad coverage. This is not very good for real-time voice services. One would rather prefer a more gradual reduction in speech quality, so-called *graceful degradation*.

7.6.4 Performance with the Proportional-Fair Scheduler

The proportional-fair scheduler is a scheduler that has been discussed in the literature. It is therefore one candidate that needs to be assessed. The packet loss rates for different load levels are shown in Figure 7.29.

The maximum, average of the packet delays and the 98 and 99 percentiles are shown in Figures 7.30 and 7.31. Further results are available in the Appendix. The figures show that the packet delay increases in a similar way as when the max-CQI scheduler is used.

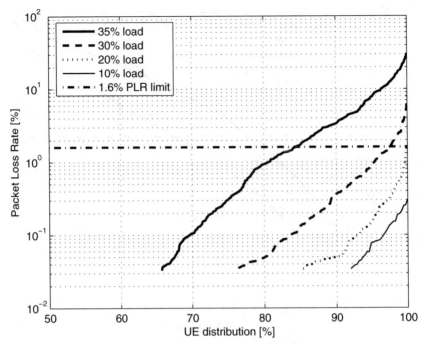

Figure 7.29: Packet loss rates for different users for different load levels with the proportional-fair scheduler. The users are sorted in increasing PLR order.

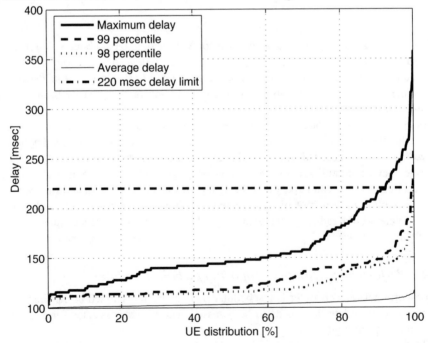

Figure 7.30: Maximum, average packet delays and 98 and 99 percentiles for 20% relative load with the proportional-fair scheduler.

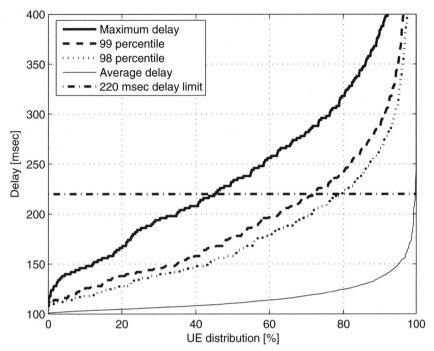

Figure 7.31: Maximum, average packet delays and 98 and 99 percentiles for 35% relative load with the proportional-fair scheduler.

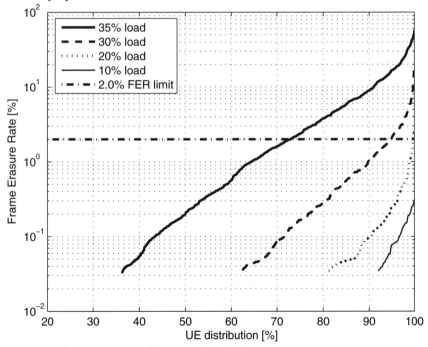

Figure 7.32: Frame erasure rates for different users for different load levels with the proportional-fair scheduler. The users are sorted in increasing FER order.

Again, one can see that the performance with the proportional-fair scheduler is similar as for the max-CQI scheduler.

Figure 7.29 suggests that the capacity limit should be slightly above 30% relative load. The packet delay statistics are however a little better than for the max-CQI scheduler and as Figure 7.32 shows, the capacity limit is slightly below 30%.

Combining the packet loss rates with the late losses introduced by long packet delays gives the frame erasure rates as shown in Figure 7.32. As can be seen in the figure, the performance for the proportional-fair scheduler shows similar on–off behavior as for the max-CQI scheduler. This makes the proportional-fair scheduler unsuitable for real-time voice.

7.6.5 Performance with the Round-Robin Scheduler

Another commonly discussed scheduler is the round-robin scheduler. The packet loss rates for different users and different load levels are shown in Figure 7.33. Figures 7.34 and 7.35 show how the packet delay statistics evolve with increasing load and Figure 7.36 shows the resulting frame erasure rate. Further results are available in the Appendix.

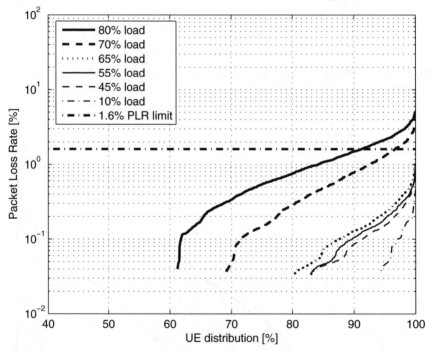

Figure 7.33: Packet loss rates for different users for different load levels with the round-robin scheduler. The users are sorted in increasing PLR order.

As can be seen, the performance is quite different from the max-CQI and the proportional-fair schedulers and shows more similarities with the delay scheduler. The satisfaction criterion is fulfilled up to a relative load of about 70% showing that this scheduler does not give as good capacity as the delay scheduler.

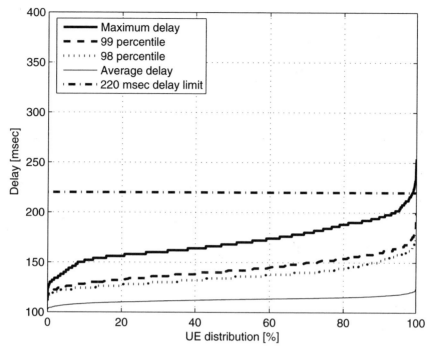

Figure 7.34: Maximum, average packet delays and 98 and 99 percentiles for 45% relative load with the round-robin scheduler.

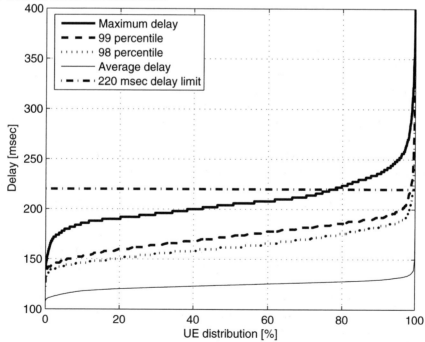

Figure 7.35: Maximum, average packet delays and 98 and 99 percentiles for 80% relative load with the round-robin scheduler.

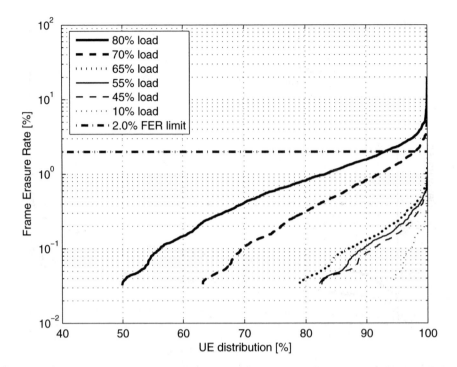

Figure 7.36: Frame erasure rates for different users for different load levels with the round-robin scheduler. The users are sorted in increasing FER order.

7.6.6   Analysis of Packet Loss Bursts

The distribution of the packet losses is also important for the speech quality. The error concealment function typically handles single and double packet losses fairly well, at least if the losses occur during steady-state signals such as vowels and background noise. But for each consecutive packet loss, the error concealment function will attenuate the signal more and more in order to avoid generating severe distortions. If the packet loss bursts are very long, the signal will be completely muted, giving an impression of a lost call for the listener.

The packet loss burst lengths are also important for the performance of the application layer redundancy. Single packet losses can be recovered by 100% redundancy while longer loss bursts either require a larger amount of redundancy or that the distance between the original frame and the redundant frame is increased. Both these solutions would however increase the end-to-end delay.

It is therefore important to also analyze how the packet losses are grouped into packet loss bursts. The users were grouped according to their respective packet loss rate:

- The group with users having less than 2% PLR is probably the most interesting group because it shows the packet loss distribution for the most normal operating conditions when application layer redundancy is not used.

- The group with users having a packet loss rate between 2% and 10% is interesting for the case when single (100%) redundancy is used.

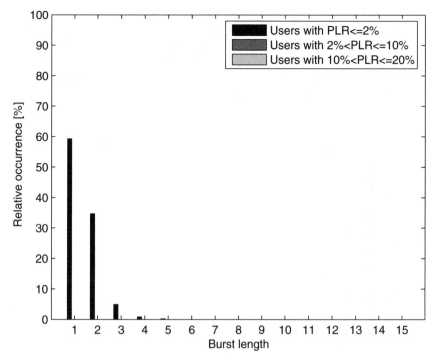

Figure 7.37: Histogram of packet loss bursts for the delay scheduler at 55% relative load.

- The group with users having a packet loss rate between 10% and 20% is interesting for the case when more than 100% application layer is used.

The analysis is performed by calculating the occurrences of packet loss burst with burst lengths of one (= single packet loss), two (= double), three (= triple) and up to 15 consecutive packet losses. The histograms of the analysis for the delay scheduler with different system loads are shown in Figures 7.37 to 7.40.

These figures show that for all users, regardless of packet loss rate, most of the packet losses are single losses or doubles. In fact, the vast majority, about 90% of the packet loss occurrences, are either single or double losses. Longer bursts are mainly three or four in a row. There are a few occurrences of long or even very long packet loss bursts, but these are both rare and occur also only for the highest relative load levels.

One concern when developing the HSPA system and the delay scheduler was that there would be a quite big difference in packet loss burst lengths between low and high load levels and also between users with different packet loss rates, i.e. between satisfied and unsatisfied users. The fact that the histograms are fairly similar for all load levels and also fairly similar for different packet loss rates shows that the scheduler manages to distribute the transmission resources quite well among the users, even for the users that are outside the satisfaction criteria.

The concentration of packet losses to singles and doubles means that application layer redundancy should work fairly well, even without the need to use excessive delay between original and redundant frames. Thereby, it seems reasonable to believe that the speech quality

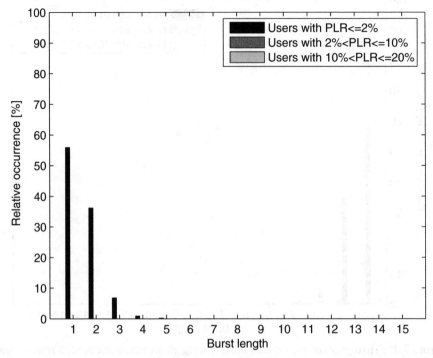

Figure 7.38: Histogram of packet loss bursts for the delay scheduler at 70% relative load.

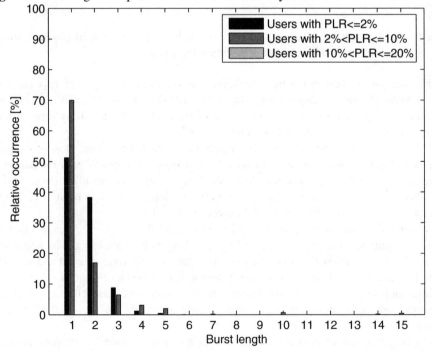

Figure 7.39: Histogram of packet loss bursts for the delay scheduler at 90% relative load.

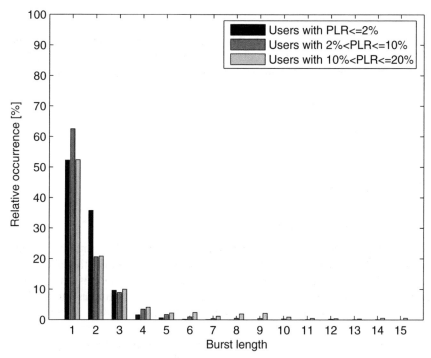

Figure 7.40: Histogram of packet loss bursts for the delay scheduler at 120% relative load.

could be significantly improved by using application layer redundancy for the users that have a packet loss rate that exceeds the satisfaction criteria.

Figures A.12–A.14 show the histograms of packet loss bursts for the max-CQI scheduler at 10% to 55% relative load levels and Figures A.17 and A.18 show the histogram of packet loss burst for the proportional-fair scheduler at 10% to 35% relative load levels. As can be seen in these figures, both the max-CQI and proportional-fair schedulers give a quite large amount of long packet loss bursts, especially as the load is increased to the capacity limit. This is further evidence that these schedulers do not work very well for real-time VoIP traffic.

Figures A.22–A.24 show the histograms of packet loss bursts for the round-robin scheduler from 10% to 80% relative load respectively. As can be seen in these figures, with the round-robin scheduler, the vast majority of the packet loss occurrences are either single, double and triple packet losses, with single losses being the dominant, as long as the system load is lower than the capacity limit. When the capacity limit is reached and exceeded, packet loss bursts with three losses in a row become dominant. This means that the round-robin scheduler is a fairly good choice, at least from a speech quality perspective and at least as long as the system load is kept below the capacity limit.

7.7   Call Setup Delays

In the earlier sections, we have focused on the system and user performance mostly from the media perspective. In this section we briefly examine another aspect of Multimedia Telephony performance, namely the call setup delays.

7.7.1 General Assumptions

The call setup is here defined to be the Multimedia Telephony session setup between two Multimedia Telephony clients implemented on mobile terminals. The Multimedia Telephony session setup delays are estimated analytically for a call between two users that:

- are served by the same operator,

- are located in the same geographical area,

- do not roam from a visited network,

- have post-paid subscriptions.

7.7.2 IMS and SIP Assumptions

It is assumed that P-CSCF, the Invited Multimedia Telephony client and the Inviting Multimedia Telephony client support and use SigComp (for more information about SigComp, see Section 4.2). It is also assumed that the SigComp implementation is suitably optimized (dynamic compression, SIP specific dictionary, etc.) for efficient compression of Multimedia Telephony session setup signaling. The resulting compression factor is assumed to be 3 : 1 on average.

No other SIP signaling is assumed to be present, i.e. there is for example no presence signaling, that interferes with the Multimedia Telephony session setup signaling.

Time for application, SIP and SigComp processing in the Multimedia Telephony client is assumed to be 100 ms. The value for the delay is approximate, but should not be difficult to reach with existing and future mobile terminals. The SIP and SigComp processing delay in the IMS core is assumed to be 80 ms for the initial SIP INVITE message and 25 ms for all other SIP messages. Again these values are only an indication of the possible delays, but should be reachable even in a highly loaded system.

It should be noted that all SIP messages pass two instances of the IMS core (the originating IMS core and the terminating IMS core), e.g. making the delay of the SIP INVITE message through the IMS node 2×80 ms $= 160$ ms. This assumption holds if the transfer delay between the IMS cores is small, i.e. the users are served by the same operator and are in the same geographical area (both users are using the same set of IMS nodes as their respective IMS cores). In the case of long distance calls or inter-operator calls, the transfer delay of the SIP messages might be larger.

Finally it is assumed that the processing and transmission between the PCRF and IMS core of a Diameter message adds a delay of 25 ms and that the processing and transmission between the PCRF and GGSN of a Diameter message also adds a delay of 25 ms.

It is assumed that the signaling related to charging happens in parallel with the final SIP 200 OK response signaling and thus does not add to the delay of the system.

7.7.3 UMTS Assumptions

For user satisfaction the mobile terminal, the RNC and the S/GGSN need to be in states that enable long battery lifetime when not sending or receiving data, but with intact PS connectivity to the IMS.

Basically there are three such state combinations:

- PMM-IDLE RRC: Idle and SM: ACTIVE;

- PMM-CONNECTED, RRC: URA_PCH and SM: ACTIVE;

- PMM-CONNECTED, RRC: Cell_PCH and SM: ACTIVE.

From an IMS session setup delay perspective, the first state combination gives the longest delay, while the last state combination gives the fastest Multimedia Telephony session setup times. As a compromise between speed and battery lifetime the second state combination was chosen to be used in this analysis. It can be assumed that the battery consumption in URA_PCH is about 3 mA h (as a comparison when transmitting in CELL_DCH the mobile terminal consumes about 300 mA h). Given a battery with 800–1000 mA h a mobile terminal should be able to camp in URA_PCH for days.

No transmission of SIP messages can be done in URA_PCH. Instead the mobile terminal and the RNC must switch to the Cell_DCH state before transmission of SIP messages can start. The state transition between URA_PCH and Cell_DCH needs two RRC procedures on the originating side and paging plus two RRC procedures on the terminating side. The two RRC procedures are: cell update and channel up switch. The technology potential of HSPA make the following delays possible.

- Cell update: 310 ms.

- Channel up switch, mobile terminal initiated: 260 ms.

- UTRAN paging: 210 ms.

- Channel up switch, UTRAN initiated: 150 ms.

As described in Chapter 3, the Multimedia Telephony communication service uses a dedicated Interactive RAB for the SIP signaling. This means that a second RAB and thus a second PDP context is needed to convey the media of the Multimedia Telephony session. It should be noted that, after the secondary PDP context has been established, the SIP signaling is still sent over the primary PDP context.

It is assumed that the signaling radio bearers that carry e.g. GPRS SM messages and RRC messages are mapped onto E-DCH and HS-DSCH. This makes it possible to do the PDP context activation in about 350 ms.

The radio conditions are assumed to be good, enabling a maximum throughput of 0.5 Mbps in the UL and 2 Mbps in the DL. The transmission time of SIP messages assumes no RLC retransmissions but it assumes that on average 10% of the MAC-hs PDUs and the MAC-e PDUs are retransmitted using H-ARQ instead. On average approximately 1 ms is added to the delay for each SIP message sent over HSPA in UL and DL.

7.7.4 Delay Calculation

The total time to send a SIP message between a Multimedia Telephony client and the IMS core (or vice versa) is given by the sum of the following:

- the time for Multimedia Telephony client processing in the mobile terminal to do e.g. SIP and SigComp message processing;

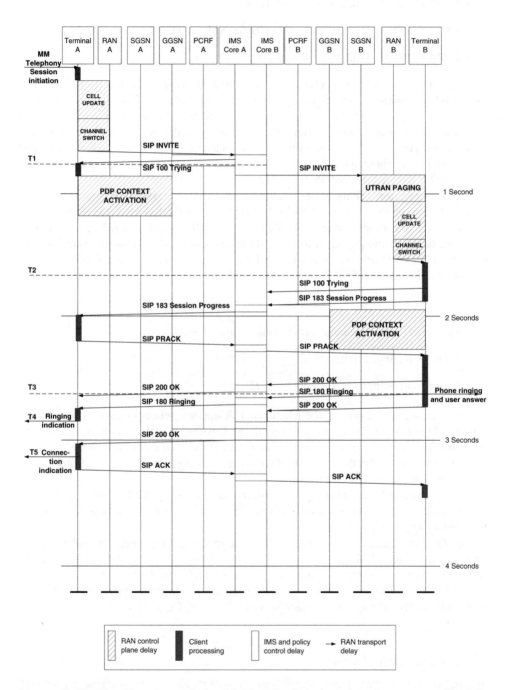

Figure 7.41: Signaling flow of the Multimedia Telephony session setup with delay components.

- the RAN transmission time of the SIP message from the mobile terminal to the P-CSCF;

- the time for SIP/SigComp processing in the IMS core and processing in the PCC (policy control) nodes.

The only exception is for the first message. For the first message in UL, time is also needed for various RAN control plane procedures:

- cell update;

- mobile terminal initiated channel up switch.

For the first message in DL, time is needed for:

- UTRAN paging;

- cell update;

- UTRAN initiated channel up switch.

The Multimedia Telephony session setup also includes PDP context activation, which is categorized as a RAN control plane procedure.

In Figure 7.41 we show the signaling flow and the sources of delay of a Multimedia Telephony session set up using mobile terminal initiated QoS (for more information about the signaling flow, see Section 4.4.1). The flow is broken down into five sections (marked T1 to T5 in the figure) and the cumulative delay for T1 to T5 is given in Table 7.8. The total time for the Multimedia Telephony session setup is about 3.1 s.

Table 7.8: The breakdown of the Multimedia Telephony session setup time.

		Cumulative Delay [ms]
T1	SIP INVITE leaves IMS core A	781
T2	SIP INVITE arrives at client B	1668
T3	SIP 180 Ringing sent from client B	2596
T4	SIP 180 Ringing arrives at client A	2808
T5	SIP 200 OK arrives at client A	3110
Total		3110

Figure 7.42 show the fraction of the total Multimedia Telephony session setup delay of the four sources of delay (client processing, RAN control plane delay, RAN transport and IMS/policy control delay). The RAN control plane is the major contributor of delay. In the proposed configuration a total of 1.25 s or 41% is spent on various RAN control plane related activities. If the mobile terminal and the RNC were in the RRC: IDLE state before the start of the session, this fraction would be even higher.

The time it takes to transport the SIP messages is only 0.25 s in the case of WCDMA HSPA. This corresponds to only 8% of the total Multimedia Telephony session setup. The SIP messages have been compressed using SigComp with an average compression ratio of 3 : 1. Due to the high peak bit rate of WCDMA HSPA almost all SIP messages in the calculations are conveyed in one transport block. Thus the delay reduction of raising the compression

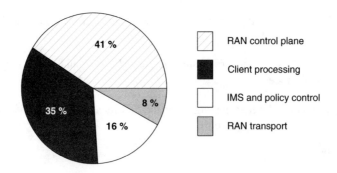

Figure 7.42: Delay breakdown for Multimedia Telephony session setup.

factor is negligible. The delay is only reduced by roughly 10 ms if the compression factor is increased to 6 : 1.

The processing time of the IMS and the PCC (the policy control) that contribute to the total Multimedia Telephony session setup delay is 0.51 s or 16% of the total Multimedia Telephony session set up.

The Multimedia Telephony client processing is a major delay component. The Multimedia Telephony client processing represents 1.1 s or 35% of the total Multimedia Telephony session setup delay. The conclusion is that it is important to have a fast and efficient implementation of the Multimedia Telephony communication service and the SIP stack in the mobile terminal.

7.7.4.1   Using Network Initiated QoS

This section presents the Multimedia Telephony session setup delay when using network initiated QoS. If network initiated QoS is used, the Multimedia Telephony session setup flow will look a bit different compared to the mobile terminal initiated QoS case. The Multimedia Telephony session setup delay is also dependent on whether the QoS control uses Early-Rx or Late-Rx signaling over the Rx interface (for more information about Early-RX and Late-Rx, see Section 4.3.2). The signaling flow with delay components for the case of Late-Rx is shown in Figure 7.43. If the same assumptions are used as for the network initiated QoS case above, the delays in Table 7.9 are valid. It can be noted that the fastest configuration analyzed is Early-Rx. The down side with Early-Rx may be that it is possible that the media transfer

Table 7.9: The breakdown of the Multimedia Telephony session setup time using network initiated QoS.

		Early-Rx Delay [ms]	Late-Rx Delay [ms]
T1	SIP INVITE leaves IMS core A	781	1227
T2	SIP INVITE arrives at client B	1768	2560
T3	SIP 180 Ringing sent from client B	1968	2760
T4	SIP 180 Ringing arrives at client A	2180	2970
T5	SIP 200 OK arrives at client A	2482	3270
Total		2482	3270

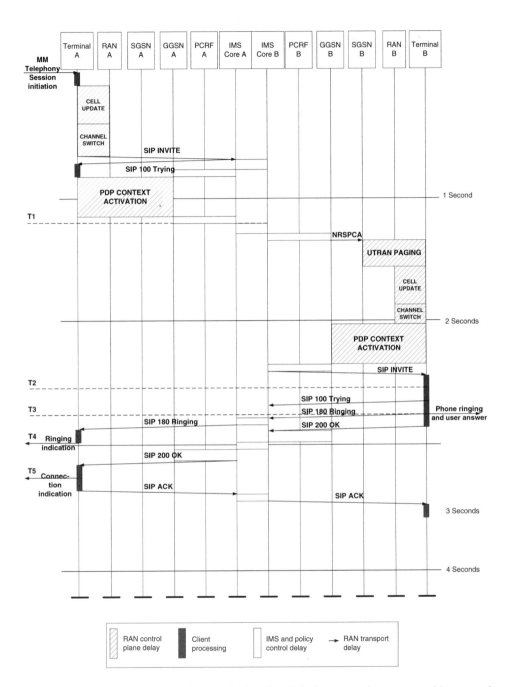

Figure 7.43: Signaling flow of the Multimedia Telephony session setup with network initiated QoS.

starts before the media RAB is set up. This may lead to audio clipping or even ghost ringing if the media RAB is for instance never set up on either the originating or terminating side. But the calculations show that audio clipping will not happen in a successful case. The Late-Rx method is a safer alternative with respect to audio clipping and ghost ringing, but gives longer Multimedia Telephony session setup delays.

Chapter 8

Other IMS Communication Services

Per Synnergren

Besides the 3GPP efforts of defining the Multimedia Telephony communication services, 3GPP and the OMA are dealing with the specification of a number of IMS communication services or service enablers that have relations to Multimedia Telephony. This chapter briefly describes the following:

- The 3GPP Circuit Switched IMS Combinational Service (CSICS) that is used to add richer media types over PS to the CS telephony service.

- The OMA Push-to-talk over Cellular (PoC) service, a half-duplex walkie-talkie style communication system using IMS as the service layer.

- The OMA Instant Messaging (IM) communication service based on the SIP and MSRP protocols that are under development.

- Presence and group list management that also act as service enablers.

8.1   3GPP CSICS

The abbreviation CSICS stands for Circuit Switched IMS Combinational Service. The 3GPP CSICS is an enabler that provides the possibility of bundling the voice service in the existing CS domain with an IMS/SIP session in the PS domain for media transfer.

Interoperability and standardization are keys for the combinational services to enter the mass market. Therefore, the CSICS concept was specified as part of 3GPP release 7; see 3GPP technical specifications 22.279 [8], 23.279 [17] and 24.279 [18]. In the standardization process, the industry view was that CSICS should use the standard based voice and data networks that to a great extent were already deployed by the operators. This was done to create a multimedia experience for the end-users without the need for expensive investments in infrastructure. Therefore, the following requirement was formulated to guide the work with CSICS in 3GPP:

IMS Multimedia Telephony over Cellular Systems   S. Chakraborty, T. Frankkila, J. Peisa and P. Synnergren
© 2007 John Wiley & Sons, Ltd

- The CS core, the PS core and the RAN are not to be impacted by CSICS. Conclusively, changes needed to deploy services based on CSICS should be restricted to the IMS elements and the mobile terminals that support CSICS for IMS.

Another goal with 3GPP CSICS was to addresses the possibility of a service driven migration to all-IP and Multimedia Telephony. By adopting CSICS it is possible to take a stepwise approach towards the long-term migration of voice traffic to IMS logic and the PS domain.

- Step 1: Keep the signaling control and traffic of the voice service in the legacy CS domain, while the signaling control and traffic of the multimedia enrichments are handled by the IMS/PS domains.

- Step 2: The signaling control for the voice service is moved to the IMS/PS domains.

- Step 3: The Multimedia Telephony scenario in which signaling control and traffic for both voice and multimedia content are handled by the IMS/PS domain.

8.1.1 CSICS Architecture

When using CSICS, the CSICS client implemented in the mobile terminal presents the CS call and the IMS session within one context to the user. To facilitate this, the following capabilities need to be supported:

- exchange of information related to the current access;

- exchange of terminal capability information;

- addition of an IMS session to an ongoing CS call;

- addition of a CS call to an ongoing IMS session.

Figure 8.1 shows a high-level end-to-end architecture of a simultaneous IMS session and CS call between two end-users belonging to the same operator. It should be pointed out that the mobile terminal needs to support simultaneous CS and PS domain access. This is a general capability for mobile terminals using the WCDMA access. A mobile terminal camping in a GSM network needs to support Dual Transfer Mode (DTM; see 3GPP technical specification 43.055 [21]) to be capable of delivering services based on 3GPP CSICS. Additionally, the mobile terminal should support the capability exchange mechanism and the capability to present the CS call and IMS session within the same context to the user. In addition, the IMS that provides the session control also needs to support the mobile terminal capability exchange mechanism. An application server may be utilized to handle the control of the IMS specific aspects of the CSICS session, for example service-based charging or volume-based charging. If service-based charging mechanisms, like charging based on the content of a multimedia message, the message type or the number of sent and/or received messages, are required, then the application server needs to be involved.

The goal of the capability exchange is to spread knowledge to the end-points (the CSICS clients) about the set of service components such as types of media that can be supported by them. The capability exchange should happen when (or shortly after) communication is established. This information can then be used to provide an indication to the users of

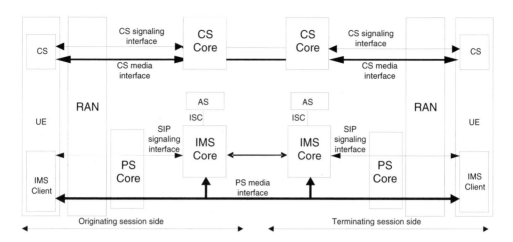

Figure 8.1: High-level architecture of the 3GPP CSICS concept.

the service components that are available; this is to encourage the use of available service components and to avoid invocation of unavailable service components.

Two types of capability exchanges may be performed. First (and optional) is the access capability exchange at CS call setup. In this capability exchange, information about whether the mobile terminal is capable of handling simultaneous CS and PS services, and whether the current radio access supports simultaneous CS and PS services, is conveyed. The access capability exchange uses CS signaling, and the network must handle the radio capability information transparently in order for this procedure to be successful. If the results are positive (i.e. the querying CSICS client finds out that simultaneous CS and PS services can occur), then the client should perform an IMS registration (if it is not already registered). The second capability exchange is the mobile terminal capability exchange. This provides input to determine the set of service components that can be invoked between two users. Here, information about which media types can be supported in the IMS/PS domain, media format parameters for supported media types, MSISDN and preferred SIP URI of the user of the mobile terminal sending capability information is exchanged. The mobile terminal capability exchange uses the SIP OPTIONS procedure. After the capability exchange procedures have been finalized, a user can add the IMS service to the ongoing CS call.

8.1.2 Interoperability with Multimedia Telephony

When a CSICS user calls a Multimedia Telephony user (or vice versa), the voice part of the call will create one IMS/SIP session where the Multimedia Telephony client and the Media Gateway Control Function (MGCF) act as the SIP end-points. The MGCF is the gateway between the CS network and the IMS domain and is in charge of converting the CS call related signaling (i.e. BICC or ISUP) to SIP signaling. However, if the CSICS user or the Multimedia Telephony user wants to enrich the call with a multimedia extension, e.g. a video stream, then a second SIP session must be established. Unlike the SIP session for the voice part this SIP session will be 'end-to-end' and it is the Multimedia Telephony client and the CSICS client that will act as the SIP end-points. So in this case of an enriched CSICS call, the involved Multimedia Telephony client will have to handle media that are related to two different SIP sessions. Thus, to interoperate with CSICS, there is a requirement that the Multimedia Telephony client needs to be able to handle media that are related to multiple SIP sessions.

8.1.3   WeShare: a 3GPP CSICS Service Example

As mentioned earlier, 3GPP CSICS is an enabler, a communication service, which can be used to packetize useful multimedia services for the end-user. One effort to create such a service is the 'Video Share definition' specifications developed by GSM Association (GSMA). In this section one service offering that complies with these specifications, the Ericsson IMS WeShare solution, is presented.

The Ericsson WeShare solution is a multimedia person-to-person service that provides the end-users with an instant way of sharing an image, video stream or some other type of stored media while speaking in a CS telephony call conversation. Two users involved in a voice conversation can at any time simply add an image, add a live video or share any multimedia file stored on the mobile terminal. The users can also share a whiteboard session during the CS telephony session. This way enriches and enhances the CS voice call by not limiting them to the capabilities of the current form of communication. This is available with a few simple keystrokes on the terminal. The following sections present the end-user features the Ericsson IMS WeShare service provides.

It should be noted that the 3GPP CSICS specification does not limit the service to the features mentioned below. With the 3GPP CSICS architecture, any kind of combinational services can be introduced quickly and cost effectively.

8.1.3.1   WeShare Image

WeShare Image allows users of the WeShare service to share images during a CS call. Figure 8.2 exemplifies the WeShare Image service. Both the calling WeShare user and the called WeShare user are both able to send and receive images during the call. The images that are shared are taken during the voice call using the in-built camera. Stored images can also be transferred but then the service is called WeShare Media File (see Section 8.1.3.3). The sender of an image can only send one image at a time. This means that the first image has to be transferred before the second image can be sent. The whole image is downloaded to the receiving mobile terminal before it is displayed on the display. The downloading state is indicated on the terminal display during the image transfer. When the image is received it will be temporarily stored in a directory on the terminal. The receiving user then gets to choose whether to store the image permanently or not.

8.1.3.2   WeShare Motion

WeShare Motion allows users of the WeShare service to share live video during a CS call. Both the calling WeShare user and the called WeShare user are able to send and receive video if they so wish. Depending on what was negotiated during WeShare session setup, both the calling WeShare user and the called WeShare user can each send video at the same time, or they can only send video in one direction at a time. Figure 8.3 exemplifies the WeShare Motion service. Both WeShare users can trigger the WeShare Motion feature independently from who started the voice call communication. The WeShare user receiving an invitation to a WeShare Motion session is able to accept or reject the video if manual answer mode

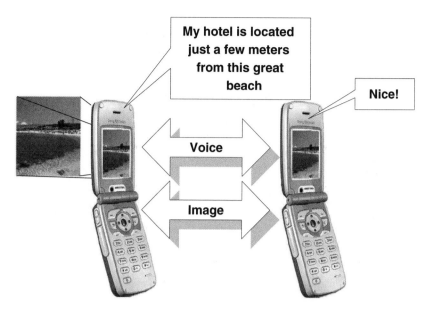

Figure 8.2: WeShare Image.

is used. The WeShare client could be preconfigured to always accept incoming invitations for a WeShare Motion session. Further, the invited WeShare user to the WeShare Motion session can at any point close down the video transfer in the WeShare Motion session without closing down the whole WeShare application. The media stream contains no voice data, as the voice content is sent via the CS connection.

8.1.3.3 WeShare Media File

WeShare Media File allows users of the WeShare service to share any stored media (e.g. image, video clip) during a CS call. Both the calling WeShare user and the called WeShare user are both able to send and receive different types of media. The media that is shared must already be stored in the mobile terminal before media transfer can start. Further, it is only possible for one WeShare user to send one media file at a time, i.e. the first media file has to be completely transferred before a second transfer of a media file can begin. Depending on media type either the media can be progressively downloaded (and be consumed while downloading) or the media transfer needs to be completed before being displayed on the terminal screen. In general, the WeShare Media File feature can transfer any types of media files (such as images, video clips, MS Office files, music, etc.), but the handling of a media file like MS Office files, music, etc. depends on the terminal capability and may require manual interaction from the WeShare user.

8.1.3.4 WeShare Whiteboard

WeShare Whiteboard allows users of the WeShare service to share a whiteboard session during a CS session. The WeShare users can draw on a blank background of configurable color, or select to share an image as the background for drawing, i.e. a map or a floor plan of a building. Both WeShare users can edit the drawing, and both WeShare users get to see

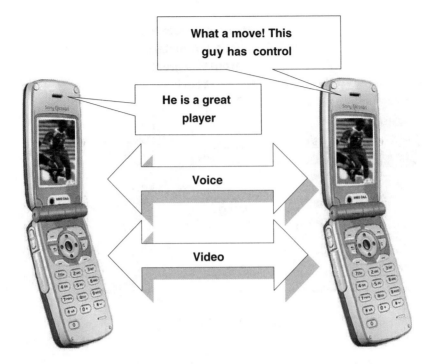

Figure 8.3: WeShare Motion.

the complete content. The involved WeShare users can individually store the content of the WeShare Whiteboard session at any time. Figure 8.4 exemplifies the WeShare Whiteboard feature.

8.2   OMA PoC

Push-to-talk over Cellular (PoC) is a quick and informal way of communicating person-to-person and with groups. With a simple push of a button, users can communicate with friends and family in much the same way as using walkie-talkies or private mobile radios. As a walkie-talkie, PoC is a half-duplex, one-way communication method between one or more participants in a communication session. But because PoC is a mobile communication service, it also enjoys the range and wide area coverage of traditional mobile services. Figure 8.5 illustrates a typical one-to-one PoC call.

1. Jennifer selects Isaac and presses the phone's push-to-talk button to start a PoC session. When she hears an audible tone she begins talking.

2. Isaac hears her voice instantly due to his terminal's automatic answer of all incoming PoC calls. Isaac presses the phone's push-to-talk button to respond to Jennifer's question.

3. Jennifer hears his voice instantly due to her terminal's automatic answer of all PoC calls.

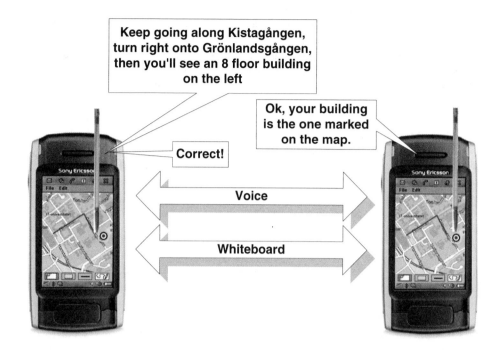

Figure 8.4: WeShare Whiteboard.

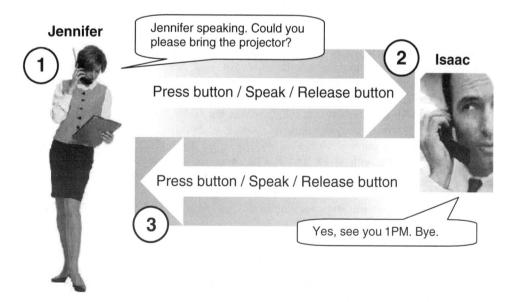

Figure 8.5: Simple PoC use case.

PoC in the consumer segment allows users to stay in touch with friends and coordinate leisure activities, such as visits to the cinema or simultaneous communication with a group of family members. In the enterprise segment, PoC can be used to share information in a group, for instance, a field technician can use it to ask colleagues for help or advice. PoC also aims for the public safety segment, in which it can be marketed as a part of a cost effective alternative to systems like TETRA. The communication service specified in the OMA PoC Release 1 specifications is a half-duplex voice communication service that can handle the management of contact lists and communication groups. The OMA PoC Release 2 specifications add a number of enhancements to the OMA PoC Release 1 specifications such as the possibility to share multimedia content with other PoC users. Both OMA PoC releases uses SIP and the IMS network to establish, modify and release the communication sessions.

8.2.1   OMA PoC Release 1 Standardization

Before the OMA PoC standardization effort began, the large telecommunication vendors Nokia and Motorola developed and commercially launched proprietary PoC technologies based on SIP/IMS. However there was an interest in the industry to develop a standardized PoC solution to counteract the fragmented PoC market. Therefore, in 2003 Ericsson, Motorola, Nokia and Siemens joined forces in an industry consortium to create a set of PoC specifications. This set of specifications was developed during 2003 and during the fall 2003 was put forward as a main contribution to the OMA PoC Release 1 standardization effort. In spite of the early work done by Ericsson, Motorola, Nokia and Siemens, the work with the OMA PoC Release 1 specifications went on for more than two years before the OMA PoC Release 1 specifications were finally approved after a phase of interoperability testing in early 2006 (the set of OMA PoC specifications contain the OMA PoC requirement document [150], the OMA PoC architecture document [149], and the OMA PoC technical specifications [142, 144, 145]). The OMA PoC Release 1 specifications are in turn based on IETF protocols and 3GPP/3GPP2 IMS mechanisms for both control and user plane. But the OMA PoC specifications do not demand the use of an IMS corresponding to the 3GPP or 3GPP2 set of specifications; in fact the OMA PoC specifications can be realized using a general SIP/IP service network. The operation of the OMA PoC communication service is identical regardless of whether it is implemented using the 3GPP or 3GPP2 version of IMS or a general SIP/IP service network, which makes the OMA PoC service identical for all access technologies providing operators with a completely access technology transparent IP multimedia communication service. However, there are some differences between mobile terminals that comply with 3GPP and 3GPP2 specifications. One such difference is the set of codecs the mobile terminals support. This was taken into consideration during the standardization effort and the OMA PoC Release 1 specifications were developed to allow full interoperability between all major mobile technologies such as GPRS, EDGE, WCDMA and CDMA2000 as specified by 3GPP and 3GPP2.

8.2.2   OMA PoC Release 1 Architecture

Figure 8.6 shows the high-level architecture of an OMA PoC Release 1 compliant system. The system includes a number of nodes and logical functions.

- The CSCF implements the S-CSCF, I-CSCF and P-CSCF functionality that provides SIP signaling, routing and registration.

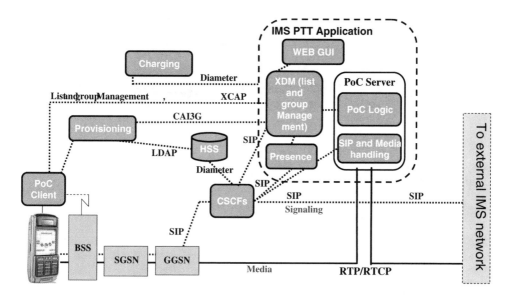

Figure 8.6: The logical architecture of OMA PoC Release 1.

- The HSS implements functionality for authentication, location query/updates and subscriber profile storage.

- The charging function post-processes the charging information, for instance delivered by the PoC server. The charging function creates the Call Data Records (CDR) and sends them towards external billing systems.

- The provisioning function is used for the provision of user related information to the PoC client.

- The PoC server (the OMA PoC AS) is a crucial part of the system. It contains two parts, the PoC logic part and the media handling part. The PoC logic part authorizes PoC users to initiate a PoC session through the use of the invited users' PoC service settings, e.g. incoming session barring (similar to the Multimedia Telephony supplementary service communication barring, see Section 4.7.7) and answer mode, as well as authorizing a user to make a group call by providing the media handling part of the group member list. The media handling part multiplies the speaker's bitstream to multiple streams for the receiving PoC users. Another important functionality of the media handling part is that it handles talk burst control, i.e. it secures that only one user speaks at a time.

- The PoC client on the mobile terminal.

- The OMA PoC network uses a group list management functionality to handle the PoC groups. The OMA PoC network may also include the presence enabler. Presence and list management are presented in Section 8.4.

In order to provide the PoC users with the possibility of doing inter-operator PoC calls the PoC server that implements the application level network functionality for the PoC service must be able to perform different roles. The two roles defined are the controlling PoC function

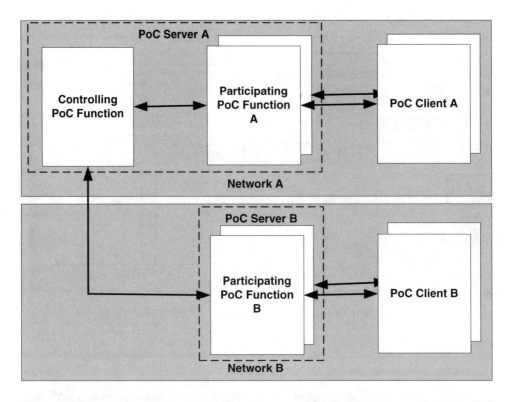

Figure 8.7: The relationship between PoC server, controlling PoC function, participating PoC functions and the PoC clients in a group session with four PoC clients.

and the participating PoC function. The PoC server can perform one of the two roles or both roles at the same time in a PoC session. The determination of the PoC server role takes place during the PoC session establishment and lasts for the duration of the whole PoC session. Figure 8.7 shows the relationship between the PoC servers, controlling PoC function, participating PoC functions and the PoC clients.

There can be only one controlling PoC function per PoC session and thus only one PoC server performing the controlling PoC function per PoC session since it is the controlling PoC function that is the 'central' intelligence in the PoC network and for instance insures that only one PoC user is talking at a given time. The controlling PoC function has N number of SIP sessions and media and talk burst control communication paths in one PoC session, where N is the number of participants in the PoC session. The controlling PoC function will not have any direct communication to the PoC client for PoC session signaling, but will interact with the PoC client via the participating PoC function for the PoC client.

There will be N participating PoC functions in the PoC session. Hence, every PoC client in the PoC session is connected to a participating PoC function so the physical unit of the PoC server may have to act as the logical unit of the participating PoC function for a multitude of PoC clients during the PoC session. A PoC server performing the participating PoC function always has a direct communication path with a PoC client and a direct communication path with the controlling PoC function (which may be located in the same physical PoC server) for PoC session signaling.

The controlling PoC function will normally route media and talk burst control messages to the PoC client via the participating PoC function for the PoC client. However, local policy in the PoC server performing the participating PoC function may allow the controlling PoC function to have a direct communication path for media and talk burst control signaling to each PoC client.

8.2.3 OMA PoC Talk Burst Control

PoC is a half-duplex, one-way communication method between participants and implies that one party speaks while the other parties listen; this requires a talk burst control mechanism, also known as floor control. The talk burst control mechanism in OMA PoC is a confirmed request/grant procedure within the PoC server performing the controlling PoC function, insuring that only one user is sending media at a given time. The talk burst control mechanism in OMA PoC forms a second control layer under the SIP layer. But since it has a very tight coupling to the media transfer this control layer is part of the media handling part of the PoC server. The Talk Burst Control Protocol (TBCP) that is used for talk burst control was developed by OMA (see [144]). The TBCP is based on RTCP APP packets which means that the protocol uses UDP for transport and may use the RTCP port negotiated at PoC session establishment. Seven control messages are required to handle four 'main' talk burst control procedures. The seven TBCP messages and the four 'main' talk burst control procedures are listed below.

- Talk burst Request – A PoC client requests that the PoC server performing the controlling PoC function shall allocate the media resources to his/her device.

- Talk burst Grant – The PoC server performing the controlling PoC function notifies the PoC client that it has been granted the floor and therefore has been granted permission to use the media resource.

- Talk burst Deny – The PoC server performing the controlling PoC function notifies a PoC client that it has been denied permission to use the media resource.

- Talk burst Taken – The PoC server performing the controlling PoC function notifies all PoC clients, except the PoC client that has been granted the floor, that the floor has been granted to another PoC client. Also the identification of the user that has been granted permission to use the media resource is communicated in the message.

- Talk burst Release – A PoC client notifies the PoC server performing the controlling PoC function that it is releasing the media resource, hence moving the PoC server performing the controlling PoC function into the 'Idle' state.

- Talk burst Idle – The PoC server performing the controlling PoC function notifies the PoC clients that no one owns the media resource, that the floor is open/available for users to request.

- Talk burst Revoke – This allows the PoC server performing the controlling PoC function to revoke the media resource from a PoC client. Typically this is used for preemption functionality, but it will also be used by the system to prevent overly long use of the floor resource.

The talk burst request procedure at session initialization. The Talk burst handler will treat the initial SIP INVITE request, derived from pressing the Push-to-talk button, as an implicit Talk burst Request and reply with Talk burst Granted ('start talking indication') or a Talk burst Deny if the floor is already taken in an ongoing session that the PoC client joins. In the case of Talk burst Granted, the Talk burst handler will send Talk burst Taken to all other users currently in the multi-party session. In the case of Talk burst Deny, if the user joins a PoC session during an ongoing talk burst the Talk burst handler will also send a Talk burst Taken message to the joining user to indicate who is talking in the PoC session. There is a slight possibility that two users can request the floor at the same time. In this case, the user of the first Talk burst Request message that reaches the Talk burst handler will get the floor. The Talk burst handler sends a Talk burst Grant message to the PoC client that got the floor and a Talk burst Deny to the PoC client that did not get the floor. It should be noted that in some cases where a PoC client rejoins a group call or is joining a so-called chat session the SIP INVITE request is not treated as an implicit Talk burst Request. But the PoC client is responded to with a TBCP message and in these cases there is a possibility that there is an ongoing PoC communication and thus the initial SIP INVITE request must be replied to with a Talk burst Taken, otherwise if the floor is idle the PoC client will get a Talk burst Idle message.

The talk burst request procedures. In the case when the invited user wants to talk, the user has to request the floor by pressing the Push-to-talk button, forcing the PoC client to send a Talk burst Request message. The Talk burst Request will be accepted if the floor is free and a Talk burst Grant message will be sent to the requesting PoC client. A Talk burst Taken message will be sent to the other PoC clients participating in the PoC session.

The talk burst release procedure. Once a PoC client sends a Talk burst Release message (when the PoC user releases the Push-to-talk button), a Talk burst Idle message will be sent to the other PoC clients participating in the PoC session. The Talk burst handler will repeatedly send Talk burst Idle messages until a Talk burst Request is received or start of a talk burst from a PoC user occurs. The Talk burst handler retransmits the Talk burst Idle packet to the PoC clients in the PoC session, according to exponential back-off in order to save resources over the air as well as in the mobile terminals. The maximum interval is configurable.

The talk burst revoke procedure. A user can control the floor for a limited time period, which is configurable. The Talk burst handler will send a stop talking indication included in a Talk burst Revoke message to a transmitting PoC client when the limit is reached. The Talk burst handler will stop transmitting the audio stream after a configurable time limit. The Talk burst Revoke message will include information about how long the user has to wait until the floor can be requested again.

Figure 8.8 exemplifies the four 'main' talk burst procedures. Beside the seven TBCP messages and four 'main' talk burst procedures there are a number of other TBCP messages and procedures defined to connect PoC clients that are using the pre-established session method (for more information on pre-established sessions, see Section 8.2.4) and to handle the optional feature, queuing of talk burst requests.

8.2.4   OMA PoC Session Establishment Methods

Maybe the most important user requirement for OMA PoC is low 'push-to-tone' latency. This is to give the 'walkie-talkie' feeling of instant communication and to be able to compete

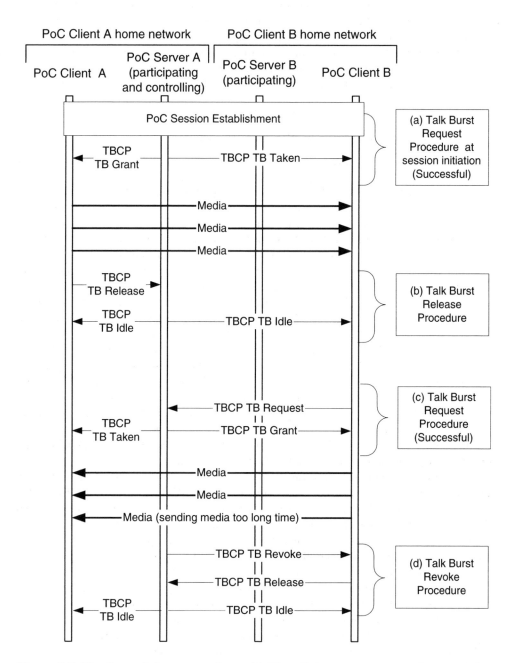

Figure 8.8: The four talk burst procedures. (a) The talk burst request procedure at session initialization. (b) The talk burst release procedure. (c) The talk burst request procedure. (d) The talk burst revoke procedure.

with commercial systems like NEXTEL's IDEN based push-to-talk system or public safety systems like TETRA. To be able to compete with these systems OMA PoC must have the technology potential of being capable of setting up a PoC session within one second. From the delay calculations made in Section 7.7 it is apparent that it is not possible to reach that kind of latency using the signaling flows used for the Multimedia Telephony communication service. Therefore, alternative methods of setting up the PoC session were developed. It could be said that there are three main OMA PoC session establishment methods, as follows.

- Method 1: On-demand session – Confirmed indication.

- Method 2: On-demand session – Unconfirmed indication.

- Method 3: Pre-established session.

Figure 8.9 shows the signaling flows that describe the different OMA PoC session establishment methods when using the auto-answer setting. Auto-answer means that the PoC client does not need any consent from its user to accept an incoming request for PoC communication. From a SIP signaling perspective auto-answer means that no SIP 180 Ringing responses are sent from the terminating PoC client when auto-answer is enabled.

The 'On-demand session – Confirmed indication' method provides a mechanism to negotiate media parameters such as IP address, ports and codecs, at the time when the PoC user wants to actually establish a PoC call. This method also confirms that the PoC server(s), all other network elements and a terminating PoC client are able and willing to receive media. The 'On-demand session – Confirmed indication' method is basically the classic three-way SIP INVITE handshake between the originating PoC client and the terminating PoC client. However, this signaling flow can be rather slow due to the fact that the terminating mobile terminal usually has to be paged and has to set up the radio connection before the SIP INVITE can be sent over the second air link. Another source of delay is transmission time of the rather large SIP INVITE message (typically 1000–1500 bytes before SIP compression) that must be sent over the air twice. The transmission time is mostly an issue when the access provides low throughput (WCDMA HSPA in poor radio conditions or GSM/GPRS/EDGE when only one timeslot (or even less) can be assigned for the SIP signaling).

When using the 'On-demand session – Unconfirmed indication' method, the originating PoC client will get a 'ready-to-speak' indication when the PoC client has successfully created a PoC session with the PoC server performing the controlling PoC function. But when the 'ready-to-speak' indication is given to the originating PoC client, the terminating PoC network should in most cases still be in the process of setting up the PoC session towards the terminating PoC client. This means that the originating PoC user will get the 'ready-to-speak' indication faster than for the 'On-demand session – Confirmed indication' method, but there is a chance that the recorded speech will never be received by the terminating PoC user, if the PoC session establishment at the terminating side fails. It could be said that the 'On-demand session – Unconfirmed indication' method tries to optimize 'push-to-tone' latency by rearranging the SIP signaling so that the originating user is given an early indication that he/she may speak before the terminating user is reached and connected.

When using the 'Pre-established session', the SIP session between the PoC client and the PoC server performing the participating PoC function is established before hand. This means that media parameters such as IP address, ports and codecs are therefore negotiated before the PoC user initiates the PoC call. When it is time to communicate, there is already a pre-established SIP session between the PoC client and the participating PoC function that

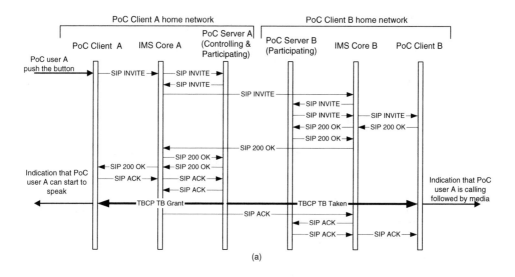

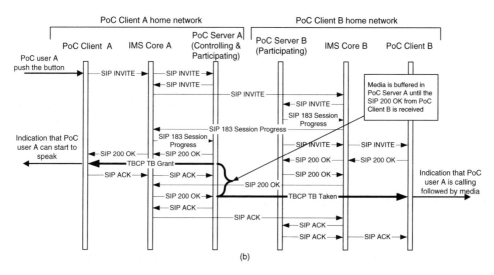

Figure 8.9: Three different methods of establishing PoC communication. (a) On-demand session – Confirmed indication. (b) On-demand session – Unconfirmed indication. (c) Pre-established session.

can be used for the PoC call. Therefore, a SIP REFER message can be used to update the pre-established session on the originating side. The size of the SIP REFER message is small compared to the SIP INVITE message used to originate an on-demand SIP session. It should be noted that in the case of a 'Pre-established session' there is no SIP session pre-established between the originating and terminating participating PoC functions and the controlling PoC function. Thus, when originating the PoC call, a SIP session is established 'on-demand' between the two participating PoC functions and the controlling PoC function. But the large SIP INVITE message is only sent in the fixed domain, i.e. in the core/service networks,

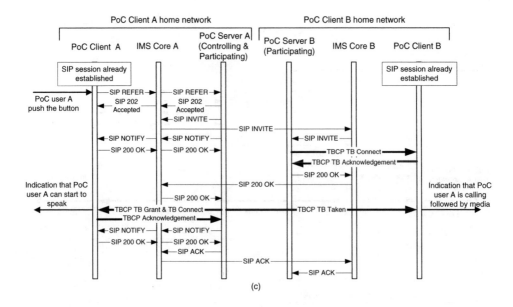

Figure 8.9: Continued.

and not over any time-consuming air link. It should also be noted that only small TBCP messages (often less than 100 bytes) need to be sent over the air for the purpose of updating the pre-established session on the terminating side. Basically, the 'Pre-established session' tries to optimize the 'push-to-tone' latency by minimizing the number of bits that need to be sent over two air links before media can be shared.

8.2.5   OMA PoC and PDP Context Establishment

The set of OMA PoC specifications do not specify when resource reservations (e.g. setting up a PDP context for the media) should be done. The reason for this is that OMA develop services for general IP access. Instead it is the responsibility of the organizations developing the underlying access to specify if resources need to be reserved and in that case how and when resources are reserved for the service. The 3GPP developed a technical report (see 3GPP TR23.979 [2]) that addresses the issue of how resource reservation for PoC should be done.

Care was taken to avoid unnecessary delays in the PoC session setup for the establishment of the media bearer. According to the 3GPP proposal, the media bearer setup should start when the PoC session setup is completed and the PoC client starts to either send or receive the first media in the PoC session. Figure 8.10 shows the 3GPP proposal for the on-demand confirmed indication case. This means that in the typical case a part of the media in the first talk burst needs to be sent over either the bearer that the SIP signaling uses or a general-purpose bearer that is available in parallel with the bearer used for SIP signaling. Thereafter the media will be transferred over the media bearer for the duration of the PoC session. To insure that this method works with acceptable quality for the PoC users, it is important that the switch of the media stream to the media PDP context is made seamless.

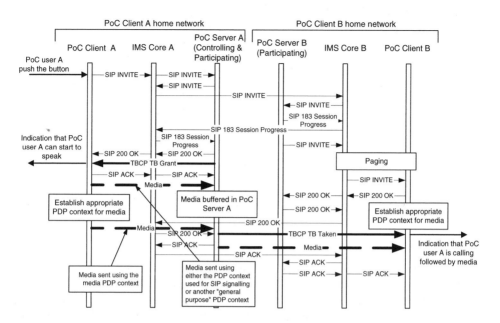

Figure 8.10: Establishment of PDP context in an on-demand session using unconfirmed indication.

8.2.6 OMA PoC Media Considerations

The OMA PoC specifications do not specify which speech codec the PoC clients should use. Instead the OMA PoC specifications refer to 3GPP and 3GPP2 specifications. In 3GPP TS26.235 [37] it is specified that the OMA PoC communication service will use either the AMR narrowband codec or the AMR wideband codec when the PoC client is implemented on a 3GPP compliant mobile terminal. On the other hand, when the PoC client is implemented on a 3GPP2 compliant mobile terminal, it is most likely that the EVRC codec is the only supported codec (see 3GPP2 S.R0100-0 [47]). Therefore, a PoC server that is to be used for inter-operator PoC calls across 3GPP and 3GPP2 networks should be able to do transcoding between the 3GPP2 and 3GPP codecs. The transcoding could be done either by the PoC server performing the controlling PoC function or by the PoC server(s) performing the participating PoC function(s).

The OMA PoC communication service uses the AMR payload format (see RFC 3267 [179]) and the EVRC payload format (see RFC 3558 [132]) to encapsulate the codec frames. When the OMA PoC communication service was introduced in the 3GPP specifications it introduced the possibility of using less restrictive payload options than the full-duplex voice service (see 3GPP TS26.236 [38]). For instance, the OMA PoC communication service allowed the packetization of several codec frames per RTP packet. These options were mainly intended to be used in networks that did not support header compression to decrease RTP/UDP/IP overhead. By packetizing several codec frames per RTP packet the RTP/UDP/IP overhead decreases according to Table 8.1.

Table 8.1: The required bandwidth for PoC media for different packetization schemes.

Number of AMR4.75 codec frames per RTP packet	Required bandwidth when IPv4 and octet-aligned mode is used [bits/s]
1	21600
2	13400
3	10667
4	9300
6	7933
8	7250
10	6840
20	6020

8.2.7   OMA PoC Release 2

As the cellular networks evolve, they will be able to provide other more advanced IP-based real-time services than the plain voice services offered in the OMA PoC Release 1 specifications. Advanced services should, in this context, be seen as services that are more demanding in terms of bit rate and delay, and that offer improved possibilities of person-to-person communication and thus increase perceived user value compared to a plain voice call. OMA is trying to make use of the new opportunities promised by the evolved cellular networks by continuing the development of the PoC communication service and for instance introducing multimedia capabilities in the PoC service offering. One example of a more advanced communication method included in the OMA PoC Release 2 specifications that promises to increase the perceived user value is PoC video sharing. PoC video sharing is here used to exemplify the evolution of OMA PoC.

The term *PoC video sharing* refers to sending and receiving of multimedia talk bursts containing video and voice to other PoC video sharing enabled mobile terminals in a PoC session. The PoC video sharing communication method can therefore be seen as a real-time person-to-person video messaging service. PoC video sharing enriches the voice communication by adding the ability to send video streams of either something you are talking about or the talker's face that enables other users to see and interpret the facial expression of the talking user.

PoC video sharing is based on the plain OMA PoC voice service. This means that the video stream of the PoC video sharing session is:

- sent using an IETF standardized protocol suite (RTP/UDP/IP);

- half-duplex;

- controlled by a media burst control mechanism (i.e. a multimedia version of the talk burst control mechanism in OMA PoC Release 1).

The right to transmit video to all PoC users in a PoC video sharing session can be requested by any of the PoC users in the PoC video sharing session, at any given time, given that no other user is currently transmitting video. Hence, concurrent bidirectional video streams are not supported in PoC video sharing; instead the video stream is always half-duplex as the voice stream in PoC. Two modes of operation are supported. The first method

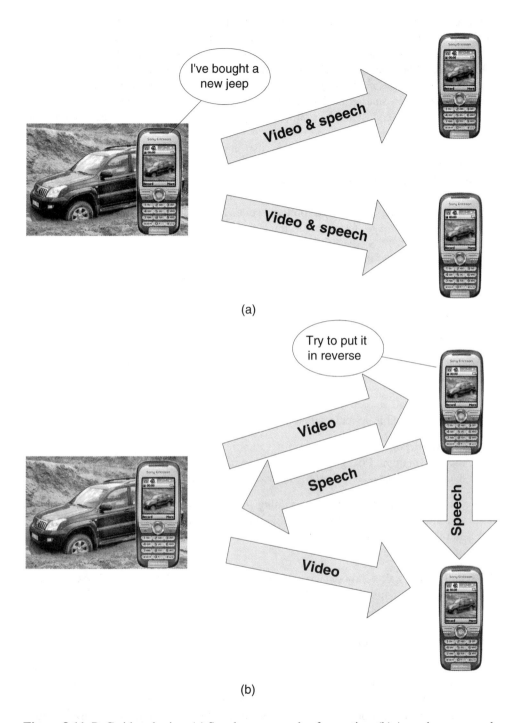

Figure 8.11: PoC video sharing. (a) Synchronous mode of operation. (b) Asynchronous mode of operation.

could be called *synchronous mode of operation*. In this mode of operation the voice and the video streams are coming from one source. During transmission of synchronized voice and video no other participant in the PoC video sharing session can be granted the right to transmit video or voice. The voice and video can be made lip-synchronized. The second method could be called *asynchronous mode of operation*. In this mode of operation only video is streamed from one PoC video sharing enabled client, while the other PoC clients in the PoC video sharing session can request and be granted the right to speak. Figure 8.11 describes the two different modes of operation.

The OMA PoC Release 2 specifications also support PoC communication methods that enable the exchange of other non-real-time media types like images, text and files such as pre-recorded video clips. Non-real-time media types are best transported using TCP/IP. Hence, the main difference compared to the PoC video sharing method is the use of TCP/IP for media transfer rather than the RTP/UDP/IP protocol suite. The application layer protocol proposed to be used for the image, text and file transfer is MSRP that also is used in Multimedia Telephony, 3GPP CSICS (see Section 8.1) and OMA Instant Messaging (see Section 8.3). It should also be noted that the OMA PoC Release 2 specifications extend the OMA PoC communication service with a large number of other enhancements beside the multimedia capabilities. The list below presents a few of the proposed enhancements.

- A 'PoC media box', which is a PoC functionality to store media burst and related information (e.g. date and time, sender identity) on behalf of a PoC user.

- Interworking functionality allowing other non-OMA push-to-talk networks to inter-work with the OMA PoC service infrastructure.

- Enhanced PoC session handling including the possibility of moderator controlled PoC sessions to enable systems with a fleet dispatcher that is in control.

- Enhanced PoC group handling, for example creation of PoC group sessions, that includes multiple PoC groups and creation of PoC group sessions based on dynamic data such as presence state of the individual PoC users.

- Quality of experience and crisis handling which calls for quality of service management and priority framework that allows the PoC service infrastructure to differentiate the end-user experience provided to individual PoC users on a subscription base and type of session request.

- Full-duplex call follow-on procedure that allows a half-duplex PoC session to be changed into a full-duplex Multimedia Telephony session or a CS call.

8.3   OMA Instant Messaging

Instant Messaging (IM) allows users to exchange information with others in near real time. The interactive nature of an IM service encourages the IM users to engage in conversations. Typically, small text messages are exchanged in IM conversations but the OMA Instant Messaging communication service supports other content types than text like images, video clips, etc., making it a multimedia IM service enabler. The OMA Instant Messaging communication service enables a user to send messages to another individual or to a group of users in which everyone can see what everyone else is sending. The service enabler supports two modes of IM communication. They are called the pager mode and IM sessions. The pager mode communication method is designed to handle brief message exchanges, while the latter is similar to a conference hosted by the network (i.e. the service can provide 'chat rooms') where the IM users join and leave the IM communication sessions over time. The OMA Instant Messaging service is equipped with the possibility to store conversations in the network. Either the messages may be stored during an active IM conversation or the storage is done when an IM user is offline and the IM user uses the storage capability as a 'IM answering machine'.

The OMA Instant Messaging communication service described here is often referred to as the OMA SIMPLE (SIP for Instant Messaging and Presence Leveraging Extensions) Instant Messaging service enabler. This is to distinguish the Instant Messaging architecture that is based on the SIP and MSRP protocols (for more information about MSRP, see [58]) from other IM solutions specified by OMA that use other protocols like for instance the Wireless Village service architecture. The OMA SIMPLE Instant Messaging communication service is described in the OMA Instant Messaging requirements document [139], the OMA SIMPLE Instant Messaging architecture document [140] and the OMA SIMPLE Instant Messaging technical specification [143] that specifies the protocol details of the service enabler.

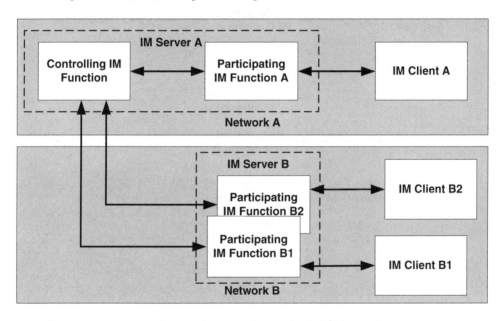

Figure 8.12: The relationship between IM server, controlling IM function, participating IM functions and IM clients in an IM group session.

8.3.1 OMA Instant Messaging Architecture

The IM architecture has similarities with the OMA PoC architecture, which is presented in Section 8.2.2. As in OMA PoC the IM server implements different roles. One reason is to be able to offer one-to-many chat sessions in a multi-operator environment. The roles used in that case are the controlling IM function and the participating IM function. Figure 8.12 show a high-level view of a multi-operator group IM communication.

But there are differences compared to PoC, one main difference being that IM does not use anything similar to the talk burst control of PoC to control the media transfer. Therefore, the controlling IM function can be optional in one-to-one IM sessions. Further, the IM Server can implement two other roles called the deferred messaging IM function and the conversation history function.

One key component of the OMA Instant Messaging communication service is deferred messaging. Deferred messaging happens when an intended recipient of an IM message is not available, either due to the IM user's current inbox setting or the IM user being 'IM offline' (i.e. not registered to the IM service). In these cases the IM messages are stored in the IM server for later delivery and thus the IM messages become deferred messages. Figure 8.13 depicts the relationship between the IM server and the deferred messaging IM function. In this case IM user B is not available for IM and the IM server activates the deferred messaging IM function to store the incoming IM message from IM user A.

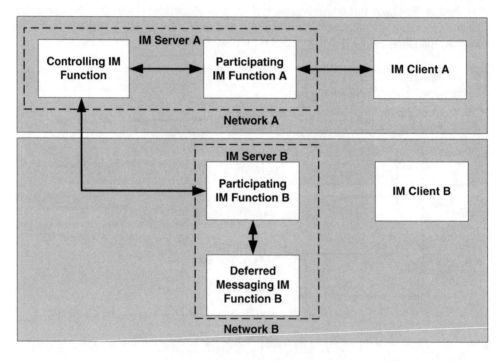

Figure 8.13: User B is not available for IM and the IM server activates the deferred messaging IM function.

The conversation history function, as the deferred messaging IM function, stores IM messages in the network. The difference is that the conversation history function stores messages from an active IM conversation that an IM user participates in. The IM server performs the conversation history function when an IM user requests the storage of an IM conversation. This can be done by an IM service setting to always store conversations for the IM user or an IM client initiated request to store a particular conversation. It should be noted that it is only the IM user that requested the storage of the conversation that can retrieve the stored IM messages.

8.3.2 Instant Messaging Modes

The OMA Instant Messaging communication service follows the 3GPP requirements (specified in 3GPP TS22.340 [28]) of an IM service and incorporates methods to do immediate messaging, session-based messaging and deferred messaging. The deferred messaging is explained above. Pager mode messaging enables immediate messaging while IM sessions provide the session-based messaging capability.

Pager mode messaging is an IM communication method that does not need the establishment of any conference-like IM sessions in the network. The pager mode messaging provides a unidirectional IM communication that can occur at any time. The characteristic of pager mode messaging is that responses and subsequent messages to a pager mode message occur independently and are unrelated to the initial pager mode message. The reason for this is that the IM server does not maintain any states when operating in the pager mode. The pager mode messaging uses either the SIP MESSAGE method (see RFC 3428 [59]) to carry messages or message transfer over MSRP. When using the SIP MESSAGE method no SIP session needs to be established. Instead the message content is inserted as a MIME attachment within the body of the SIP MESSAGE request message. Figure 8.14 shows an IM flow using the SIP MESSAGE method. The drawback with the SIP MESSAGE based pager mode messaging approach is that there is a limitation on the size of the instant messages (1300 bytes) that can be sent. The size restriction is especially limiting for multimedia IM messages. To solve this problem the second MSRP-based pager mode messaging method was developed. By inserting the content of the message in MSRP messages, the pager mode messaging can handle arbitrarily large messages. But the usage of MSRP needs the establishment of an SIP session between the interested parties. Figure 8.14 shows the signaling and media flow of an MSRP-based messaging event. It can be noted that MSRP needs a TCP connection for media transfer. The SIP session is in general only used to send one message then it is torn down. Pager mode messaging handles group messages (both ad-hoc groups and predefined groups). One-to-many communication is possible regardless of whether the SIP MESSAGE method or the MSRP method is used to carry the content of the messages.

The IM session mode is sometimes referred to as a 'chat', i.e. the joining of an IM session is similar to the concept of joining a 'chat room'. The IM session joins together the IM users for a period of time and the messages that are exchanged in the 'chat room' are associated together in the context of the IM session. The IM session requires a central point of control that provides conference-like capabilities. This central point is the controlling IM function. The media transfer in an IM session is always done by MSRP regardless of size (a difference from pager mode messaging that sends everything less than 1300 bytes in SIP MESSAGE messages).

8.3.3 OMA Instant Messaging Media Types

When applying the OMA Instant Messaging communication service on 3GPP enabled terminals then an interoperable baseline set of media types must be supported by the mobile terminal. This set of media types is defined in 3GPP TS26.141 [29] and contains quite a large number of media types. But it should be pointed out that the mobile terminal may also support media types other than the ones in the baseline set. The design criterion used when developing this baseline set of media types was to maximize the technology reuse of other services like MMS and the streaming service defined in 3GPP. The IM service should be able to provide the service of sending/receiving the following media types in a 3GPP environment.

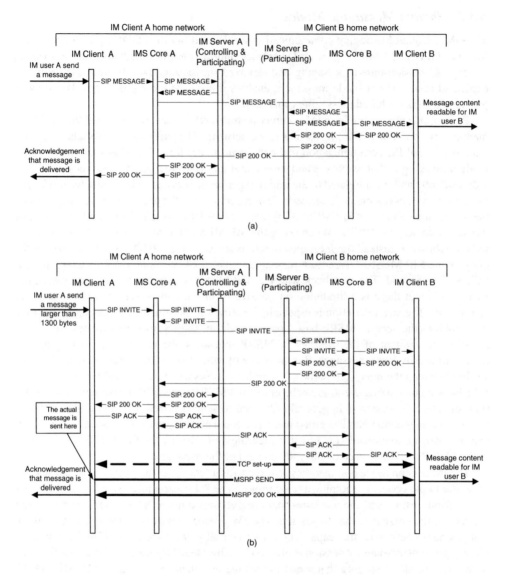

Figure 8.14: IM flows. (a) Pager mode messaging using the SIP MESSAGE method. (b) Pager mode messaging using MSRP.

- **Text**: Plain text, any character encoding that contains a subset of Unicode logical characters can be used.

- **Still image**: The JPEG compression and JFIF file format are used.

- **Bitmap graphics**: The GIF and PNG formats are used.

- **Speech**: The AMR narrowband and AMR wideband speech codecs are used.

- **Audio**: The audio codecs AAC+ and extended AMR-WB are supported.

- **Video**: Certain profiles of the video codecs H.263, MPEG-4 and H.264 are supported.

- **Synthetic audio**: The Scalable Polyphony MIDI content format is supported.

- **Vector graphics**: The Scalable Vector Graphics (SVG) Tiny format and the EMCAScript are supported.

- **Media synchronization**: A subset of SMIL 2.0 is used for media synchronization.

8.4 Presence and List Management

Presence and list management are service enablers that aim to facilitate communication and to support communication services like Multimedia Telephony and OMA PoC. Both service enablers are specified by the PAG group in OMA. OMA actually specifies two different presence solutions. The one discussed here is the presence solution based on SIP, which is often referred to as OMA Presence Simple. The set of SIP-based presence specifications contains the OMA Presence Simple requirements document [138], the OMA Presence Simple architecture document [146] and the OMA Presence Simple technical specifications [147, 148, 151]. The list management is referred to as XML Document Management (XDM). The set of OMA XDM specifications include the OMA XDM requirement document [154], the OMA XDM architecture document [153] and the OMA XDM technical specifications [155, 152, 141].

8.4.1 Presence Simple

Presence is a service enabler that allows a user to subscribe to presence information and thus see other user's availability and willingness to communicate (Figure 8.15). The presence state of a user can, depending on implementation, hold information on whether the user is online or not, current mood or geographical location. Presence is not a new service. Several presence implementations designed for Internet usage exist and have done so for a number of years. The first 'block buster' application that used presence was ICQ that was released in its first version in 1996. But unfortunately, the majority of the presence solutions were, and still are, proprietary and incompatible with each other. Therefore, standardization bodies like IETF and OMA started to work on standardizing presence solutions. The SIMPLE working group in IETF developed several RFCs including RFC 3856 [165], RFC 3857 [166] and RFC 3858 [167], which forms the basis for the OMA Presence Simple solution. Figure 8.16 give a technical overview of the OMA Presence Simple service enabler.

In the figure, a user (the watcher) subscribes to the presence service by sending a SIP SUBSCRIBE message to the presence server that acts as an IMS application server. After the subscription phase is done, the presence server notifies the watcher about the presence state of the users (the presentities) in the watcher's presence group. Usually the presence states of the presentities are offered to the watcher via the so-called buddy list. The notification is done by the transmission of a SIP NOTIFY message from the presence server to the watcher's presence client. When a presentity changes his/her presence state the presence client needs to update the stored state in the presence server. This is done by transmitting a SIP PUBLISH message to the presence server indicating the new presence state. The watcher is then notified about the changed presence state of the presentity by the reception of a new SIP NOTIFY message. The notification may be sent immediately after the presence state of the presentity

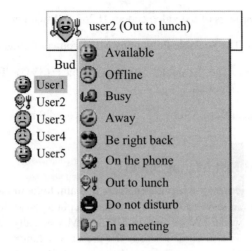

Figure 8.15: Presence, the availability and willingness to communicate.

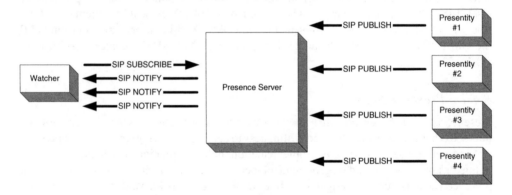

Figure 8.16: High-level view of the SIP-based presence service.

has changed in the presence server or after a certain time or when a certain event triggers the transmission of the SIP NOTIFY. This depends on the presence model used. The most common presence models are as follows:

- **Push**: This model updates the watcher's presence state immediately after a change of a presence state is published. The push method gives good interactivity but may create a lot of traffic.

- **Throttling**: This model tries to group together state changes during a certain time period. The presence server notifies the watcher about all presence state changes of all the presentities in the watcher's presence group after a 'presence update interval timer' expiry. The interactivity of throttling depends on the presence update interval. The amount of traffic created by presence is somewhat under the operator's control by the presence update interval timer setting.

- **Pull**: In this model the watcher has to manually request the presence server to update the presence states of the watcher's presence group. By polling the presence server for presence information, there is a trade-off between limiting the presence traffic and keeping the local presence information in the watcher's presence client accurate.

When deciding which presence model to use care should be taken about the type of underlying access that the presence service runs on. The WCDMA networks are in general well suited for services like streaming that create a continuous data transfer after a certain amount of time. Presence generates traffic with a quite different characteristic than streaming. Changes of a presentity's state generate traffic that the presence server multiplies to all watchers subscribing to the presence state of this particular presentity. These messages can easily be many but they can be considered small in size. The troublesome part is that they have a large overhead in terms of establishing radio connections between the mobile terminal and the base station. The time to establish and release a radio connection can easily be measured as several seconds while the transmission of the presence message can be done in a few milliseconds due to the high peak bit rate of for instance WCDMA HSPA. Thus the percentage of the channel utilization of presence will in general be very low. If presence updates happen with a very high frequency, there may be issues with for instance code limitations in a WCDMA system and the presence service will become costly in terms of capacity. However, the different presence models give the system architect a tool to limit the frequency of the presence messages if the desire is to minimize the presence service usage of network resources.

The amount of traffic that presence creates is not only dependent on the presence model. Also the number of presence states and type of presence information available may impact the traffic volume. The presence information can be grouped into a number of categories.

- **Usage information**: A basic feature is to provide information on whether the presentity is registered to the service or not. The typical states are *online* and *offline*.

- **Availability information**: This category provides information about the possibility to reach the presentity with e.g. a phone call. Typical state may be *I am active*, *busy*, *do-not-disturb*. More detailed states could be *out to lunch* or *in a meeting*.

- **Mood information**: The current mood could be published as presence information. By publishing the mood, further communication between the watcher and the presentity is encouraged. Examples of mood states are *happy*, *angry* or *sad*.

- **Geographical information**: In its simplest case, a presentity choses between pre-determined presence states such as *at work* or *at home*. In more advanced presence solutions the published geographical information may be automatically generated by a GPS or some similar positioning system.

- **Reachability**: Presence information may also contain information about the communication methods the presentity's mobile terminal supports or the presentity subscribes to. This could allow the watcher's mobile terminal to indicate what communication methods are possible for communication with a presentity. The presence information could also contain information about preferred communication methods by for instance defining priorities for the different supported communication methods.

Presence changes will in some cases be automatic. For instance when a Multimedia Telephony client sends or receives an invitation to a Multimedia Telephony session the presence client may publish the availability information *in a call* automatically. If the presence client has connection to the in-built calendar in the mobile terminal it could automatically publish *in a meeting* whenever the calendar indicates that there is a meeting.

Presence services add value for the end-user and are today widely used on the Internet. There is a large probability that presence services will become popular in the telecommunication sector when they are deployed in larger scale in mobile terminals. Presence is also interesting from a radio network planning perspective. The traffic characteristics of presence is challenging since presence updates may happen quite often, leading to numerous establishments and releases of radio connections during a day for each user. But the traffic volumes sent/received may be quite small since in general only one message limited in size is sent and one is received for each presence transaction. The presence traffic itself is in general not time critical so the presence service should make good use of QoS not to disturb e.g. a real-time service like a Multimedia Telephony basic voice call.

8.4.2 List Management

The OMA XML Document Management (OMA XDM) defines a common mechanism that makes user-specific service-related information accessible to the services that need the information. The intention is to store the user-specific service-related information in the network in a logical repository called the XDM server where it is accessed and manipulated (created, changed, deleted, etc.). For communication services it is foreseen that information like contact lists, group lists and access lists are to be created and stored in the XDM server:

- **Contact list**: The contact entries in the 'phone book' that are stored in the network for easy access from whatever XDM enabled device is used for communication.

- **Group list**: The defined communication groups used in e.g. OMA PoC communication. The XDM is the enabler that is used to create, change, store and delete the PoC groups.

- **Access list**: The access rules, that is who is allowed or not allowed to reach a specific user via a specific service.

Manipulation of XML documents is achieved using the XCAP protocol (see [168]) over the interface between the XDM client and the XDM server. The XCAP protocol use HTTP methods for the communication between e.g. the XDM entities. Hence, the XDM client uses the HTTP GET method to read information from the XDM server and the HTTP PUT method to create or modify information in the XDM server. Since XCAP resources are also HTTP resources, it is easy to realize a system in which users and administrators can manage the contact lists and groups directly from the mobile terminal and via a web portal.

Chapter 9

Summary

Per Synnergren, Janne Peisa

In this chapter we summarize the key concepts and performance of the standardized IMS Multimedia Telephony communication service. The Multimedia Telephony communication service was standardized in 3GPP release 7. The standardization effort was made to position IMS-based Multimedia Telephony as the standardized operator-controlled way to realize interoperable IP-based voice and multimedia communication that has the same quality, reliability and security as the legacy CS telephony service.

Multimedia Telephony is an all-IP and SIP-based communication solution that uses the IMS as the control plane. The IMS is access independent and offers smooth interoperation with wireline terminals; this is achieved by using SIP, which is an IETF defined session control capability. SIP establishes, modifies and terminates multimedia sessions. It can be used to invite new users to an existing session to create a conference or it can be used to create brand new sessions alongside existing ones. SIP is independent of the type of multimedia session it handles. SIP, therefore, uses the same mechanisms whether it is used to set up a video conference, voice calls or shared whiteboards sessions. The mechanism used to describe the session is SDP. SDP carries the session description, which is distributed among the potential participants of an IMS session via SIP.

The access independence of IMS is a prime driver in the integration of wireline and wireless technologies. IMS focuses around convergence of components and services. The component convergence means that the IMS core (CSCFs, HSS, etc.) are applicable to both fixed and mobile networks. Service convergence aims to enable users to access the same set of services on the same address over fixed broadband access or mobile access and at the same time.

Maybe the most apparent thing that differentiates IMS Multimedia Telephony from CS telephony from an end-user perspective is the increased service flexibility. Multimedia Telephony enables users to enhance the communication sessions by adding/dropping media components at any time during the session. The Multimedia Telephony specifications mention a core set of media capabilities that are included in a Multimedia Telephony service offering. The media capabilities are:

- voice,

- video,

- text,

- file sharing between two clients,

- sharing of content that is played out directly at reception (images, video clips, audio clips).

The voice, video and text offer a full-duplex, real-time communication service, while the other components are not real-time.

The Multimedia Telephony client is not limited by the core set of media capabilities mentioned above. Instead, a Multimedia Telephony service offering may allow the end-users to communicate or interact with other users using other media components like whiteboard sharing or various types of gaming. However, Multimedia Telephony should be able to replace CS telephony and it is therefore natural that Multimedia Telephony supports a usage model similar to traditional CS telephony. Hence, Multimedia Telephony service offerings include the support of supplementary services that behave almost identically to the supplementary services offered in ISDN/PSTN. In the Multimedia Telephony specifications the following set of supplementary services has been included:

- Originating Indication Presentation (OIP) and Originating Indication Restriction (OIR),

- Terminating Indication Presentation (TIP) and Terminating Indication Restriction (TIR),

- Communication DIVersion (CDIV),

- Communication Hold (HOLD),

- Communication Barring (CB),

- Message Waiting Indication (MWI),

- CONFerence (CONF),

- Explicit Communication Transfer (ECT).

Multimedia Telephony uses the IMS as the service platform and thus inherits the service requirements of IMS. One such requirement that is needed for fast uptake of Multimedia Telephony is interworking with other networks. Multimedia Telephony will support interworking between PS and CS services and with PSTN/ISDN. This is to support the anticipated usage model of Multimedia Telephony: one user connecting to any other user, regardless of operator and access technology.

Multimedia Telephony will only be well received by the end-users if the quality of the offered service is as good as or even better than for the CS telephony service the end-users are accustomed to. IMS addresses this issue by the support of negotiable QoS for IMS sessions. To achieve sufficient QoS for each type of supported media, the QoS framework used for Multimedia Telephony can assign different QoS levels to the individual media components used in a session. This means that media components with different QoS levels get differential treatment in the radio access and in the transport network to secure a good end-user perception of the service.

The operators want a service that can provide surplus value for the end-user and also have the potential to lower costs. The IMS architecture will reduce long-term cost of ownership compared to the CS core network and other IP systems as the operators add more services into their network. But it is also important that the realization of Multimedia Telephony is cost effective over the connected access networks. Hence, Multimedia Telephony must provide a capacity similar to or better than CS telephony over the air interface.

The main technical innovations allowing a high capacity enhanced telephony service is the introduction of the High Speed Packet Access and similar methods in other systems. Even though the IMS Multimedia Telephony communication service is access independent, and can be used over all current cellular technologies as well as the fixed lines, the High Speed Packet Access provides an unusually fertile deployment opportunity for Multimedia Telephony.

The HSPA enables efficient support for packet data applications. It is based on the requirement to let applications transmit at highest possible data rate when possible, allowing the flexibility needed for the many different combinations of Multimedia Telephony while still maintaining the requirement for high capacity for 'simple' services such as pure Voice over IP.

The key ingredients of HSPA are the High Speed Downlink Packet Access (HSDPA) and the Enhanced UpLink (EUL). HSDPA provides a single shared data channel for all users in the downlink. The allocation of the resources for each user is done based on the service requirements and channel quality for each 2 ms transmission time interval. The data rate of the channel is adjusted to the channel conditions of individual users, and any possible errors in the adjustment as well as transmission errors on the radio are corrected using Hybrid ARQ. Unlike HSDPA, which provides a shared resource, the Enhanced Uplink provides a dedicated uplink channel for each user. However, it is still possible to do fast resource assignment via Node B controlled scheduling. Similarly to the HSDPA, the Enhanced Uplink provides Hybrid ARQ and short transmission time interval.

The IMS allows, among other things, the coupling of the radio access realization and service via the QoS framework. This makes it possible to optimize system performance for the offered media components, and allows high capacity for selected services such as a Multimedia Telephony basic voice call. The combination of the IMS support for coupling the application to the radio characteristics and the overall high performance and flexibility of HSPA results in a system in which the different service realizations, such as Multimedia Telephony video call, can be supported, and the system capacity can be optimized for the most used media combinations. Thus IMS and HSPA at the same time allow the Multimedia Telephony service to act as both the replacement for the CS telephony service and the enhancement of the traditional service.

For media processing, it is necessary to operate in a unique environment, especially due to the usage of the shared channels of HSPA. The requirement of high capacity together with the HARQ used in the HSPA channels creates relatively large jitter (of the order of 100 ms), as well as occasional packet losses. The amount of jitter is directly coupled to the system load, and this jitter needs to be compensated using advanced de-jittering methods.

The main principle of advanced de-jittering algorithms is adaptive de-jittering, which allows the usage of a small jitter buffer when the jitter is low, and only switching to large jitter buffer when the jitter increases. It is also necessary to adapt the buffer level during the call. The most promising technique for this is the adaptive playout, in which the actual speech samples are expanded or reduced to allow quick adaptation in the de-jittering level.

In addition to the de-jittering, the limited air interface capacity also forces the applications to be prepared for sudden changes in the available data rate. Typically the applications should be prepared to change the application data rate to match the currently available data rate.

For Multimedia Telephony application, this means the ability to detect the changes in the current data rate, and the ability to reduce rate by either dropping complete service components (such as video) or reducing the rate of the used video/speech codec to better match the available data rate.

In this book we show by simulation results that the actual performance of Multimedia Telephony can be as good as or even better than the performance of the CS telephony service in terms of the end-user perceived quality and system capacity. Especially this is true if the requirements on the speech quality and delay for Multimedia Telephony basic voice call are equal to the requirements on the CS telephony service. However, many different service realizations are possible:

- It is possible to introduce new codecs (such as AMR wideband) to enhance speech quality and still maintain relatively high system capacity.

- It is possible to deploy service with reduced interactiveness (e.g. larger mouth-to-ear delay) and significantly increased capacity.

- It is possible to dynamically switch between different quality and capacity based e.g. on system load.

As Multimedia Telephony is planned to replace CS telephony, it is important to provide at least as good capacity and coverage for the voice component of the Multimedia Telephony as for the CS voice. The performance evaluations in Chapter 7 show that the capacity and coverage depend a lot on the following:

- Protocol overhead. It is vital to have an efficient header compression mechanism for IP/UDP/RTP headers.

- The actual air interface design. The improvements made for the air interface as part of the High Speed Packet Access are important for the voice capacity as well.

- Downlink scheduler. In order to match the CS capacity, the downlink scheduler needs to be optimized for voice service. It is especially important to avoid schedulers heavily favoring the channel conditions.

- Optimized configuration. The overall system parameters need to be optimized for voice service in order to reach the maximum capacity.

- Maximum allowed scheduling delay. Longer scheduling delay corresponds directly to better capacity and coverage. Thus it is important to reduce both processing and transmission delays in the complete system, and to allow for longer scheduling and retransmission delays on the air interface.

The two main reasons for the improved capacity are the introduction of the fast link adaptation in the downlink and the usage of the HARQ in both uplink and downlink. The link adaptation allows the system to use the channel (on average) with improved link quality. This translates immediately to capacity increase. Similarly the HARQ enables early stopping of the transmission in cases when the link was better than expected, which also increases capacity.

The Multimedia Telephony standardization effort did not aim to specify a complete autonomous service. Instead Multimedia Telephony is viewed as an IMS communication service for which all client-to-network and network-to-network interfaces are specified. The standard does not define a specific application. In the end an end-user application might utilize several different IMS communication services to offer the end-users a unique service offering. An IMS communication service can be proprietary, but in standards today there exist a few IMS communication services:

- Multimedia Telephony in 3GPP;

- CSICS in 3GPP;

- PoC in OMA;

- instant messaging in OMA;

- presence and list management in OMA;

- WeShare.

3GPP CSICS allows users to enhance mobile CS telephony with packet switched multimedia sessions over the IMS. A typical service offering built using the CSICS communication service is adding the possibility for users to share images, video and video clips during a mobile CS telephony call. OMA PoC is a 'walkie-talkie' style one-to-one and one-to-many communication method where one person speaks at a time. The PoC communication service is being evolved to support multimedia communications as well. OMA Instant Messaging is an enabler that provides common functionality to support a messaging service in the IMS environment. Presence provides status information like availability and mood of other users to help a user to select the most appropriate time and method for communication. List management give the users the possibility to build contact lists and groups for e.g. PoC to support the new type of communication methods that IMS brings. A commercial Multimedia Telephony service offering will most probably in most cases be bundled with presence and list management and it has to support interworking with 3GPP CSICS for rapid uptake of the new IMS services.

To summarize this book, the technical realization and performance results demonstrated in the different chapters show that Multimedia Telephony over HSPA has the potential to become competitive in terms of functionality, capacity and media quality compared to legacy systems for voice communications. With a realization using the flexible HSPA bearer, Multimedia Telephony over HSPA has the potential to become a competitive and future-proof service offering.

Appendix

Additional Simulation Results

In this appendix, we show additional simulation results that can be used to assess the media performance in more detail. For easier comparison, some figures from Chapter 7 are duplicated. All load levels are normalized with respect to the maximum capacity of the reference circuit switched UTRAN system.

A.1   Delay Scheduler

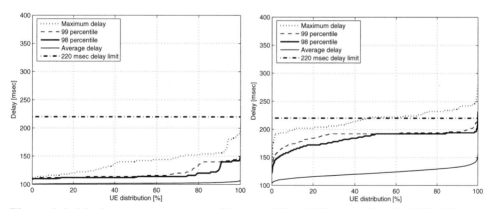

Figure A.1: Maximum, average packet delays and 98 and 99 percentiles for 10% (left) and 55% (right) relative load with the delay scheduler.

IMS Multimedia Telephony over Cellular Systems   S. Chakraborty, T. Frankkila, J. Peisa and P. Synnergren
© 2007 John Wiley & Sons, Ltd

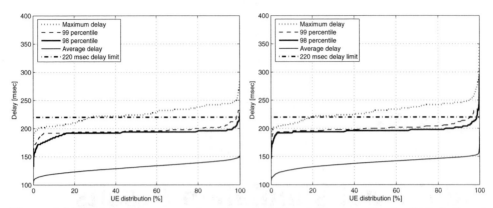

Figure A.2: Maximum, average packet delays and 98 and 99 percentiles for 70% (left) and 90% (right) relative load with the delay scheduler.

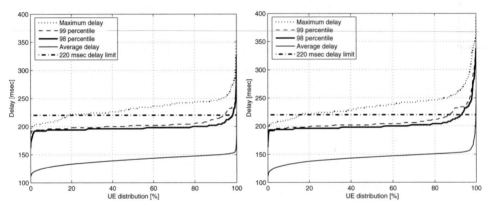

Figure A.3: Maximum, average packet delays and 98 and 99 percentiles for 100% (left) and 110% (right) relative load with the delay scheduler.

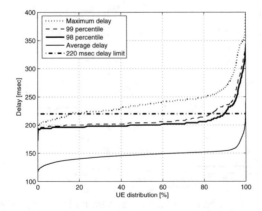

Figure A.4: Maximum, average packet delays and 98 and 99 percentiles for 120% relative load with the delay scheduler.

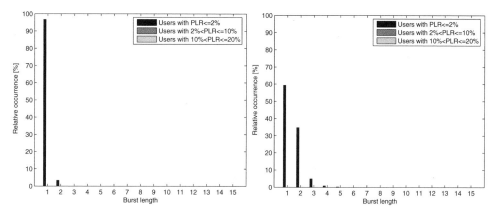

Figure A.5: Histogram of packet loss bursts for the delay scheduler at 10% (left) and 55% (right) relative load.

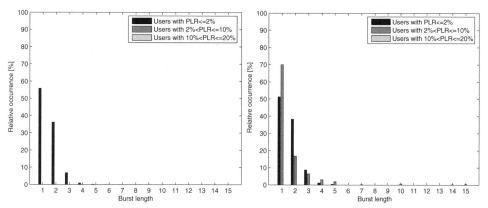

Figure A.6: Histogram of packet loss bursts for the delay scheduler at 70% (left) and 90% (right) relative load.

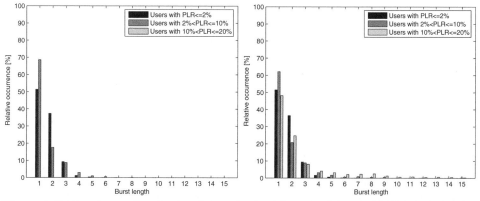

Figure A.7: Histogram of packet loss bursts for the delay scheduler at 100% (left) and 110% (right) relative load.

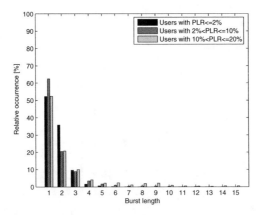

Figure A.8: Histogram of packet loss bursts for the delay scheduler at 120% relative load.

A.2   Max-CQI Scheduler

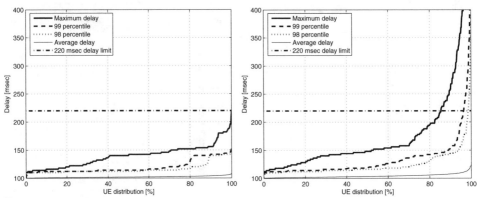

Figure A.9: Maximum, average packet delays and 98 and 99 percentiles for 10% (left) and 20% (right) relative load with the max-CQI scheduler.

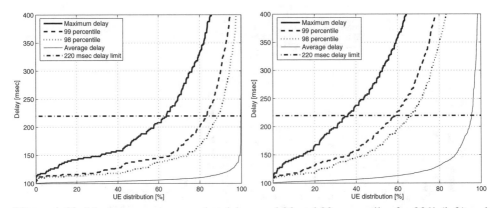

Figure A.10: Maximum, average packet delays and 98 and 99 percentiles for 30% (left) and 35% (right) relative load with the max-CQI scheduler.

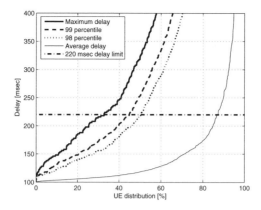

Figure A.11: Maximum, average packet delays and 98 and 99 percentiles for 55% load with the max-CQI scheduler.

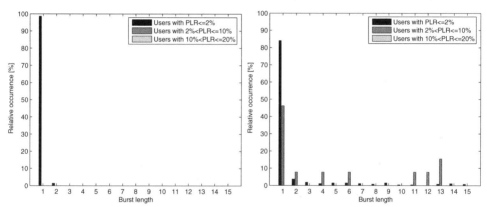

Figure A.12: Histogram of packet loss bursts for the max-CQI scheduler at 10% (left) and 20% (right) relative load.

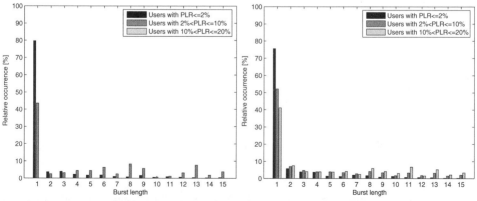

Figure A.13: Histogram of packet loss bursts for the max-CQI scheduler at 30% (left) and 35% (right) relative load.

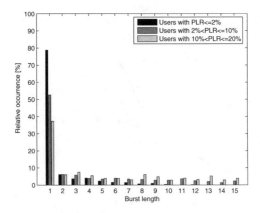

Figure A.14: Histogram of packet loss bursts for the max-CQI scheduler at 55% relative load.

A.3 Proportional-Fair Scheduler

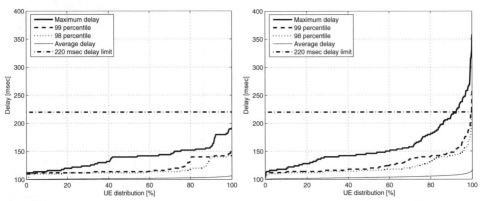

Figure A.15: Maximum, average packet delays and 98 and 99 percentiles for 10% (left) and 20% (right) relative load with the proportional-fair scheduler.

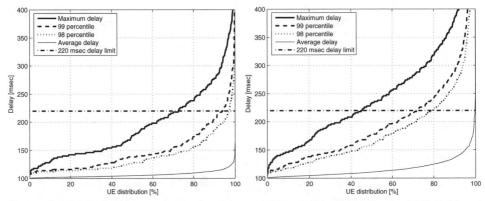

Figure A.16: Maximum, average packet delays and 98 and 99 percentiles for 30% (left) and 35% (right) relative load with the proportional-fair scheduler.

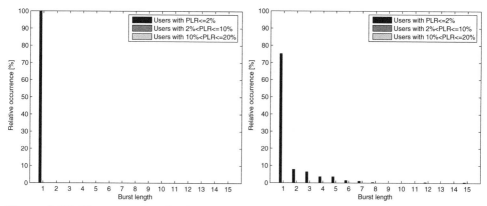

Figure A.17: Histogram of packet loss bursts for the proportional-fair scheduler at 10% (left) and 20% (right) relative load.

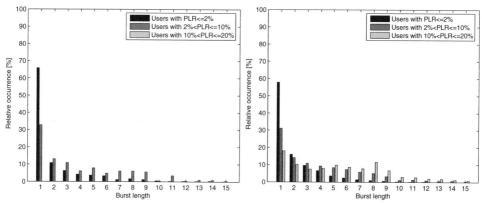

Figure A.18: Histogram of packet loss bursts for the proportional-fair scheduler at 30% (left) and 35% (right) relative load.

A.4   Round-Robin Scheduler

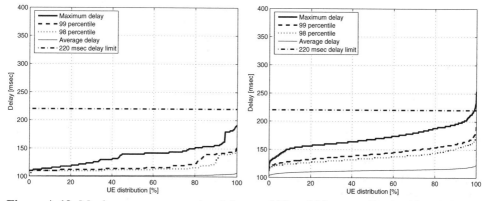

Figure A.19: Maximum, average packet delays and 98 and 99 percentiles for 10% (left) and 45% (right) relative load with the round-robin scheduler.

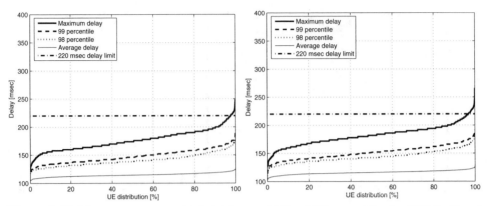

Figure A.20: Maximum, average packet delays and 98 and 99 percentiles for 55% (left) and 65% (right) relative load with the round-robin scheduler.

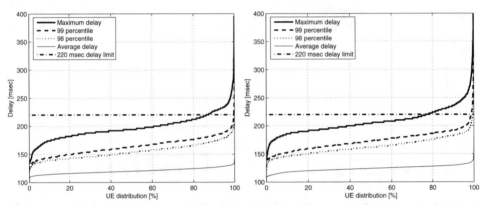

Figure A.21: Maximum, average packet delays and 98 and 99 percentiles for 70% (left) and 80% (right) relative load with the round-robin scheduler.

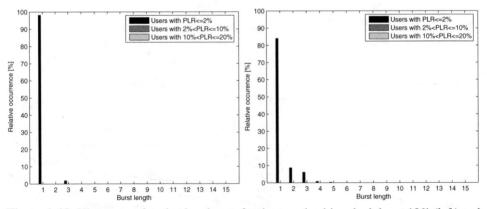

Figure A.22: Histogram of packet loss bursts for the round-robin scheduler at 10% (left) and 45% (right) relative load.

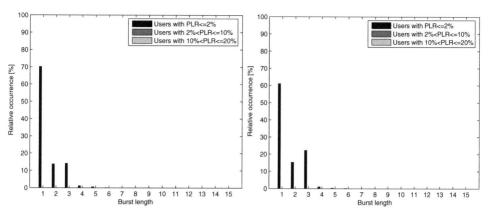

Figure A.23: Histogram of packet loss bursts for the round-robin scheduler at 55% (left) and 65% (right) relative load.

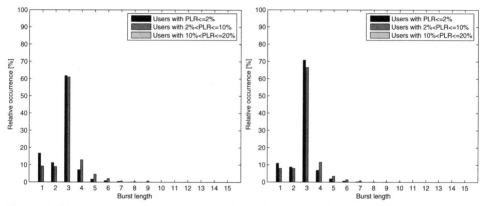

Figure A.24: Histogram of packet loss bursts for the round-robin scheduler at 70% (left) and 80% (right) relative load.

References

[1] 3GPP. 3G security; Network Domain Security (NDS); IP network layer security. TS 33.210, 3rd Generation Partnership Project (3GPP), December 2005.

[2] 3GPP. 3GPP enablers for Open Mobile Alliance (OMA) Push-to-talk over Cellular (PoC) services; Stage 2. TR 23.979, 3rd Generation Partnership Project (3GPP), June 2005.

[3] 3GPP. AMR speech codec; Error concealment of lost frames. TS 26.091, 3rd Generation Partnership Project (3GPP), January 2005.

[4] 3GPP. AMR speech codec; General description. TS 26.071, 3rd Generation Partnership Project (3GPP), January 2005.

[5] 3GPP. Cellular text telephone modem; General description. TS 26.226, 3rd Generation Partnership Project (3GPP), January 2005.

[6] 3GPP. Codec for circuit switched Multimedia Telephony service; Modifications to H.324. TS 26.111, 3rd Generation Partnership Project (3GPP), January 2005.

[7] 3GPP. Codec for circuit switched Multimedia Telephony service; Terminal implementor's guide. TR 26.911, 3rd Generation Partnership Project (3GPP), October 2005.

[8] 3GPP. Combining circuit switched (CS) and IP Multimedia Subsystem (IMS) services; Stage 1. TS 22.279, 3rd Generation Partnership Project (3GPP), December 2005.

[9] 3GPP. End-to-end Quality of Service (QoS) concept and architecture. TS 23.207, 3rd Generation Partnership Project (3GPP), October 2005.

[10] 3GPP. Enhanced Full Rate (EFR) speech processing functions; General description. TS 46.051, 3rd Generation Partnership Project (3GPP), January 2005.

[11] 3GPP. Generic access to the A/Gb interface; Stage 2. TS 43.318, 3rd Generation Partnership Project (3GPP), February 2005.

[12] 3GPP. Inband Tandem Free Operation (TFO) of speech codecs; Service description; Stage 3. TS 28.062, 3rd Generation Partnership Project (3GPP), December 2005.

[13] 3GPP. Signalling flows for the IP multimedia call control based on Session Initiation Protocol (SIP) and Session Description Protocol (SDP); Stage 3. TS 24.228, 3rd Generation Partnership Project (3GPP), December 2005.

[14] 3GPP. Speech codec speech processing functions; Adaptive Multi-Rate – Wideband (AMR-WB) speech codec; General description. TS 26.171, 3rd Generation Partnership Project (3GPP), January 2005.

[15] 3GPP. 3G security; Access security for IP-based services. TS 33.203, 3rd Generation Partnership Project (3GPP), June 2006.

[16] 3GPP. 3G security; Security architecture. TS 33.102, 3rd Generation Partnership Project (3GPP), January 2006.

[17] 3GPP. Combining circuit switched (CS) and IP Multimedia Subsystem (IMS) services; Stage 2. TS 23.279, 3rd Generation Partnership Project (3GPP), June 2006.

[18] 3GPP. Combining Circuit Switched (CS) and IP Multimedia Subsystem (IMS) services; Stage 3. TS 24.279, 3rd Generation Partnership Project (3GPP), June 2006.

[19] 3GPP. Common test environments for User Equipment (UE) conformance testing. TS 34.108, 3rd Generation Partnership Project (3GPP), June 2006.

[20] 3GPP. Conferencing using the IP Multimedia (IM) Core Network (CN) subsystem; Stage 3. TS 24.147, 3rd Generation Partnership Project (3GPP), March 2006.

[21] 3GPP. Dual Transfer Mode (DTM); Stage 2. TS 43.055, 3rd Generation Partnership Project (3GPP), July 2006.

[22] 3GPP. General Packet Radio Service (GPRS); Service description; Stage 2. TS 23.060, 3rd Generation Partnership Project (3GPP), June 2006.

[23] 3GPP. IMS Multimedia Telephony – Media handling and interaction. TS 26.114, 3rd Generation Partnership Project (3GPP), September 2006.

[24] 3GPP. IMS Multimedia Telephony service and supplementary services; Stage 3. TS 24.173, 3rd Generation Partnership Project (3GPP), September 2006.

[25] 3GPP. Internet Protocol (IP) multimedia call control protocol based on Session Initiation Protocol (SIP) and Session Description Protocol (SDP); Stage 3. TS 24.229, 3rd Generation Partnership Project (3GPP), June 2006.

[26] 3GPP. Interworking between the IP Multimedia (IM) Core Network (CN) subsystem and circuit switched (CS) networks. TS 29.163, 3rd Generation Partnership Project (3GPP), June 2006.

[27] 3GPP. IP Multimedia core network Subsystem (IMS) Multimedia Telephony service and supplementary services; Stage 1. TS 22.173, 3rd Generation Partnership Project (3GPP), June 2006.

[28] 3GPP. IP Multimedia Subsystem (IMS) messaging; Stage 1. TS 22.340, 3rd Generation Partnership Project (3GPP), January 2006.

[29] 3GPP. IP Multimedia System (IMS) messaging and presence; Media formats and codecs. TS 26.141, 3rd Generation Partnership Project (3GPP), March 2006.

[30] 3GPP. Media Gateway Control Function (MGCF) – IM Media Gateway (IM-MGW) Mn interface. TS 29.332, 3rd Generation Partnership Project (3GPP), June 2006.

[31] 3GPP. Medium Access Control (MAC) protocol specification. TS 25.321, 3rd Generation Partnership Project (3GPP), June 2006.

[32] 3GPP. Messaging using the IP Multimedia (IM) Core Network (CN) subsystem; Stage 3. TS 24.247, 3rd Generation Partnership Project (3GPP), March 2006.

[33] 3GPP. Network Domain Security; Authentication Framework (NDS/AF). TS 33.310, 3rd Generation Partnership Project (3GPP), January 2006.

[34] 3GPP. Numbering, addressing and identification. TS 23.003, 3rd Generation Partnership Project (3GPP), June 2006.

[35] 3GPP. Overall high level functionality and architecture impacts of flow based charging; Stage 2. TS 23.125, 3rd Generation Partnership Project (3GPP), March 2006.

[36] 3GPP. Packet Data Convergence Protocol (PDCP) specification. TS 25.323, 3rd Generation Partnership Project (3GPP), June 2006.

[37] 3GPP. Packet switched conversational multimedia applications; Default codecs. TS 26.235, 3rd Generation Partnership Project (3GPP), March 2006.

[38] 3GPP. Packet switched conversational multimedia applications; Transport protocols. TS 26.236, 3rd Generation Partnership Project (3GPP), March 2006.

[39] 3GPP. Policy and charging control architecture. TS 23.203, 3rd Generation Partnership Project (3GPP), September 2006.

[40] 3GPP. Quality of Service (QoS) concept and architecture. TS 23.107, 3rd Generation Partnership Project (3GPP), March 2006.

[41] 3GPP. Radio Link Control (RLC) protocol specification. TS 25.322, 3rd Generation Partnership Project (3GPP), June 2006.

[42] 3GPP. Security aspects of early IP Multimedia Subsystem (IMS). TR 33.978, 3rd Generation Partnership Project (3GPP), March 2006.

[43] 3GPP. Service requirements for the Internet Protocol (IP) Multimedia core network Subsystem (IMS); Stage 1. TS 22.228, 3rd Generation Partnership Project (3GPP), June 2006.

[44] 3GPP. Services and service capabilities. TS 22.105, 3rd Generation Partnership Project (3GPP), June 2006.

[45] 3GPP. Speech codec list for GSM and UMTS. TS 26.103, 3rd Generation Partnership Project (3GPP), March 2006.

[46] 3GPP. Typical examples of Radio Access Bearers (RABs) and Radio Bearers (RBs) supported by Universal Terrestrial Radio Access (UTRA). TR 25.993, 3rd Generation Partnership Project (3GPP), June 2006.

[47] 3GPP2. Push-to-talk over Cellular (PoC) system requirements. TS S.R0100-0, 3rd Generation Partnership Project 2 (3GPP2).

[48] B. Aboba and M. Beadles. The Network Access Identifier. RFC 2486, Internet Engineering Task Force, January 1999.

[49] B. S. Atal and M. R. Schroeder. Stochastic coding of speech signals at very low bit rates. In *IEEE International Conference on Communication (ICC 1984)*, pages 1610–1613, May 1984.

[50] M. Baugher, D. McGrew, M. Naslund, E. Carrara and K. Norrman. The Secure Real-time Transport Protocol (SRTP). RFC 3711, Internet Engineering Task Force, March 2004.

[51] C. Bormann, C. Burmeister, M. Degermark, H. Fukushima, H. Hannu, L.-E. Jonsson, R. Hakenberg, T. Koren, K. Le, Z. Liu, A. Martensson, A. Miyazaki, K. Svanbro, T. Wiebke, T. Yoshimura and H. Zheng. RObust Header Compression (ROHC): Framework and four profiles: RTP, UDP, ESP, and uncompressed. RFC 3095, Internet Engineering Task Force, July 2001.

[52] C. Bormann, L. Cline, G. Deisher, T. Gardos, C. Maciocco, D. Newell, J. Ott, G. Sullivan, S. Wenger and C. Zhu. RTP Payload Format for the 1998 Version of ITU-T Rec. H.263 Video (H.263+). RFC 2429, Internet Engineering Task Force, October 1998.

[53] G. Camarillo. The Binary Floor Control Protocol (BFCP). Internet-Draft draft-camarillo-xcon-bfcp-00, Internet Engineering Task Force, May 2004. Work in progress.

[54] G. Camarillo, G. Eriksson, J. Holler and H. Schulzrinne. Grouping of media lines in the Session Description Protocol (SDP). RFC 3388, Internet Engineering Task Force, December 2002.

[55] G. Camarillo and M. A. García-Martín. *The 3G IP Multimedia Subsystem (IMS)*, 2nd Edition. John Wiley & Sons, 2006.

[56] G. Camarillo and P. Kyzivat. Update to the Session Initiation Protocol (SIP) preconditions framework. RFC 4032, Internet Engineering Task Force, March 2005.

[57] G. Camarillo, W. Marshall and J. Rosenberg. Integration of resource management and Session Initiation Protocol (SIP). RFC 3312, Internet Engineering Task Force, October 2002.

[58] B. Campbell. The Message Session Relay Protocol. Internet-Draft draft-ietf-simple-message-sessions-10, Internet Engineering Task Force, February 2005. Work in progress.

[59] B. Campbell, J. Rosenberg, H. Schulzrinne, C. Huitema and D. Gurle. Session Initiation Protocol (SIP) extension for instant messaging. RFC 3428, Internet Engineering Task Force, December 2002.

[60] S. Casner and V. Jacobson. Compressing IP/UDP/RTP headers for low-speed serial links. RFC 2508, Internet Engineering Task Force, February 1999.

[61] M. Degermark. CRTP over cellular radio links. Internet-Draft draft-degermark-crtp-cellular-01, Internet Engineering Task Force, December 1999. Work in progress.

[62] M. Degermark, B. Nordgren and S. Pink. IP header compression. RFC 2507, Internet Engineering Task Force, February 1999.

[63] EBU. The relative timing of the sound and vision components of a television signal. Technical Report EBU R37-2006, European Broadcasting Union (EBU), February 2006.

[64] B. Campbell, R. Mahy and C. Jennings (Eds). The Message Session Relay Protocol. Internet-Draft draft-ietf-simple-message-sessions-14, Internet Engineering Task Force, February 2006. Work in progress.

[65] ETSI. Telecommunications and Internet converged Services and Protocols for Advanced Networking (TISPAN); Extensible Markup Language (XML) Configuration Access Protocol (XCAP) over the Ut interface for manipulating NGN PSTN/ISDN simulation services; Protocol specification. TS 183 011, European Telecommunications Standards Institute (ETSI), 2006.

[66] ETSI. Telecommunications and Internet converged Services and Protocols for Advanced Networking (TISPAN); PSTN/ISDN simulation services; Anonymous Communication Rejection (ACR) and Communication Barring (CB); Protocol specification. TS 183 011, European Telecommunications Standards Institute (ETSI), 2006.

[67] ETSI. Telecommunications and Internet converged Services and Protocols for Advanced Networking (TISPAN); PSTN/ISDN simulation services; Communication Diversion (CDIV); Protocol specification. TS 183 004, European Telecommunications Standards Institute (ETSI), 2006.

[68] ETSI. Telecommunications and Internet converged Services and Protocols for Advanced Networking (TISPAN); PSTN/ISDN simulation services; Communication HOLD (HOLD); Protocol specification. TS 183 010, European Telecommunications Standards Institute (ETSI), 2006.

[69] ETSI. Telecommunications and Internet converged Services and Protocols for Advanced Networking (TISPAN); PSTN/ISDN simulation services; Conference (CONF); Protocol specification. TS 183 005, European Telecommunications Standards Institute (ETSI), 2006.

[70] ETSI. Telecommunications and Internet converged Services and Protocols for Advanced Networking (TISPAN); PSTN/ISDN simulation services; Explicit Communication Transfer (ECT); Protocol specification. TS 183 029, European Telecommunications Standards Institute (ETSI), 2006.

[71] ETSI. Telecommunications and Internet converged Services and Protocols for Advanced Networking (TISPAN); PSTN/ISDN simulation services; Message Waiting Indication (MWI); Protocol specification. TS 183 006, European Telecommunications Standards Institute (ETSI), 2006.

[72] ETSI. Telecommunications and Internet converged Services and Protocols for Advanced Networking (TISPAN); PSTN/ISDN simulation services; Originating Identification Presentation (OIP) and Originating Identification Restriction (OIR); Protocol specification. TS 183 007, European Telecommunications Standards Institute (ETSI), 2006.

[73] ETSI. Telecommunications and Internet converged Services and Protocols for Advanced Networking (TISPAN); PSTN/ISDN simulation services; Terminating Identification Presentation (TIP) and Terminating Identification Restriction (TIR); Protocol specification. TS 183 008, European Telecommunications Standards Institute (ETSI), 2006.

[74] ETSI/TISPAN. TISPAN web page, http://www.tispan.org. Technical report, ETSI/TISPAN, September 2006.

[75] F. Rui, F. Persson, M. Nordberg, S. Wänstedt, M. Qingyu and G. Xinyu. Coverage study for VoIP over enhanced uplink. In *IEEE Vehicular Technology Conference*, 2006.

[76] P. Faltstrom. E.164 number and DNS. RFC 2916, Internet Engineering Task Force, September 2000.

[77] WiMAX forum. WiMAX forum web page, http://www.wimaxforum.org. Technical report, WiMAX forum, September 2006.

[78] S. Frankel, R. Glenn and S. Kelly. The AES-CBC cipher algorithm and its use with IPsec. RFC 3602, Internet Engineering Task Force, September 2003.

[79] T. Friedman, R. Caceres and A. Clark. RTP Control Protocol Extended Reports (RTCP XR). RFC 3611, Internet Engineering Task Force, November 2003.

[80] M. Garcia-Martin, C. Bormann, J. Ott, R. Price and A. B. Roach. The Session Initiation Protocol (SIP) and Session Description Protocol (SDP) static dictionary for Signaling Compression (SigComp). RFC 3485, Internet Engineering Task Force, February 2003.

[81] S. Garg and M. Kappes. Can I add a VoIP call? In *IEEE International Conference on Communications*, vol. 2, pages 779–783, 2003.

[82] J. D. Gibson. *The Mobile Communications Handbook*. CRC Press Inc., 1996.

[83] GSMA. The GSM Assocation web page, http://www.gsmworld.com. Technical report, GSM Assocation, September 2006.

[84] M. Handley and V. Jacobson. SDP: Session Description Protocol. RFC 4566, Internet Engineering Task Force, July 2006.

[85] H. Hannu, J. Christoffersson, S. Forsgren, K.-C. Leung, Z. Liu and R. Price. Signaling Compression (SigComp) – Extended operations. RFC 3321, Internet Engineering Task Force, January 2003.

[86] D. Harkins and D. Carrel. The Internet Key Exchange (IKE). RFC 2409, Internet Engineering Task Force, November 1998.

[87] M. Hata. Empirical formula for propagation loss in land mobile radio services. *IEEE Transactions on Vehicular Technology*, VT-29, 1980.

[88] G. Hellstrom and P. Jones. RTP payload for text conversation. RFC 4103, Internet Engineering Task Force, June 2005.

[89] H. Holma and A. Toskala. *WCDMA for UMTS*. John Wiley & Sons, 2002.

[90] ISO/IEC. Information Technology – Coding of moving pictures and associated audio for digital storage media at up to about 1.5 Mbit/s – Part 2: Video. International Standard ISO/IEC 11172-2, 1993.

[91] ISO/IEC. Information Technology – Generic coding of moving pictures and associated audio information – Part 2: Video. International Standard ISO/IEC 13812-2, 2000.

[92] ISO/IEC. Information Technology – Universal Multiple-Octet Coded Character Set (UCS) – Part 1: Architecture and basic multilingual plane. Technical Specification INCITS/ISO/IEC 10646-1-2000, 2000.

[93] ISO/IEC. Information Technology – Coding of audio-visual objects – Part 2: Visual. International Standard ISO/IEC 14496-2, 2004.

[94] ISO/IEC. Information Technology – Coding of audio-visual objects – Part 10: Advanced video coding. International Standard ISO/IEC 14496-10, 2005.

[95] ITU. 7 kHz audio-coding within 64 kbit/s. Technical Report ITU-T Rec. G.722, International Telecommunications Union (ITU), 1988.

[96] ITU. Pulse Code Modulation (PCM) of voice frequencies. Technical Report ITU-T Rec. G.711, International Telecommunications Union (ITU), 1988.

[97] ITU. Specification for an intermediate reference system. Technical Report ITU-T Rec. P.48, International Telecommunications Union (ITU), 1988.

[98] ITU. 40, 32, 24, 16 kbit/s Adaptive Differential Pulse Code Modulation (ADPCM). Technical Report ITU-T Rec. G.726, International Telecommunications Union (ITU), 1990.

[99] ITU. Coding of speech at 8 kbit/s using Conjugate-Structure Algebraic-Code-Excited Linear-Prediction (CS-ACELP). Technical Report ITU-T Rec. G.729, International Telecommunications Union (ITU), March 1996.

[100] ITU. Methods for subjective determination of transmission quality. Technical Report ITU-T Rec. P.800, International Telecommunications Union (ITU), August 1996.

[101] ITU. Subjective performance assessment of telephone-band and wideband digital codecs. Technical Report ITU-T Rec. P.830, International Telecommunications Union (ITU), February 1996.

[102] ITU. Protocol for multimedia application text conversation. Technical Report ITU-T Rec. T.140, International Telecommunications Union (ITU), February 1998.

[103] ITU. Text chat application entity. Technical Report ITU-T Rec. T.134, International Telecommunications Union (ITU), February 1998.

[104] ITU. Application of the E-model: A planning guide. Technical Report ITU-T Rec. G.108, International Telecommunications Union (ITU), September 1999.

[105] ITU. Signalling system No. 7 – ISDN user part functional description. Technical Report ITU-T Rec. Q.761, International Telecommunications Union (ITU), 1999.

[106] ITU. Operational and interworking requirements for DCEs operating in the text telephone mode. Technical Report ITU-T Rec. V.18, International Telecommunications Union (ITU), November 2000.

[107] ITU. Bearer Independent Call Control protocol (Capability Set 2); Functional description. Technical Report ITU-T Rec. Q.1902.1, International Telecommunications Union (ITU), 2001.

[108] ITU. Perceptual evaluation of speech quality (PESQ): An objective method for end-to-end speech quality assessment of narrow-band telephone networks and speech codecs. Technical Report ITU-T Rec. P.862, International Telecommunications Union (ITU), February 2001.

[109] ITU. One-way transmission time. Technical Report ITU-T Rec. G.114, International Telecommunications Union (ITU), May 2003.

[110] ITU. The E-model, a computational model for use in transmission planning. Technical Report ITU-T Rec. G.107, International Telecommunications Union (ITU), March 2003.

[111] ITU-R. Studio encoding parameters of digital television for standard 4:3 and wide-screen 16:9 aspect ratios. Recommendation BT.601, International Telecommunication Union, October 1995.

[112] ITU-T. Video codec for audiovisual services at $p \times 64$ kbit/s. Recommendation H.261, International Telecommunication Union, March 1993.

[113] ITU-T. Information technology – Generic coding of moving pictures and associated audio information: Video. Recommendation H.262, International Telecommunication Union, February 2000.

[114] ITU-T. Multiplexing protocol for low bit rate multimedia communication. Recommendation H.223, International Telecommunication Union, July 2001.

[115] ITU-T. Gateway control protocol: Version 2. Recommendation H.248, International Telecommunication Union, May 2002.

[116] ITU-T. Packet-based multimedia communication systems. Recommendation H.323, International Telecommunication Union, July 2003.

[117] ITU-T. Narrow-band visual telephone systems and terminal equipment. Recommendation H.320, International Telecommunication Union, March 2004.

[118] ITU-T. Advanced video coding for generic audiovisual services. Recommendation H.264, International Telecommunication Union, March 2005.

[119] ITU-T. Terminal for low bit-rate multimedia communication. Recommendation H.324, International Telecommunication Union, September 2005.

[120] ITU-T. Video coding for low bit rate communication. Recommendation H.263, International Telecommunication Union, May 2005.

[121] ITU-T. Control protocol for multimedia communication. Recommendation H.245, International Telecommunication Union, May 2006.

[122] V. Jakobson. Compressing TCP/IP headers for low-speed serial links. RFC 1144, Internet Engineering Task Force, February 1990.

[123] N. S. Jayant. High-quality coding of telephone speech and wideband audio. In *IEEE Communications Magazine*, pages 10–20, January 2000.

[124] C. Jennings, J. Peterson and M. Watson. Private extensions to the Session Initiation Protocol (SIP) for asserted identity within trusted networks. RFC 3325, Internet Engineering Task Force, November 2002.

[125] F. Johansson and T. Johansson. Mobile IPv4 extension for carrying network access identifiers. RFC 3846, Internet Engineering Task Force, June 2004.

[126] I. Johansson. VoIP inband signaling using shim extensions. Internet-Draft draft-johansson-avt-rtp-shim-00, Internet Engineering Task Force, July 2006. Work in progress.

[127] I. Johansson, T. Frankkila and P. Synnergren. Bandwidth efficient AMR operation for VoIP. In *IEEE Speech Coding Workshop*, 2002.

[128] S. Kent and R. Atkinson. IP Encapsulating Security Payload (ESP). RFC 2406, Internet Engineering Task Force, November 1998.

[129] S. Kent and R. Atkinson. Security architecture for the Internet Protocol. RFC 2401, Internet Engineering Task Force, November 1998.

[130] Y. Kikuchi, T. Nomura, S. Fukunaga, Y. Matsui and H. Kimata. RTP payload format for MPEG-4 audio/visual streams. RFC 3016, Internet Engineering Task Force, November 2000.

[131] R. Kumar and M. Mostafa. Conventions for the use of the Session Description Protocol (SDP) for ATM Bearer Connections. RFC 3108, Internet Engineering Task Force, May 2001.

[132] A. Li. RTP payload format for Enhanced Variable Rate Codecs (EVRC) and Selectable Mode Vocoders (SMV). RFC 3558, Internet Engineering Task Force, July 2003.

[133] S. Lin and D. J. Costello Jr. *Error Control Coding: Fundamentals and Applications*. Prentice-Hall, 1983.

[134] C. Madson and R. Glenn. The use of HMAC-MD5-96 within ESP and AH. RFC 2403, Internet Engineering Task Force, November 1998.

[135] C. Madson and R. Glenn. The use of HMAC-SHA-1-96 within ESP and AH. RFC 2404, Internet Engineering Task Force, November 1998.

[136] R. Mahy. A message summary and message waiting indication event package for the Session Initiation Protocol (SIP). RFC 3842, Internet Engineering Task Force, August 2004.

[137] A. Niemi, J. Arkko and V. Torvinen. Hypertext Transfer Protocol (HTTP) digest authentication using Authentication and Key Agreement (AKA). RFC 3310, Internet Engineering Task Force, September 2002.

[138] OMA. Presence SIMPLE requirements, Candidate Version 1.0 – 06 Oct 2005. Requirements Document OMA-RD-Presence_SIMPLE-V1_0-20051006-C, Open Mobile Alliance (OMA), October 2005.

[139] OMA. Instant Messaging requirements, Candidate Version 1.0 – 6 Jun 2006. Requirements Document OMA-RD_IM-V1_0-20060606-D, Open Mobile Alliance (OMA), June 2006.

[140] OMA. Instant Messaging using SIMPLE architecture, Draft Version 1.0.0 – 17 May 2006. Architecture Document OMA-AD-IM_SIMPLE-V1_0_0-20060517-D, Open Mobile Alliance (OMA), May 2006.

[141] OMA. OMA Management Object for XML Document Management, Approved Version 1.0 – 12 Jun 2006. Technical Specification OMA-TS-XDM_MO-V1_0-20060612-A, Open Mobile Alliance (OMA), June 2006.

[142] OMA. OMA PoC control plane, Approved Version 1.0 – 09 Jun 2006. Technical Specification OMA-TS-PoC-ControlPlane-V1_0-20060609-A, Open Mobile Alliance (OMA), June 2006.

[143] OMA. OMA-TS-IM_SIMPLE, Draft Version 1.0 – 10 04 2006. Technical Specification OMA-TS-IM_SIMPLE-V1_0_0-20060410-D, Open Mobile Alliance (OMA), April 2006.

[144] OMA. PoC user plane, Approved Version 1.0 – 09 Jun 2006. Technical Specification OMA-TS_PoC-UserPlane-V1_0-20060609-A, Open Mobile Alliance (OMA), June 2006.

[145] OMA. PoC XDM specification, Approved Version 1.0 – 09 Jun 2006. Technical Specification OMA-TS-PoC_XDM-V1_0-20060609-A, Open Mobile Alliance (OMA), June 2006.

[146] OMA. Presence SIMPLE architecture document, Candidate Version 1.0 – 10 Jan 2006. Architecture Document OMA-AD-Presence_SIMPLE-V1_0-20060110-C, Open Mobile Alliance (OMA), January 2006.

[147] OMA. Presence SIMPLE specification, Candidate Version 1.0 – 18 Apr 2006. Technical Specification OMA-TS-Presence_SIMPLE-V1_0-20060418-C, Open Mobile Alliance (OMA), April 2006.

[148] OMA. Presence XDM specification, Candidate Version 1.0 – 18 Apr 2006. Technical Specification OMA-TS-Presence_SIMPLE_XDM-V1_0-20060418-C, Open Mobile Alliance (OMA), April 2006.

[149] OMA. Push to Talk over Cellular (PoC) – Architecture, Approved Version 1.0 – 09 Jun 2006. Architecture Document OMA-AD-PoC-V1_0-20060609-A, Open Mobile Alliance (OMA), June 2006.

[150] OMA. Push to Talk over Cellular requirements, Approved Version 1.0 – 09 Jun 2006. Requirements Document OMA-RD-PoC-V1_0-20060609-A, Open Mobile Alliance (OMA), June 2006.

[151] OMA. Resource List Server (RLS) XDM specification, Candidate Version 1.0 – 18 Apr 2006. Technical Specification OMA-TS-Presence_SIMPLE_RLS_XDM-V1_0-20060418-C, Open Mobile Alliance (OMA), April 2006.

[152] OMA. Shared XDM specification, Approved Version 1.0 – 12 Jun 2006. Technical Specification OMA-TS-XDM_Shared-V1_0-20060612-A, Open Mobile Alliance (OMA), June 2006.

[153] OMA. XML Document Management architecture, Approved Version 1.0 – 12 Jun 2006. Architecture Document OMA-AD-XDM-V1_0-20060612-A, Open Mobile Alliance (OMA), June 2006.

[154] OMA. XML Document Management requirements, Approved Version 1.0 – 12 Jun 2006. Requirements Document OMA-RD-XDM-V1_0-20060612-A, Open Mobile Alliance (OMA), June 2006.

[155] OMA. XML Document Management (XDM) specification, Approved Version 1.0 – 12 Jun 2006. Technical Specification OMA-TS-XDM_Core-V1_0-20060612-A, Open Mobile Alliance (OMA), June 2006.

[156] S. Parkvall, E. Englund, P. Malm, T. Hedberg, M. Persson and J. Peisa. Wcdma evolved – high-speed packet-data services. *Ericsson Review*, (2):56–65, 2003.

[157] J. Peisa and E. Englund. TCP performance over HS-DSCH. In *Proceedings of Vehicular Technology Conference*, Spring 2002.

[158] R. Pereira and R. Adams. The ESP CBC-mode cipher algorithms. RFC 2451, Internet Engineering Task Force, November 1998.

[159] C. Perkins, I. Kouvelas, O. Hodson, V. Hardman, M. Handley, J. C. Bolot, A. Vega-Garcia and S. Fosse-Parisis. RTP payload for redundant audio data. RFC 2198, Internet Engineering Task Force, September 1997.

[160] J. Peterson. A privacy mechanism for the Session Initiation Protocol (SIP). RFC 3323, Internet Engineering Task Force, November 2002.

[161] R. Price, C. Bormann, J. Christoffersson, H. Hannu, Z. Liu and J. Rosenberg. Signaling Compression (SigComp). RFC 3320, Internet Engineering Task Force, January 2003.

[162] R. Cox, W. B. Kleijn and P. Kroon. Robust CELP coders for noisy backgrounds and noisy channels. In *IEEE International Conference on Acoustics, Speech, and Signal Processing (ICASSP)*, pages 739–742, 1989.

[163] R. G. Cole and J. H. Rosenbluth. Voice over IP performance monitoring. *Computer Communication Review*, 31(2):9–24, April 2001.

[164] J. Rosenberg. The Session Initiation Protocol (SIP) UPDATE method. RFC 3311, Internet Engineering Task Force, October 2002.

[165] J. Rosenberg. A presence event package for the Session Initiation Protocol (SIP). RFC 3856, Internet Engineering Task Force, August 2004.

[166] J. Rosenberg. A watcher information event template-package for the Session Initiation Protocol (SIP). RFC 3857, Internet Engineering Task Force, August 2004.

[167] J. Rosenberg. An Extensible Markup Language (XML) based format for watcher information. RFC 3858, Internet Engineering Task Force, August 2004.

[168] J. Rosenberg. The Extensible Markup Language (XML) Configuration Access Protocol (XCAP). Internet-Draft draft-ietf-simple-xcap-07, Internet Engineering Task Force, June 2005. Work in progress.

[169] J. Rosenberg and H. Schulzrinne. An RTP payload format for generic forward error correction. RFC 2733, Internet Engineering Task Force, December 1999.

[170] J. Rosenberg and H. Schulzrinne. An offer/answer model with Session Description Protocol (SDP). RFC 3264, Internet Engineering Task Force, June 2002.

[171] J. Rosenberg and H. Schulzrinne. Reliability of provisional responses in Session Initiation Protocol (SIP). RFC 3262, Internet Engineering Task Force, June 2002.

[172] J. Rosenberg, H. Schulzrinne, G. Camarillo, A. Johnston, J. Peterson, R. Sparks, M. Handley and E. Schooler. SIP: Session Initiation Protocol. RFC 3261, Internet Engineering Task Force, June 2002.

[173] J. Rosenberg, H. Schulzrinne, and P. Kyzivat. Indicating user agent capabilities in the Session Initiation Protocol (SIP). RFC 3840, Internet Engineering Task Force, August 2004.

[174] M. R. Schroeder and B. S. Atal. Code-excited linear prediction (CELP): High quality at very low bit rates. In *IEEE International Conference on Acoustics, Speech, and Signal Processing (ICASSP)*, pages 937–941, March 1985.

[175] H. Schulzrinne. The tel URI for telephone numbers. RFC 3966, Internet Engineering Task Force, December 2004.

[176] H. Schulzrinne and S. Casner. RTP profile for audio and video conferences with minimal control. RFC 3551, Internet Engineering Task Force, July 2003.

[177] H. Schulzrinne, S. Casner, R. Frederick and V. Jacobson. RTP: a transport protocol for real-time applications. RFC 3550, Internet Engineering Task Force, July 2003.

[178] J. Sjoberg. Real-time Transport Protocol (RTP) payload format and file storage format for the Adaptive Multi-Rate (AMR) and Adaptive Multi-Rate Wideband (AMR-WB) audio codecs. Internet-Draft draft-ietf-avt-rtp-amr-bis-01, Internet Engineering Task Force, January 2005. Work in progress.

[179] J. Sjoberg, M. Westerlund, A. Lakaniemi and Q. Xie. Real-time Transport Protocol (RTP) payload format and file storage format for the Adaptive Multi-Rate (AMR) and Adaptive Multi-Rate Wideband (AMR-WB) audio codecs. RFC 3267, Internet Engineering Task Force, June 2002.

[180] R. Steinmetz. Human perception of jitter and media synchronization. *IEEE Journal on Selected Areas in Communications*, 14(1), 1996.

[181] TIA. TDMA cellular/PCS radio interface enhanced full-rate voice codec. Technical Report ANSI/TIA/EIA 136-410, Telecommunications Industry Association (TIA), November 1999.

[182] A. Vaha-Sipila. URLs for telephone calls. RFC 2806, Internet Engineering Task Force, April 2000.

[183] K. Nyberg and V. Niemi. *UMTS Security*. John Wiley & Sons, 2003.

[184] S. Wänstedt, M. Ericson, K. Sandlund, M. Nordberg and T. Frankkila. Realization and performance evaluation of IMS Multimedia Telephony for HSPA. In *The 17th Annual IEEE International Symposium on Personal, Indoor and Mobile Radio Communications (PIMRC'06)*, 2006.

[185] S. Wenger, U. Chandra, M. Westerlund and B. Burman. Codec control messages in the Audio-Visual Profile with Feedback (AVPF). Internet-Draft draft-wenger-avt-avpf-ccm-04, Internet Engineering Task Force, May 2006. Work in progress.

[186] S. Wenger, M. M. Hannuksela, T. Stockhammer, M. Westerlund and D. Singer. RTP payload format for H.264 Video. RFC 3984, Internet Engineering Task Force, February 2005.

Index